Ecology for Nonecologists

FRANK R. SPELLMAN

Government Institutes
An imprint of
The Scarecrow Press, Inc.
Lanham, Maryland • Toronto • Plymouth, UK
2008

 **Government Institutes**

Published in the United States of America
by Government Institutes, an imprint of The Scarecrow Press, Inc.
A wholly owned subsidiary of
The Rowman & Littlefield Publishing Group, Inc.
4501 Forbes Boulevard, Suite 200
Lanham, Maryland 20706
http://govinst.scarecrowpress.com

Estover Road
Plymouth PL6 7PY
United Kingdom

British Library Cataloguing in Publication Information Available

Library of Congress Cataloging-in-Publication Data

Spellman, Frank R.
    Ecology for nonecologists / Frank R. Spellman.
        p. cm.
    Includes bibliographical references and index.
    ISBN-13: 978-0-86587-197-7 (pbk. : alk. paper)
    ISBN-10: 0-86587-197-3 (pbk. : alk. paper)
    1. Ecology–Popular works. I. Title.
QH541.13.S64 2008
577–dc22   2007032773

♾ ™ The paper used in this publication meets the minimum requirements of
American National Standard for Information Sciences—Permanence of
Paper for Printed Library Materials, ANSI/NISO Z39.48-1992.
Manufactured in the United States of America.

For Revonna M. Bieber

Contents

Figures

Preface

Following the large, flagship footprints provided by *Toxicology for Nontoxicologists* and *Chemistry for Nonchemists*, the purpose of this book is to present basic concepts in ecology with the focus primarily on the interrelationship (the ecology) of biota (life forms) in our environment. In addition, this book fills the gap between general introductory science texts and the more advanced environmental science books used in graduate courses. *Ecology for Nonecologists* fills this gap by covering the basics of ecology relative to current conditions and issues. While primarily designed as an information source and presented in simple, straightforward, easy-to-understand English, *Ecology for Nonecologists* provides a levelheaded look at a very serious discipline, one based on years of extensive research.

Ecology is a multidisciplinary field that incorporates aspects of biology, chemistry, physics, geology, geography, meteorology, pedology, agriculture, fisheries, forestry, and many other fields. Books on the subject are typically geared toward professionals in these fields. This makes undertaking a study of ecology daunting to those without this specific background. However, this complexity also indicates ecology's broad scope of impact. Because ecology affects us, it is important to understand some basic concepts of the discipline.

Although this book is primarily intended for professionals in the environmental field, such as managers, geologists, biologists, environmental scientists, and everyday practitioners, it is also designed for a more general audience: the curious, those with a love to learn anything and everything. It is hoped that this book will enable readers to broaden their outlook on the environment and, in particular, to aid in understanding it. Krebs (1972) defines ecology as the place "where organisms are found, how many occur there, and why"—with the emphasis on "why."

Ecology for Nonecologists—again, suitable for use by both the technical practitioner in the field, by students in the classroom, and by others—emphasizes basic concepts, definitions, and descriptions, all presented with a touch of prose, poetry, and irony. To ensure correlation with modern practice and design, illustrative problems are presented in terms of commonly used ecological parameters. Here is all the information you need to make technical and personal decisions about ecology. It is important to remember, however, that left to her own devices, Mother Nature can perform wonders—but overload her and there might be hell to pay.

Each chapter ends with a Chapter Review Test to help evaluate your mastery of the main concepts presented. Before going on to the next chapter, take the

Review Test, compare your answers to the key provided in the appendix, and review the pertinent information for any problems you missed. If you miss many items, review the whole chapter.

✔   *Note*: The checkmark displayed in various locations throughout this text indicates or emphasizes an important point or points to study carefully.

Again, this text is accessible to those who have no experience with ecology. If you work through the text systematically, you will be surprised at how easily you acquire an understanding of ecology—adding a critical component to your professional knowledge.

Notes

Krebs, C. H. 1972. *Ecology: The experimental analysis of distribution and abundance*. New York: Harper & Row.

Part I

FUNDAMENTAL ECOLOGY

Rolling hills, Shenandoah National Park, Virginia.
Photograph by Frank R. Spellman

CHAPTER 1

Introduction

> The "control of nature" is a phrase conceived in arrogance, born of the Neanderthal age of biology and the convenience of man.
>
> —Rachel Carson (1962)

Topics

Setting the Stage
History of Ecology
Levels of Organization
Ecosystem
Summary of Key Terms
Chapter Review Questions

Setting the Stage

What is ecology? Why is it important? Why study it? These are all simple, straightforward questions. However, providing simple answers is not that easy. Notwithstanding the inherent difficulty with explaining any complex science in simple, straightforward terms, such an undertaking is nevertheless the purpose of this text. In short, the task of this text is to outline basic information that explains the functions and values of ecology and its interrelationships with other sciences, including ecology's direct impact on our lives. In doing so, I hope to not only dispel the common misconception that ecology is too difficult for the average person to understand but also to instill the concept of ecology as an asset that can be learned and cherished.

WHAT IS ECOLOGY?

Ecology can be defined in various ways. For example, ecology, or ecological science, is commonly defined in the literature as the scientific study of the distribution and abundance of living organisms and how the distribution and abundance are affected by interactions between the organisms and their environment. The term *ecology* was coined in 1866 by the German biologist Haeckel and loosely means "the study of the household [of nature]." Odum (1983) explains that the word *ecology* is derived from the Greek *oikos*, meaning home. Ecology, then, means the study of organisms at home. Ecology is the study of the relation of an organism or a group of organisms to their environment. In a broader sense, ecology is the study of the relation of organisms or groups to their environment.

✔ *Important Point*: No ecosystem can be studied in isolation. If we were to describe ourselves, our histories, and what made us the way we are, we could not leave the world around us out of our description! So it is with streams: They are directly tied in with the world around them. They take their chemistry from the rocks and dirt beneath them as well as for a great distance around them (C. Cave, cited in Spellman 1996, 1).

Charles Darwin explained ecology in a famous passage in his *Origin*, a passage that helped establish the science of ecology. A "web of complex relations" binds all living things in any region, Darwin writes. Adding or subtracting even a single species causes waves of change that race through the web, "onwards in ever-increasing circles of complexity." The simple act of adding cats to an English village would reduce the number of field mice. Killing mice would benefit the bumblebees, whose nest and honeycombs the mice often devour. Increasing the number of bumblebees would benefit the heartsease and red clover, which are fertilized almost exclusively by bumblebees. So adding cats to the village could result in adding flowers. For Darwin, the whole of the Galapagos archipelago argues this fundamental lesson. The volcanoes are much more diverse in their ecology than their biology. The contrast suggests that in the struggle for existence, species are shaped at least as much by the local flora and fauna as by the local soil and climate. "Why else would the plants and animals differ radically among islands that have the same geological nature, the same height, and climate" (Darwin 1998).

Probably the best way to understand ecology—to get a really good "feel" for it—or to get to the heart of what ecology is all about is to read the following by Rachel Carson (1962):

> We poison the caddis flies in a stream and the salmon runs dwindle and die. We poison the gnats in a lake and the poison travels from link to link of the food chain and soon the birds of the lake margins become victims. We spray our elms and the following springs are silent of robin song, not because we sprayed the robins directly but because the poison traveled, step by step, through the now familiar elm leaf-earthworm-robin cycle. These are matters of record, observable, part of the visible world around us. They reflect the web of life—or death—that scientists know as ecology.

As Rachel Carson points out, what we do to any part of our environment has an impact upon other parts. In other words, there is an interrelationship between the parts that make up our environment. Probably the best way to state this interrelationship is to define ecology definitively—that is, to define it as it is used in this text: "Ecology is the science that deals with the specific interactions that exist between organisms and their living and nonliving environment" (Tomera 1989).

When environment was mentioned in the preceding and as it is discussed throughout this text, it (the environment) includes everything important to the organism in its surroundings. The organism's environment can be divided into four parts:

1. Habitat and distribution—its place to live
2. Other organisms—whether friendly or hostile
3. Food
4. Weather—light, moisture, temperature, soil, etc.

There are four major subdivisions of ecology:

- Behavioral ecology
- Population ecology (autecology)
- Community ecology (synecology)
- Ecosystem ecology

Behavioral ecology is the study of the ecological and evolutionary basis for animal behavior. *Population ecology* (or autecology) is the study of the individual organism or a species. It emphasizes life history, adaptations, and behavior. It is the study of communities, ecosystems, and biosphere. An example of autecology would be when biologists study the ecology of the salmon throughout the entire lifetime of the salmon. *Community ecology* (or synecology), on the other hand, is the study of groups of organisms associated together as a unit and deals with the environmental problems caused by mankind. For example, the effect of discharging phosphorous-laden effluent into a stream involves several organisms. The activities of human beings have become a major component of many natural areas. As a result, it is important to realize that the study of ecology must involve people. *Ecosystem ecology* is the study of how energy flow and matter interact with biotic elements of ecosystems (Odum 1971).

✔ *Important Point*: Ecology is generally categorized according to complexity; the primary kinds of organisms under study (plant, animal, insect ecology); the biomes principally studied (forest, desert, benthic, grassland, etc.); the climatic or geographic area (e.g., artic or tropics); and/or the spatial scale (macro or micro) under consideration. Specialized branches of ecology are shown in figure 1.1.

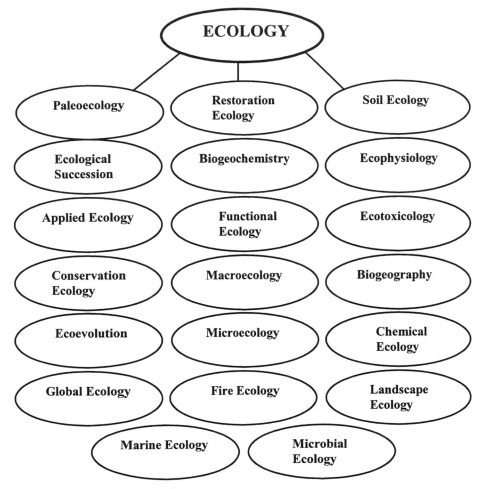

Figure 1.1. Branches of ecology

WHY IS ECOLOGY IMPORTANT?

Ecology, in its true sense, is a holistic discipline that does not dictate what is right or wrong. Instead, ecology is important to life on Earth simply because it makes us aware, to a certain degree, of what life on Earth is all about. Ecology shows us that each living organism has an ongoing and continual relationship with every other element that makes up our environment. Simply, ecology is all about interrelationships, intraspecific and interspecific, and on how important it is to maintain these relationships—to ensure our very survival.

At this point in this discussion, there are literally countless examples that could be used to point out the importance of ecology and interrelationships. However, to not only demonstrate the importance of ecology but to also point out that an ecological principle can be a double-edged sword, depending on point of view (ecological problems along with pollution can be a judgment call; that is, they are a matter of opinion), a famous parable adapted from Peter Marshall's *Mr. Jones: Meet the Master* (1950) is used here to demonstrate that many misconceptions exist in ecology.

After reviewing Marshall's parable and the restoration of the spring, the average person might say to himself or herself, "Gee, all is well with the town again." The residents had swans, irrigation, hydropower, and pretty views, and they seem to be pleased that the stream was restored to its "normal" state. The trained ecologist, however, would take a different view of this same stream. The ecologist would go beyond the hype (as portrayed in the popular media, including literature) about what a healthy stream is. For example, the trained ecologist would know that a perfectly clean stream, clear of all terrestrial plant debris (woody debris and leaves) would not be conducive to ensuring diverse, productive invertebrates and fish, would not preserve natural sediment and water regimes, and would not ensure overall stream health (Dolloff and Webster 2000).

WHY STUDY ECOLOGY?

Does anyone really need to be an ecologist or a student of ecology to appreciate the following words of Will Carleton (1845–1912) from his classic poem, *Autumn Days*?

The Keeper of the Spring

This is the story of the keeper of the spring. He lived high in the Alps above an Austrian town and had been hired by the town council to clear debris from the mountain springs that fed the steam that flowed through the town. The man did his work well and the village prospered. Graceful swans floated in the stream. The surrounding countryside was irrigated. Several mills used the water for power. Restaurants flourished for townspeople and for a growing number of tourists.

Years went by. One evening at the town council meeting, someone questioned the money being paid to the keeper of the spring. No one seemed to know who he was or even if he was still on the job high up in the mountains. Before the evening was over, the council decided to dispense with the old man's services.

Weeks went by and nothing seemed to change. Then autumn came. The trees began to shed their leaves. Branches broke and fell into the pools high up in the mountains. Down below the villagers began to notice the water becoming darker. A foul odor appeared. The swans disappeared. So did the tourists. Soon disease spread through the town.

When the town council reassembled, they realized that they had made a costly error. They found the old keeper of the spring and hired him back again. Within a few weeks, the stream cleared up and life returned to the village as they had known it before.

> Sweet and smiling are thy ways,
> Beauteous, gold Autumn days.

Moreover, does anyone need to study ecology to observe and to relish and to feel and/or to sense the real thing: Nature's annual color palette in full kaleidoscope display—where those "yellow, mellow, ripened days are sheltered in a golden coating"? Those clear and sunny days and cool and crisp nights of autumn provide an almost irresistible lure to those of us (ecologists and non-ecologists alike) who enjoy the outdoors. To take in the splendor and delight of autumn's color display, many head for the hills, the mountains, countryside, lakes, streams, and recreation areas of the national forests. The more adventurous horseback ride or backpack through Nature's glory on trails winding deep into forest tranquility—just being out-of-doors in those golden days rivals any thrill in life. Even those of us fettered to the chains of city life are often exposed to city streets with those columns of life ablaze in color.

No, one need not study ecology to witness, appreciate, or understand the enchantment of autumn's annual color display—summer extinguished in a blaze of color. It is a different story, however, for those involved in trying to understand all of the complicated actions—and even more complicated interactions—involving pigments, sunlight, moisture, chemicals, temperatures, site, hormones, length of daylight, genetic traits, and so on that make for a perfect autumn color display (U.S. Department of Agriculture [USDA] 1999). This is the work of the ecologist—to probe deeper and deeper into the basics of Nature, constantly seeking answers. To find the answers, the ecologist must be a synthesis scientist; that is, he or she must be well versed in botany, zoology, physiology, genetics, and other disciplines such as geology, physics, and chemistry.

Earlier, an adaptation of Peter Marshall's parable was used to make the point that a clean stream and other downstream water bodies can be a good thing, depending on one's point of view—pollution is a judgment call. It was also pointed out that this view might not be shared by a trained ecologist, especially a stream ecologist. The ecologist knows, for example, that terrestrial plant debris is not only a good thing but that it is absolutely necessary. Why? Consider the following explanation (Spellman 1996):

> In a stream, there are two possible sources of primary energy: in-stream photosynthesis by algae, mosses, and higher aquatic plants and imported organic matter from streamside vegetation (e.g., leaves and other parts of vegetation). Simply put, a significant portion of the food that is eaten grows right in the stream, like algae, diatoms, nymphs and larvae, and fish. This food that originates from within the stream is called autochthonous (Benfield, 1996).
>
> Most food in a stream, however, comes from outside the stream. This is especially the case in small, heavily wooded streams, where there is normally insufficient light to support substantial in-stream photosynthesis so energy pathways are sup-

ported largely by imported energy. A large portion of this imported energy is provided by leaves. Worms drown in floods and get washed in. Leafhoppers and caterpillars fall from trees. Adult mayflies and other insects mate above the stream, lay their eggs in it, and then die in it. All of this food from outside the stream is called allochthonous.

🖝 *Note*: Anytime the author, a stream ecologist, thinks or writes about streams and allochthonous inputs, haunting wet refrains mingle with his thoughts. For example, consider the following by Henry David Thoreau (*Walking: Winter Walk* 1862):

> When every stream in its penthouse
> Goes gurgling on its way,
> And in his gallery the mouse
> Nibbleth the meadow of hay.

Another important reason to study and learn ecology can be garnered from another simple stream ecology example.

History of Ecology

The chronological development of most sciences is clear and direct. Listing the progressive stages in the development of biology, math, chemistry, and physics is a relatively easy, straightforward process. The science of ecology is different. Having only gained prominence in the latter part of the 20th century, ecology is generally spoken of as a new science. However, ecological thinking at some level has been around for a long time, and the principles of ecology have developed gradually and more like a multistemmed bush than a tree with a single trunk (Smith 1996).

Smith and Smith (2006) point out that one can argue that ecology goes back to Aristotle or perhaps his friend and associate Theophrastus, both of whom had interest in the relations between organisms and the environment and in many species of animals. Theophrastus described interrelationships between animals and between animals and their environment as early as the fourth century BCE (Ramalay 1940).

Modern ecology has its early roots in plant geography (i.e., plant ecology, which developed earlier than animal ecology) and natural history. The early plant geographers (ecologists) included Carl Ludwig Willdenow (1765–1812) and Friedrich Alexander von Humboldt (1769–1859). Willdenow was one of the first phytogeographers; he was also a mentor to von Humboldt. Willdenow, for whom the perennial vine Willdenow's spikemoss (*Selaginella willdenowii*) is named, developed the notion, among many others, that plant distribution patterns changed over time. Von Humboldt, considered by many to be the father of ecology, further

Leaf Processing in Streams

Autumn leaves entering streams are nutrition-poor because trees absorb most of the sugars and amino acids (nutrients) that were present in the green leaves (Suberkoop, Godshalk, and Klug 1978). Leaves falling into streams may be transported short distances but usually are caught by structures in the streambed to form leaf packs. These leaf packs are then processed in place by components of the stream communities in a series of well-documented steps (see figure 1.2; Peterson and Cummins 1974).

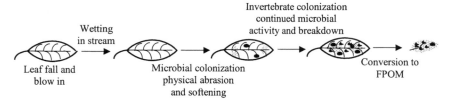

Figure 1.2.   The processing or "conditioning" sequence for a medium-fast deciduous tree leaf in a temperate stream

Adapted from J. D. Allen, *Stream Ecology: Structure and Functions of Running Waters* (Tandem: Chapman & Hall, 1996), p. 114.

Within 24 to 48 hours of entering a stream, many of the remaining nutrients in leaves leach into the water. After leaching, leaves are composed mostly of structural materials like nondigestible cellulose and lignin. Within a few days, fungi (especially Hyphomycetes), protozoa, and bacteria process the leaves by microbial processing (see figure 1.2; Barlocher and Kendrick 1975). Two weeks later, microbial conditioning leads to structural softening of the leaf and, among some species, fragmentation. Reduction in particle size from whole leaves (coarse particulate organic matter) to fine particulate organic matter is accomplished mainly through the feeding activities of a variety of aquatic invertebrates collectively known as "shredders" (Cummins 1974; Cummins and Klug 1979). Shredders (stoneflies, for example) help to produce fragments shredded from leaves but not ingested and fecal pellets, which reduce the particle size of organic matter. The particles then are collected (by mayflies, for example) and serve as a food resource for a variety of micro- and macroconsumers. Collectors eat what they want and send even smaller fragments downstream. These tiny fragments may be filtered out of the water by a true fly larva (i.e., a filterer). Leaves may also be fragmented by a combination of microbial activity and physical factors such as current and abrasion (Benfield, Jones, and Patterson 1977; Paul, Benfield, and Cairns 1978).

Leaf-pack processing by all the elements mentioned above (i.e., leaf species, microbial activity, physical and chemical features of the stream) is important. However, the most important point is that these integrated ecosystem processes convert whole leaves into fine particles, which are then distributed downstream and used as an energy source by various consumers.

The bottom line on allochthonous material in a stream: Insects that have fallen into a stream are ready-to-eat, and may join leaves, exuviae, copepods, dead and dying animals, rotifers, bacteria, and dislodged algae and immature insects in their float (drift) downstream to a waiting hungry mouth.

developed many of the Willldenow's notions, including the notion that barriers to plant dispersion were not absolute (Smith 2007).

Another scientist who is considered a founder of plant ecology was Johannes E. B. Warming (1841–1924). Warming studied the tropical vegetation of Brazil. He is best known for working on the relations between living plants and their surroundings. He is also recognized for his flagship text on plant ecology, *Plantesamfund* (1895). In addition, he wrote *A Handbook of Systematic Botany* (1878).

Meanwhile, other naturalists were assuming important roles in the development of ecology. First and foremost amongst the naturalists was Charles Darwin. While working on his origin of species, Darwin came across the writings of Thomas Malthus (1766–1834). Malthus advanced the principle that populations grow in a geometric fashion, doubling at regular intervals until they outstrip the food supply—ultimately resulting in death and thus restraining population growth (Smith and Smith 2006). Darwin, in his autobiography (1876), stated,

> In October 1838, that is, fifteen months after I had begun my systematic inquiry, I happened to read for amusement Malthus on *Population* and being well prepared to appreciate the struggle for existence which everywhere goes on from long-continued observation of the habits of animals and plants it at once struck me that under these circumstances favorable variations would tend to be preserved, and unfavorable ones to be destroyed. The results of this would be the formation of a new species. Here, then I had at last got a theory by which to work.

During the period Darwin was formulating his theory of the origin of species, Gregor Mendel (1822–1884) was studying the transmission of characteristics from one generation of pea plants to another. Mendel's plus Darwin's work provided the foundation for population genetics, the study of evolution and adaptation.

Time marched on; the preceding work of chemists Lavoisier (he lost his head during the French Revolution) and Horace B. de Saussere and the Austrian geologist Eduard Suess, who proposed the term *biosphere* in 1875, together with the later studies mentioned above, all set the foundations of the advanced work that continues today.

✔ *Important Point*: The Russian geologist, Vladimir Vernadsky, detailed the idea of biosphere in 1926.

Several forward strides in animal ecology, independent of plant ecology, were made during the 19th century that enabled the 20th-century scientists R. Hesse, Charles Elton, Charles Adams, and Victor Shelford to refine the discipline.

Smith and Smith (2006) point out that many early plant ecologists were "concerned with observing the patterns of organisms in nature, attempting to understand how patterns were formed and maintained by interactions with the physical environment." Instead of looking for patterns, Frederic E. Clements (1874–

Family Picnic Hosts Insect Intruders

On one of their late August holiday outings, a family of 18 picnickers from a couple of small rural towns visited a local stream that coursed its way alongside and through one of the towns. This annual outing was looked upon with great anticipation for it was that one time each year when aunts, uncles, and cousins came together as a one big family. The streamside setting was perfect for such an outing, but historically, until quite recently, the stream had been posted "DANGER—NO SWIMMING—CAMPING or FISHING!"

Because the picnic area was such a popular location for picnickers, swimmers, and fishermen over the years, several complaints about the polluted stream were filed with the County Health Department. The Health Department finally took action to restore the stream to a relatively clean condition: sanitation workers removed debris and old tires, and plugged or diverted end-of-pipe industrial outfalls upstream of the picnic area. After two years of continuous worker-aided stream cleanup and the stream's natural self-purification process, the stream was given a clean bill of health by the Health Department. The danger postings were removed.

Once the stream had been declared clean, fit for use by swimmer and fisher, with postings removed, it did not take long for the word to get out. Local folks and others alike made certain, at first opportunity, that they flocked to the restored picnic and swimming and fishing site alongside the stream.

During most visits to the restored picnic-stream area, visitors, campers, fishermen, and others were pleased with their cleaned-up surroundings. However, during late summer, when the family of 18 and several others visited the restored picnic-stream area, they found themselves swarmed over by thousands of speedy dragonflies and damselflies, especially near the bank of the stream. Soon they found the insects too much to deal with so they stayed clear of the stream. To themselves and to anyone who would listen, the same complaint was heard over and over again: "What happened to our nice clean stream? With all those nasty bugs, the stream is polluted again." So, when August arrived with its hordes of dragonfly-type insects, the picnickers, campers, swimmers, and fishermen avoided the place until the insects departed, and the human visitors thought the stream was clean again.

However, there is one local family that does not avoid the stream-picnic area in August; on the contrary, August is one of their favorite times to visit, camp, swim, take in nature, and fish—they usually have most of the site to themselves. The family is led by a local

---

1945) sought a system of organizing nature. Conducting his studies on vegetation in Nebraska, he postulated that the plant community behaves as a complex organism that grows and develops through stages, resembling the development of an individual organism, to a mature (climax) stage. Clements's theory of vegetation was criticized significantly by Arthur Tansley, a British ecologist, and others.

A new direction in ecology was given a boost in 1913 when Victor Shelford stressed the interrelationship of plants and animals. He conducted early studies on succession in the Indiana dunes and on experimental *physiological ecology*. Because of his work, ecology became a science of communities. His *Animal Communities in Temperate America* (1913) was one of the first books to treat ecology as a separate science. The well-known and influential biology educator Eugene P. Odum was one of Shelford's students.

The *New World Encyclopedia* (Ecology 2007) points out that in 1935, Tansley coined the term *ecosystem*—the interactive system established between a group of living creatures (biocoenosis), and the environment in which it lives (biotype). Tansley's ecosystem concept was adopted by Odum. Along with his brother, How-university professor of ecology and she knows the truth about the picnic-stream area and the dragonflies and other insects. She knows that dragonflies and damselflies are macroinvertebrate indicator organisms; they only inhabit, grow, and thrive in and around streams that are clean and healthy—when dragonflies and damselflies are around, they indicate nonpolluted water. Further, the ecology professor knows that dragonflies are valued as predators, friends, and allies in waging war against flies and controlling populations of harmful insects, such as mosquitoes. In regards to mosquitoes, dragonflies take the wrigglers in the water, and the adults, on swiftest wings (25–35 mph), hover over streams and ponds laying their eggs.

The ecology professor's husband, an amateur poet, also understood the significance of the presence of the indicator insects and had no problem sharing the same area with them. He also viewed the winged insects differently, with the eye of a poet. He knew that the poets have been lavish in their attention to the dragonflies and have paid them delightful tributes. James Witcomb Riley (1849–1916) says,

> Till the dragon fly, in light gauzy armor
> burnished bright,
> Came tilting down the waters in a wild,
> bewildered flight.

But perhaps James Russell Lowell's (1819–1891) poem, *The Fountain of Youth*, gives us the perfect description of these insects:

> In summer-noon flushes
> When all the wood hushes,
> Blue dragon-flies knitting
> To and fro in the sun,
> With sidelong jerk flitting,
> Sink down on the rushes.
> And, motionless sitting,
> Hear it bubble and run,
> Hear its low inward singing
> With level wings swinging
> On green tasseled rushes,
> To dream in the sun.

ard Odum, Eugene P. Odum wrote a textbook that (starting in 1953) educated multiple generations of biologists and ecologists in North American (including the author of this text). Eugene P. Odum is often called the "father of modern ecosystem ecology."

Human ecology began in the 1920s. About the same time, the study of populations split into two fields: *population ecology* and *evolutionary ecology*. Closely associated with population ecology and evolutionary ecology is *community ecology*. At the same time, physiological ecology arose. Later, natural history observations spawned behavioral ecology (Smith and Smith, 2006).

The history of ecology has been tied to advances in biology, physics, and chemistry that have spawned new areas of study in ecology, such as landscape, conservation, restoration, and global ecology. At the same time, ecology was ripe with conflicts and opposing camps. Smith (1996) notes that the first major split in ecology was between plant ecology and animal ecology, which even lead to a controversy over the term *ecology*, with botanists dropping the initial "o" from oecology, the spelling in use at the time, and zoologists refusing to use the term *ecology* at all, because of its perceived affiliation with botany. Other historical schisms were between organismal and individualist ecology, holism versus reductionism, and theoretical versus applied ecology (Ecology 2007).

Levels of Organization

Odum explains that "the best way to delimit modern ecology is to consider the concept of levels of organization" (Odum 1983). Levels of organization can be simplified as follows:

Organs → Organism → Population → Communities → Ecosystem → Biosphere

In this relationship, organs form an organism; organisms of a particular species form a population; populations occupying a particular area form a community; communities, interacting with nonliving or abiotic factors, separate in a natural unit to create a stable system known as the ecosystem (the major ecological unit); and the part of Earth in which an ecosystem operates is known as the biosphere. Tomera (1989) points out that "every community is influenced by a particular set of abiotic factors." The abiotic part of the ecosystem is represented by inorganic substances such as oxygen, carbon dioxide, several other inorganic substances, and some organic substances.

The physical and biological environment in which an organism lives is referred to as its habitat. For example, the habitat of two common aquatic insects, the "backswimmer" (*Notonecta*) and the "water boatman" (*Corixa*) is the littoral zone of ponds and lakes (shallow, vegetation-choked areas). (See figure 1.3; Odum 1983.)

Within each level of organization of a particular habitat, each organism has a special role. The role the organism plays in the environment is referred to as its niche. A niche might be that the organism is food for some other organism or is a

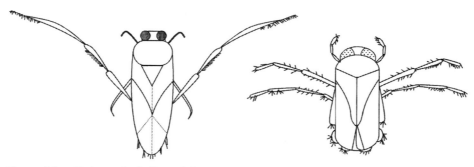

Figure 1.3. *Notonecta* (left) and *Corixa* (right)

Adapted from Eugene P. Odom, *Basic Ecology* (Saunders College Publishing, 1983), p. 402.

predator of other organisms. Odum refers to an organism's niche as its "profession" (Odum 1971). In other words, each organism has a job or role to fulfill in its environment. Although two different species might occupy the same habitat, "niche separation based on food habits" differentiates between two species. Such niche separation can be seen by comparing the niches of the water backswimmer and the water boatman. The backswimmer is an active predator, while the water boatman feeds largely on decaying vegetation (Odum 1983).

✔ *Important Note*: In order for an ecosystem to exist, a dynamic balance must be maintained among all biotic and abiotic factors—a concept known as homeostasis.

Ecosystem

As mentioned, *ecosystem*, a contraction of "ecological" and "system," is a term introduced by Tansley to denote an area that includes all organisms therein and their physical environment. Specifically, an ecosystem is defined as a geographic area and includes all the living organisms, their physical surroundings, and the natural cycles that sustain them. All of these elements are interconnected (U.S. Fish and Wildlife Service [USFWS] 2007). Simply, the ecosystem is the major ecological unit in nature. Elements of an ecosystem may include flora, fauna, lower life forms, water, and soil (Ecosystem 2007). "There is a constant interchange of the most various kinds within each system, not only between the organisms but between the organic and the inorganic" (Tansley 1935). Living organisms and their nonliving environment are inseparably interrelated and interact upon each other to create a self-regulating and self-maintaining system. To create a self-regulating and self-maintaining system, ecosystems are homeostatic, that is, they resist any change through natural controls. These natural controls are important in ecology. This is especially the case since it is people through their complex activities who tend to disrupt natural controls. Tansley regarded the ecosystem as not only the organism complex but also the whole complex of physical factors forming what we call the environment. It was first applied to levels of biological organization represented by units such community and the biome. Odum (1952) and Evans (1956) expanded the extent of the concept to include other levels of organization (USDA 1982).

✔ *Important Point*: According to the *Concise Oxford Dictionary of Ecology* (1994), "natural" is commonly defined as being "present in or produced by nature . . . with relatively little modification by humans."

✔ *Important Point*: Modern usage of the term *ecosystem* derives from the work done by Raymond Lindeman. Lindeman's central concepts were that of functional organization and ecological energy efficiency ratios. This approach is connected to ecological energetics and might also be thought of as environmental rationalism. It was subsequently applied by H. T. Odum in founding the transdiscipline known as *systems ecology* (Lindeman 1942).

As stated earlier, an ecosystem encompasses both the living and nonliving factors in a particular environment. The living or biotic part of the ecosystem is formed by two components: autotrophic and heterotrophic. The autotrophic (self-nourishing) component does not require food from its environment but can manufacture food from inorganic substances. For example, some autotrophic components (plants) manufacture needed energy through photosynthesis. Heterotrophic components, on the other hand, depend upon autotrophic components for food (Porteous 1992).

The nonliving or abiotic part of the ecosystem is formed by three components: inorganic substances, organic compounds (which link biotic and abiotic parts), and climate regime. Figure 1.4 is a simplified diagram showing a few of the living and nonliving components of an ecosystem found in a freshwater pond.

An ecosystem is a cyclic mechanism in which biotic and abiotic materials are constantly exchanged through biogeochemical cycles. Biogeochemical cycles are defined as follows: *bio* refers to living organisms and *geo* to water, air, rocks, or solids; *chemical* is concerned with the chemical composition of the Earth. Biogeochemical cycles are driven by energy, directly or indirectly, from the sun. They will be discussed later.

The simplified freshwater pond shown in figure 1.4 depicts an ecosystem where biotic and abiotic materials are constantly exchanged. Producers construct organic substances through photosynthesis and chemosynthesis. Consumers and

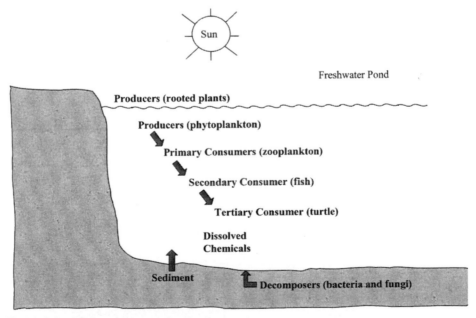

Figure 1.4.   Major components of a freshwater pond ecosystem

decomposers use organic matter as their food and convert it into abiotic components. That is, they dissipate energy fixed by producers through food chains. Fix, in this instance, means to change into a stable compound or to make into some other available form of energy to the consumers and decomposers. The abiotic part of the pond in figure 1.4 is formed of inorganic and organic compounds dissolved and in sediments such as carbon, oxygen, nitrogen, sulfur, calcium, hydrogen, and humic acids. The biotic part is represented by producers such as rooted plants and phytoplanktons. Fish, crustaceans, and insect larvae make up the consumers. Detrivores, which feed on organic detritus, are represented by mayfly nymphs. Decomposers make up the final abiotic part. They include aquatic bacteria and fungi, which are distributed throughout the pond.

As stated earlier, an ecosystem is a cyclic mechanism. From a functional viewpoint, an ecosystem can be analyzed in terms of several factors. The factors important in this study include biogeochemical cycles, energy, and food chains; these factors are discussed in detail later.

TYPES OF ECOSYSTEMS

Individual ecosystems consist of physical, chemical, and biological components. As mentioned, the physical and chemical components are known as abiotic (non-living components of the environment) factors that influence living organisms in both terrestrial and aquatic ecosystems. The abiotic factors are:

For terrestrial ecosystems
- Sunlight
- Temperature
- Precipitation
- Wind
- Latitude
- Altitude
- Fire frequency
- Soil

For aquatic ecosystems
- Light penetration
- Water currents
- Dissolved nutrient concentrations
- Suspended solids
- Salinity

The biotic (living environment) factors making up an ecosystem include the producers, consumers, and decomposers.

Abiotic and biotic factors combine to make up the following types of terrestrial and aquatic ecosystems:

- Estuaries
- Swamps and marshes
- Tropical rain forest
- Temperate forest
- Northern coniferous forest (taiga)
- Savanna
- Tundra (artic and alpine)
- Desert scrub

- Agricultural land
- Woodland and shrubland
- Temperate grassland
- Lake and streams
- Continental shelf
- Opean ocean
- Extreme desert

To illustrate one way in which the ecosystem classification is used, a real-world example from the United States Department of Agriculture is provided.

EXAMPLE ECOSYSTEM: AGROECOSYSTEM MODEL

What are the basic components of agroecosystems [agricultural ecosystems]? Just as natural ecosystems they can be thought of as including the processes of primary production, consumption, and decomposition interacting with abiotic environmental components and resulting in energy flow and nutrient cycling. Economic, social, and environment factors must be added to this primary concept because of the human element that is so closely involved with agroecosystem creation and maintenance.

Agroecosystem Characteristics

Agricultural ecosystems (referred to as agroecosystems) have been described by Odum (1984) as domesticated ecosystems. He states that they are in many ways intermediate between natural ecosystems (such as grasslands and forests) and fabricated ecosystems (cities).

Agroecosystems are solar powered (as are natural systems) but differ from natural systems in that

1. there are auxiliary energy sources that are used to enhance productivity; these sources are processed fuels along with animal and human labor;
2. species diversity is reduced by human management in order to maximize yield of specific foodstuffs (plant or animal);
3. dominant plant and animal species are under artificial rather than natural selection; and,
4. control is external and goal-oriented rather than internal via subsystem feedback as in natural ecosystems.

Agroecosystems do not happen without human intervention in the landscape. Therefore, creation of these ecosystems (and maintenance of them as well) is necessarily concerned with the (human) economic goals of production, productivity, and conservation. Agroecosystems are controlled, by definition, by management of ecological processes.

Crossley, et al. (1984) addressed the possible use of agroecosystem as a unifying and in many ways clarifying concept for proper management of managed landscape units. All ecosystems are open, that is, they exchange biotic and abiotic elements with other ecosystems. Agroecosystems are extremely open—with major exports of primary and secondary production (plant and animal production) as well as increased opportunity for loss of nutrient elements. Because modern agroecosystems are entirely dependent on human intervention, they would not persist but for that intervention. It is for this reason that they are sometimes referred to as artificial systems as opposed to natural systems that do not require intervention to persist through space and time.

Definitions of agroecosystems often include the entire support base of energy and material subsidies, seeds, and chemicals, and even a sociopolitical-economic matrix in which management decisions are made. Crossley (1984) stated that while this is logical, he preferred to designate the individual field as the agroecosystem because it is consistent with designating an individual forest catchment or lake as an ecosystem. He envisions the *farm system* as consisting of a set of agroecosystems—field with similar or different crops—together with support mechanisms and socioeconomic factors contributing to their management.

Agroecosystems retain most if not all the functional properties of natural ecosystems—nutrient conservation mechanisms, energy storage and use patterns, and regulation of biotic diversity.

Ecosystem Pattern and Process

Throughout the United States, the landscape consists of patches of natural ecosystems scattered (or imbedded) in a matrix of different agroecosystems and fabricated ecosystems. In fact, about three-quarters of the land area of the United States (USDA, 1982) is occupied by agroecosystems.

The pattern created by this interspersion incorporates elements of the variability of structure and separation of functions among the various ecosystems. Pattern variables quantify the structure and relationships between systems; process implies functional relationships between and within the biotic and abiotic ecosystem components. Within agroecosystems, processes include:

• enhanced productivity of producers through fertilization
• improved productivity through selective breeding
• management of pests with various control methods
• management of various aspects of the hydrologic cycle
• landforming

Agricultural Ecology

Simply stated, agricultural ecology is the study of agricultural ecosystems and their components as they function within themselves and in the context of the landscapes that contain them. Application of this knowledge can lead to development of more sustainable agricultural ecosystems in harmony with their larger ecosystem and ecoregion. (USDA 2007)

Summary of Key Terms

Abiotic factor—the nonliving part of the environment composed of sunlight, soil, mineral elements, moisture, temperature, topography, minerals, humidity, tide, wave action, wind, and elevation.

✔   *Important Note*: Every community is influenced by a particular set of abiotic factors. While it is true that the abiotic facts affect the community members, it is also true that the living (biotic factors) may influence the abiotic factors. For example, the amount of water lost through the leaves of plants may add to the moisture content of the air. Also, the foliage of a forest reduces the amount of sunlight that penetrates the lower regions of the forest. The air temperature is therefore much lower than in nonshaded areas (Tomera 1989).

Autotroph—green plants that fix energy of the sun and manufacture food from simple, inorganic substances.

Biogeochemical cycles—cyclic mechanisms in all ecosystems by which biotic and abiotic materials are constantly exchanged.

Biotic factor (community)—the living part of the environment composed of organisms that share the same area, are mutually sustaining, interdependent, and constantly fixing, utilizing, and dissipating energy.

Community—in an ecological sense, community includes all the populations occupying a given area.

Consumers and decomposers—energy dissipated and fixed by the producers through food chains or webs. The available energy decreases by 80–90 percent during transfer from one trophic level to another.

Ecology—the study of the interrelationship of an organism or a group of organisms and their environment.

Ecosystem—the community and the nonliving environment functioning together as an ecological system.

Environment—everything that is important to an organism in its surroundings.

Heterotrophs—animals that use food stored by the autotroph, rearrange it, and finally decompose complex materials into simple inorganic compounds. Heterotrophs may be carnivorous (meat eaters), herbivorous (plant eaters), or omnivorous (plant and meat eaters).

Homeostasis—a natural occurrence during which an individual population or an entire ecosystem regulates itself against negative factors and maintains an overall stable condition.

Niche—the role that an organism plays in its natural ecosystem, including its activities, resource use, and interaction with other organisms.

Pollution—an adverse alteration to the environment by a pollutant.

Chapter Review Questions

✔   *Important Note*: Answers to chapter review questions are found in the appendix.

1.1   Another word for environmental interrelationship is _____.
1.2   Define autecology.

Label the Pond exercise:

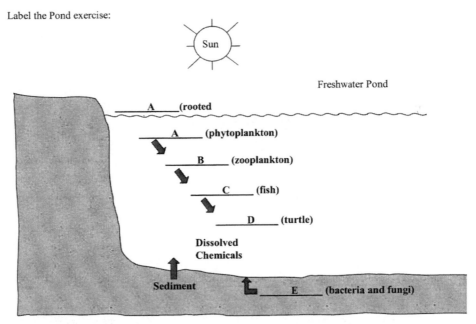

Figure 1.5.   Exercise: Major components of a freshwater pond ecosystem

1.3   Define synecology.

1.4   What is a niche?

1.5   Define pollution.

1.6   List eight abiotic factors found in a typical ecosystem.

1.7   Label the Pond Exercise: Identify the lettered components in figure 1.5 below. Insert the correct descriptions from the list to correspond with the correct letter.

primary consumers   A. _____

secondary consumers   B. _____

producers   C. _____

tertiary consumers   D. _____

decomposers   E. _____

Cited References and Recommended Reading

Barlocher, R., and Kendrick, L. 1975. Leaf conditioning by microorganisms. *Oecologia* 20:359–62.

Benfield, E. F. 1996. Leaf breakdown in streams ecosystems. In *Methods in stream ecology*, ed. F. R. Hauer and G. A. Lambertic, 579–90. San Diego: Academic Press.

Benfield, E. F., Jones, D. R., and Patterson, M. F. 1977. Leaf pack processing in a pastureland stream. *Oikos* 29:99–103.

Benjamin, C. L., Garman, G. R., and Funston, J. H. (1997). *Human biology*. New York: McGraw-Hill.

Carson, R. 1962. *Silent spring*. Boston: Houghton Mifflin.

Clements, E. S. 1960. *Adventures in ecology*. New York: Pageant Press.

Crossley, D. A., Jr., House, G. J., Snider, R. M., Snider, R. J., and Stinner, B. R. 1984. The positive interactions in agroecosystems. In *Agricultural ecosystems*, ed. R. Lowrance, B. R. Stinner, and G. J. House. New York: John Wiley & Sons.

Cummins, K. W. 1974. Structure and function of stream ecosystems. *Bioscience* 24:631–41.

Cummins, K. W., and Klug, M. J. 1979. Feeding ecology of stream invertebrates. *Annual Review of Ecology and Systematics* 10:631–41.

Darwin, C. 1998. *The origin of species*, ed. G. Suriano. New York: Grammercy.

Dolloff, C. A., and Webster, J. R. 2000. Particulate organic contributions from forests to streams: Debris isn't so bad. In *Riparian management in forests of the continental eastern United States*, ed. E. S. Verry, J. W. Hornbeck, and C.A. Dolloff. Boca Raton, FL: Lewis.

Ecology. 2007. *New World Encyclopedia*. http://www.newworldencyclopedia.org/preview/Ecology (accessed February 10, 2007).

Ecosystem. 2007. *Wikipedia*. http://en.wikipedia.org/wiki/Ecosystem (accessed February 11, 2007).

Evans, F. C. 1956. Ecosystem as the basic unit in ecology. *Science* 23:1127–28.

Krebs, C. H. 1972. *Ecology: The experimental analysis of distribution and abundance*. New York: Harper & Row.

Lindeman, R. L. 1942. The trophic-dynamic aspect of ecology. *Ecology* 23:399–418.

Margulis, L., and Sagan, D. 1997. *Microcosmos: Four billion years of evolution from our microbial ancestors*. Berkeley: University of California Press.

Marshall, P. 1950. *Mr. Jones, meet the master*. New York: Fleming H. Revel.

Odum, E. P. 1952. *Fundamentals of ecology* (1st ed.). Philadelphia: W. B. Saunders.

Odum, E. P. 1971. *Fundamentals of ecology* (3rd ed.). Philadelphia: W. B. Saunders.

Odum, E. P. 1983. *Basic ecology*. Philadelphia: Saunders College.

Odum, E. P. 1984. Properties of agroecosystems. In *Agricultural ecosystems*, ed. R. Lowrance, B. R. Stinner, and G. J. House. New York: John Wiley & Sons.

Odum, E. P., and Barrett, G. W. 2005. *Fundamentals of ecology* (5th ed.). Belmont, CA: Thomson Brooks/Cole.

Paul, R. W., Jr., Benfield, E. F., and Cairns, J., Jr. 1978. Effects of thermal discharge on leaf decomposition in a river ecosystem. *Verhandlugen der Internationalen Vereinigung fur Thoeretsche and Angewandte Limnologie* 20:1759–66.

Peterson, R. C., and Cummins, K. W. 1974. Leaf processing in woodland streams. *Freshwater Biology* 4:345–68.

Porteous, A. 1992. *Dictionary of environmental science and technology*. New York: John Wiley & Sons.

Ramalay, F. 1940. The Growth of a science. *University of Colorado Studies* 26:3–14.

Smith, C. H. 2007. Karl Ludwig Willldenow. *Western Kentucky University*. http://www.wku.edu/~smithch/chronob/WILL1765.htm (accessed February 9, 2007).

Smith, R. L. 1996. *Ecology and field biology*. New York: HarperCollins College.

Smith, T. M., and Smith, R. L. 2006. *Elements of ecology* (6th ed.). San Francisco: Pearson, Benjamin Cummings.

Spellman, F. R. 1996. *Stream ecology and self-purification*. Lancaster, PA: Technomic.

Suberkoop, K., Godshalk, G. L., and Klug, M. J. 1976. Changes in the chemical composition of leaves during processing in a woodland stream. *Ecology* 57:720–27.

Tansley, A. G. 1935. The use and abuse of vegetational concepts and terms. *Ecology* 16:284–307.

Tomera, A. N. 1989. *Understanding basic ecological concepts*. Portland, ME: J. Weston Walch.

U.S. Department of Agriculture (USDA). 1982. *Agricultural statistics 1982*. Washington, D.C.: U.S. Government Printing Office.

U.S. Department of Agriculture (USDA). 1999. *Autumn colors—How leaves change color.*
http://www.na.fs.fed.us/spfo/pubs/misc/autumn/autumn_colors.htm (accessed February
8, 2007).

U.S. Department of Agriculture (USDA). 2007. *Agricultural ecosystems and agricultural ecology.*
http://www.nrcs.usda.gov/technical/ECS/agecol/ecosystem.html (accessed February 11,
2007).

U.S. Fish and Wildlife Service (USFWS). 2007. *Ecosystem conservation.* http://www.fws.gov/
ecosystems/ (accessed February 11, 2007).

Running waters, Shenandoah National Park, Virginia.
Photograph by Frank R. Spellman

CHAPTER 2

Biogeochemical Cycles

Some time ago I heard of an old man down on a hill farm in the South, who sat on his front porch as a newcomer to the neighborhood passed by. The newcomer to make talk said, "Mister, how does the land lie around here?" The old man replied, "Well—I don't know about the land alying; it's these real estate people that do the lying."

In a very real sense the land does not lie; it bears a record of what men write on it. In a larger sense a nation writes its record on the land, and civilization writes its record on the land—a record that is easy to read by those who understand the simple language of the land.

—W. C. Lowdermilk, 1953

Topics

Nutrient Cycles

All matter [i.e., carbon, nitrogen, oxygen, or molecules (water)] cycles; it is neither created nor destroyed. Because the Earth is essentially a closed system with respect to matter, it can be said that all matter on Earth cycles. Ecologists study the flow of nutrients in ecosystems.

Ecosystem elements, such as streams—and their larger cousins, rivers—are complex ecosystems that take part in the physical and chemical cycles (biogeochemical cycles) that shape our planet and allow life to exist. A *biogeochemical cycle* is composed of bioelements (chemical elements that cycle through living organisms) and occurs when there is an interaction between the biological and physical exchanges of bioelements. In a biogeochemical cycle, nutrient cycling and recycling through ecosystems results from the actions of geology, meteorology, and living things. Various nutrient biogeochemical cycles include (Spellman 1996):

- water cycle
- carbon cycle
- oxygen cycle
- nitrogen cycle
- phosphorus cycle
- sulfur cycle

✔ *Important Note*: Contrary to an incorrect assumption, energy does not cycle through an ecosystem, chemicals do. The inorganic nutrients cycle through more than the organisms; however, they also enter into the oceans, atmosphere, and even rocks. Since these chemicals cycle through both the biological and the geological world, we call the overall cycles "biogeochemical cycles."

Each chemical has its own unique cycle, but all of the cycles do have some things in common. Reservoirs are those parts of the cycle where the chemical is held in large quantities for long periods of times (e.g., in the oceans for water and in rocks for phosphorous). In exchange pools, on the other hand, the chemical is held for only a short time (e.g., the atmosphere, a cloud). The length of time a chemical is held in an exchange pool or a reservoir is termed its residence time. The biotic community includes all living organisms. This community may serve as an exchange pool (although for some chemicals such as carbon, bound in certain tree species for a thousand years, it may seem more like a reservoir) and also serve to move chemicals (bioelements) from one stage of the cycle to another. For instance, the trees of the tropical rain forest bring water up from the forest floor to be transpired into the atmosphere. Likewise, coral organisms take carbon from the water and turn it into limestone rock. The energy for most of the transportation of chemicals from one place to another is provided either by the sun or by the heat released from the mantle and core of the Earth (Spellman 1996).

✔ *Important Point:* Water is exchanged between the hydrosphere, lithosphere, atmosphere, and biosphere. The oceans are large reservoirs that store water; they ensure thermal and climatic stability.

In addition to these exchanges of nutrients (losses and gains), Abedon (1997) points out examples of other losses and gains:

- Minerals can be lost from ecosystems by the action of rain.
- Nutrients can also be carried into ecosystems by the action of wind or migrating animals.

- Movement of salmon up rivers is an example of how nutrients might be delivered into an upstream ecosystem (e.g., from the oceans back to terrestrial forests).
- A consequence of ecosystem disruption is an impaired ability to recycle nutrients, which leads to nutrient loss and long-term ecosystem impoverishment.
- In general, a disturbed habitat probably loses (rather than recycles) nutrients to a much greater degree than an undisturbed habitat where the actions of human activities are not necessarily rapidly or readily reversible. A common consequence of human disturbance of ecosystems and the associated irreversible loss of nutrients is desertification.

In the case of chemical elements that cycle through living things—that is, bioelements (Illinois State Water Survey [ISWS] 2005):

- All bioelements reside in compartments or defined spaces in nature.
- A compartment contains a certain quantity, or pool, of bioelements.
- Compartments exchange bioelements. The rate of movement of bioelements between two compartments is called the flux rate.
- The average length of time a bioelement remains in a compartment is called the mean residence time (MRT).
- The flux rate and pools of bioelements together define the nutrient cycle in an ecosystem.
- Ecosystems are not isolated from one another, and bioelements come into an ecosystem through meteorological, geological, or biological transport mechanisms:
  - meteorological (e.g., deposition in rain and snow, atmospheric gases)
  - ecological (e.g., surface and subsurface drainage)
  - biological (e.g., movement of organisms between ecosystems)

As a result, biogeochemical cycles can be:

- local
- global

Smith (1974) categorizes biogeochemical cycles into two types: the gaseous and the sedimentary. Gaseous cycles include the carbon and nitrogen cycles. The main pool (or sink) of nutrients in the gaseous cycle is the atmosphere and the ocean. The sedimentary cycles include the sulfur and phosphorous cycles. The main sink for sedimentary cycles is soil and rocks of the Earth's crust.

Between 20 and 40 elements of the Earth's 92 naturally occurring elements are ingredients that make up living organisms. The chemical elements carbon, hydrogen, oxygen, nitrogen, and phosphorus are critical in maintaining life as we know it on Earth. Odum (1971) points out that of the elements needed by living organisms to survive—oxygen, hydrogen, carbon, and nitrogen—are needed in larger quantities than are some of the other elements. The point is—no matter what particular elements are needed to sustain life, these elements exhibit definite biogeochemical cycles. These biogeochemical cycles will be discussed in detail later. For now it is important to cover the life-sustaining elements in greater detail.

The elements needed to sustain life are products of the global environment. The global environment consists of three main subdivisions, as shown in figure 2.1.

1. Hydrosphere—includes all the components formed of water bodies on the Earth's surface.
2. Lithosphere—comprises the solid components, such as rocks.
3. Atmosphere—is the gaseous mantle that envelopes the hydrosphere and lithosphere.

To survive, organisms require inorganic metabolites from all three parts of the biosphere. For example, the hydrosphere supplies water as the exclusive source of needed hydrogen. Essential elements such as calcium, sulfur, and phosphorus are provided by the lithosphere. Finally, oxygen, nitrogen, and carbon dioxide are provided by the atmosphere.

Within the biogeochemical cycles, all the essential elements circulate from the environment to organisms and back to the environment. Because of the critical importance of elements in sustaining life, it may be easily understood why the biogeochemical cycles are readily and realistically labeled nutrient cycles.

Through these biogeochemical or nutrient cycles, nature processes and re-processes the critical life-sustaining elements in definite inorganic-organic cycles. Some cycles, such as carbon, are more perfect than others; that is, there is no loss

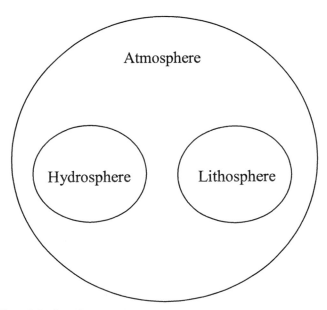

Figure 2.1. The global environment

of material for long periods of time. One major point to keep in mind: energy (to be explained later) flows "through" an ecosystem, but nutrients are cycled and recycled.

Humans need most of these recycled elements to survive. Because we need almost all the elements in our complex culture, we have speeded up the movement of many materials so that the cycles tend to become imperfect or what Odum (1971) calls acyclic. Odum goes on to explain that our environmental impact on phosphorus demonstrates one example of a somewhat imperfect cycle.

> We mine and process phosphate rock with such careless abandon that severe local pollution results near mines and phosphate mills. Then, with equally acute myopia we increase the input of phosphate fertilizers in agricultural systems without controlling in any way the inevitable increase in run-off output that severely stresses our waterways and reduces water quality through eutrophication. (Odum 1971)

As related above, in agricultural ecosystems, we often supply necessary nutrients in the form of fertilizer to increase plant growth and yield. In natural ecosystems, however, these nutrients are recycled naturally through each trophic level. For example, the elemental forms are taken up by plants. The consumers ingest these elements in the form of organic plant material. Eventually, the nutrients are degraded to the inorganic form again. The following pages present and discuss the nutrient cycles for water, carbon, nitrogen, phosphorus, and sulfur.

Water Cycle

Simply, the water cycle describes how water moves through the environment and identifies the links between groundwater, surface water, and the atmosphere (see figure 2.2). As illustrated in figure 2.2 and the Water Cycle Box Model below, water is taken from the Earth's surface to the atmosphere by evaporation from the surface of lakes, rivers, streams, and oceans. This evaporation process occurs when the sun heats water. The sun's heat energizes surface molecules, allowing them to break free of the attractive force binding them together and then evaporate and rise as invisible vapor in the atmosphere. Water vapor is also emitted from plant leaves by a process called *transpiration*. Every day, an actively growing plant transpires five to ten times as much water as it can hold at once; for example, one acre of corn gives off 3,000–4,000 gallons (11,400–15,100 liters) of water per day, and a large oak tree can transpire 40,000 gallons (151,000 liters) per year (U.S. Geological Survey [USGS] 2005). As water vapor rises, it cools and eventually condenses, usually on tiny particles of dust in the air. When it condenses, it becomes a liquid again or turns directly into a solid (ice, hail, or snow). These water particles then collect and form clouds. The atmospheric water formed in clouds eventually falls to Earth as precipitation. The precipitation can contain contami-

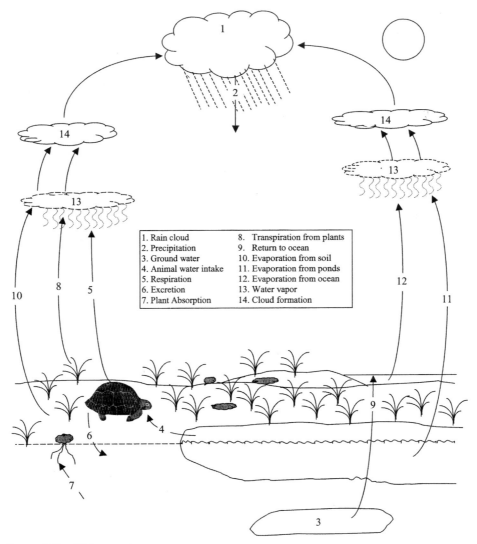

1. Rain cloud
2. Precipitation
3. Ground water
4. Animal water intake
5. Respiration
6. Excretion
7. Plant Absorption
8. Transpiration from plants
9. Return to ocean
10. Evaporation from soil
11. Evaporation from ponds
12. Evaporation from ocean
13. Water vapor
14. Cloud formation

Figure 2.2. Water cycle

nants from air pollution. The precipitation may fall directly onto surface waters, be intercepted by plants or structures, or fall onto the ground. Most precipitation falls in coastal areas or in high elevations. Some of the water that falls in high elevations becomes runoff water, the water that runs over the ground (sometimes collecting nutrients from the soil) to lower elevations to form streams, lakes, and fertile valleys.

Evapotranspiration is the sum of evaporation and plant transpiration. In regards to evapotranspiration and the water cycle, evapotranspiration is a significant water loss from a watershed. Types of land use and vegetation significantly affect evapotranspiration, and therefore the amount of water leaving a watershed. Evapotranspiration cannot be measured directly but may be estimated by creating an equation of the water balance of a watershed. The equation balances the change in water stored in the basin (S) with inputs and exports:

$$S = P - ET - Q - D \qquad (2.1)$$

where

S = water stored in watershed
P = input in precipitation
ET = missing flux
Q = stream flow
D = groundwater recharge

The water we see is known as *surface water*. Surface water can be broken down into five categories: oceans, lakes, rivers (streams), estuaries, and wetlands.

WATER RESERVOIRS, FLUXES, AND RESIDENCE TIMES

Earth's water reservoirs include the atmosphere, oceans, land surfaces (lakes and rivers), land subsurfaces (groundwater), and ice (glaciers). Water cycle fluxes include precipitation, evaporation, transpiration included in evaporation, surface runoff, subsurface runoff, infiltration, springs, and human use. Water cycle residence times are accumulated while in the atmosphere, ocean, streams and rivers, and groundwater. Table 2.1 lists quantities for water reservoirs, fluxes, and residence times.

Because the amount of rain and snow remains almost constant but population and usage per person are both increasing rapidly, water is in short supply. In the United States alone, water usage is four times greater today than it was in 1900. In the home, this increased use is directly related to an increase in the number of bathrooms, garbage disposals, home laundries, and lawn sprinklers. In industry, usage has increased 13 times since 1900.

More than 170,000 small-scale suppliers provide drinking water to approximately 200 million or more Americans by more than 60,000 community water supply systems, and to nonresidential locations, such as schools, factories, and campgrounds. The rest of Americans are served by private wells. The majority of the drinking water used in the United States is supplied from groundwater. Untreated water drawn from groundwater and surface waters, and used as a drinking water supply, can contain contaminants that pose a threat to human health.

Carbon Cycle

Carbon, which is an essential ingredient of all living things, is one of the most abundant elements in the solar system; it is the basic building block of the large organic molecules necessary for life. Inorganic forms of carbon (carbon dioxide,

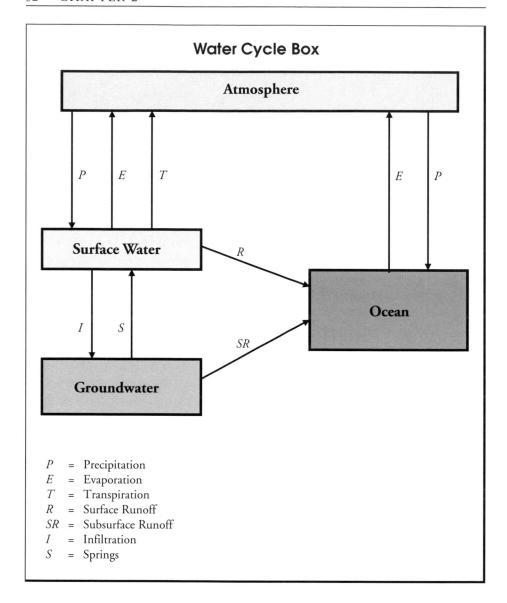

bicarbonate, and carbonate) strongly affect the acidity of soils and natural waters, the heat-insulating capability of the atmosphere, and the rates of such key natural processes as photosynthesis, weathering, and biomineralization. In reduced form, carbon provides the elemental backbone for myriad organic molecules that comprise living organisms, soil humus, and fossil fuels (National Science Foundation [NSF] 2000). Carbon is cycled into food chains from the atmosphere, as shown in figure 2.3 and the Carbon Cycle Box Model.

The carbon cycle (see figure 2.3) is based on carbon dioxide, which makes up only a small percentage of the atmosphere. From figure 2.3, it can be seen that

Table 2.1.   Water Reservoirs, Fluxes and Residence Times

Reservoirs		km^3	%
Atmosphere		12,700	0.001
Ocean		1,230,000,000	97.2
Land surface			
Lakes		123,000	0.009
Rivers and streams		1,200	0.0001
Land subsurface (groundwater)		4,000,000	0.31
Ice (glaciers)		28,600,000	2.15

Fluxes			km^3/yr
Precipitation		total	496,000
Land			111,000
Ocean			385,000
Evaporation (includes transpiration)		total	496,000
Land			71,000
Ocean			425,000
Surface runoff			26,000
Subsurface runoff		liquid	12,000
Ice			2,000
Infiltration			14,000
Springs			2,000
Human Use		total	3,000

Residence	Residence Time
Atmosphere	0.03 year or 9 days
Ocean	2,900 years
Streams and rivers	0.05 year or 17 days
Groundwater	330 years

green plants obtain carbon dioxide (CO_2) from the air and, through photosynthesis—described by Asimov as the "most important chemical process on Earth"—it produces the food and oxygen that all organisms live on (Asimov 1989). Part of the carbon produced remains in living matter and the other part is released as CO_2 in cellular respiration. Miller points out that the carbon dioxide released by cellular respiration in all living organisms is returned to the atmosphere (Miller 1988).

✔ *Important Note*: About a tenth of the estimated 700 billion tons of carbon dioxide in the atmosphere is fixed annually by photosynthetic plants. A fur-

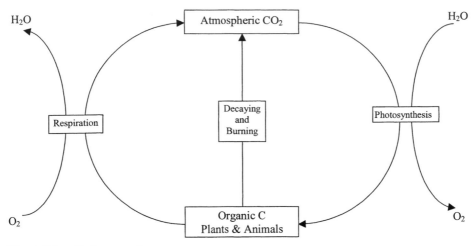

Figure 2.3. Carbon cycle

ther trillion tons are dissolved in the ocean, more than half in the photosynthetic layer.

Some carbon is contained in buried dead animal and plant materials. Much of these buried plant and animal materials were transformed into fossil fuels. Fossil fuels, coal, oil, and natural gas contain large amounts of carbon. When fossil fuels are burned, stored carbon combines with oxygen in the air to form carbon dioxide, which enters the atmosphere (Moran, Morgan, and Wiersma 1986).

In the atmosphere, carbon dioxide acts as a beneficial heat screen as it does not allow the radiation of Earth's heat into space. This balance is important. The problem is that as more carbon dioxide from burning is released into the atmosphere, this balance is altered. Odum (1983) warns that the recent increases in consumption of fossil fuels "coupled with the decrease in the 'removal capacity' of the green belt is beginning to exceed the delicate balance." Massive increases of carbon dioxide into the atmosphere tend to increase the possibility of global warming. The consequences of global warming "would be catastrophic . . . and the resulting climatic change would be irreversible" (Abrahamson 1988).

Nitrogen Cycle

Nitrogen is important to all life because it is a necessary nutrient. Nitrogen in the atmosphere or in the soil can go through many complex chemical and biological changes, be combined into living and nonliving material, and return back to the soil or air in a continuing cycle. This is called the nitrogen cycle (Killpack and Buchholz 1993). The nitrogen cycle consists of the following processes and various states:

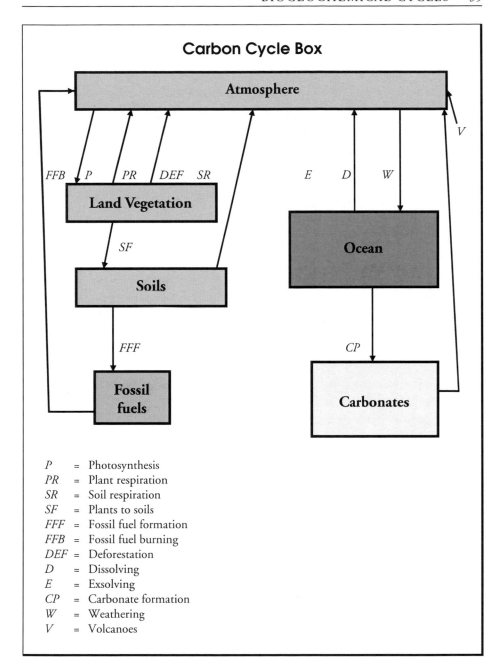

Carbon Cycle Box

P	= Photosynthesis
PR	= Plant respiration
SR	= Soil respiration
SF	= Plants to soils
FFF	= Fossil fuel formation
FFB	= Fossil fuel burning
DEF	= Deforestation
D	= Dissolving
E	= Exsolving
CP	= Carbonate formation
W	= Weathering
V	= Volcanoes

NITROGEN CYCLE

Processes   Immobilization
Fertilizer   Nitrification
Volatilization   Biological fixation
Animal wastes   Mineralization
Organic matter   Denitrification
Crop uptake   *States*
                       N_2—elemental nitrogen is a gaseous form of nitrogen
                       NH_3—ammonia is a gaseous form of nitrogen
                       NO—nitric oxide is a gaseous for of nitrogen
                       $NH_4{}^+$—ammonium is attracted to soil particles
                       N_2O—nitrous oxide is a gaseous form of nitrogen
                       NO_3—nitrate is not attracted to soil particles

IMPORTANT NITROGEN CYCLE TERMINOLOGY

- *Limiting nutrient*—amount of an element necessary for plant life is in short supply.
- *Nitrogen fixation*—chemical conversion from N_2 to NH_3 (ammonia) or NO_3 (nitrate)
- *Denitrification*—chemical conversion from nitrate (NO_3) back to N_2.

Today, we readily apply nitrogen in the form of fertilizer to plants to ensure proper growth. Before fertilizer containing nitrogen was commercially available, how did agriculture survive? Early farmers had to rely on natural regeneration of fixed nitrogen:

- annual floods—bring fresh sediments (e.g., Nile Valley)
- slash/burn agriculture—once the soil nutrients are depleted, move on to a new place
- crop rotation—certain crops (e.g., soybeans) are good at fixing nitrogen, others (e.g. corn) use it up; plant on alternate years

The atmosphere contains 78 percent, by volume, of nitrogen. Moreover, as stated above, nitrogen is an essential element for all living matter and constitutes 1–3 percent dry weight of cells, yet nitrogen is not a common element on Earth. Although it is an essential ingredient for plant growth, it is chemically very inactive; before it can be incorporated by the vast majority of the biomass, it must be "fixed" (Porteous 1992).

Price describes the nitrogen cycle as an example "of a largely complete chemical cycle in ecosystems with little leaching out of the system." From the water/wastewater specialist's point of view, nitrogen and phosphorous are both commonly considered as limiting factors for productivity. Of the two, nitrogen is harder to control but is found in smaller quantities in wastewater.

As stated earlier, nitrogen gas makes up about 78 percent of the volume of the Earth's atmosphere. As such, it is useless to most plants and animals. Fortunately, nitrogen gas is converted into compounds containing nitrate ions, which are taken up by plant roots as part of the nitrogen cycle, shown in simplified form in figure 2.4 and in box model form below.

Aerial nitrogen is converted into nitrates mainly by microorganisms, bacteria, and blue-green algae. Lightning also converts some aerial nitrogen gas into forms that return to the Earth as nitrate ions in rainfall and other types of precipitation. From figure 2.4, it can be seen that ammonia plays a major role in the nitrogen cycle. Excretion by animals and anaerobic decomposition of dead organic matter by bacteria produce ammonia. Ammonia, in turn, is converted by nitrification bacteria into nitrites and then into nitrates. This process is known as nitrification. Nitrification bacteria are aerobic. Bacteria that convert ammonia into nitrites are known as nitrite bacteria (*Nitrosococcus* and *Nitrosomonas*); they convert nitrites into nitrates and nitrate bacteria (*Nitrobacter*). In wastewater treatment, ammonia is produced in the sludge digester, nitrates in the aerobic sewage treatment process.

Nitrogen reservoirs and quantities in millions of metric tons are:

• Atmosphere: 4,000,000,000
• Land plants: 3500
• Soils: 9500

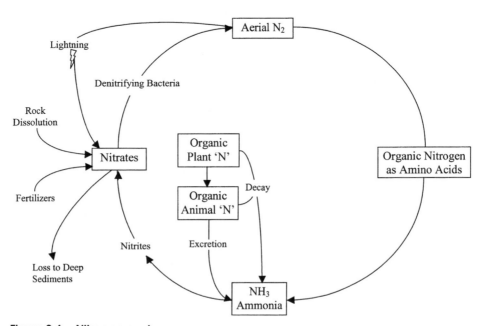

Figure 2.4. Nitrogen cycle

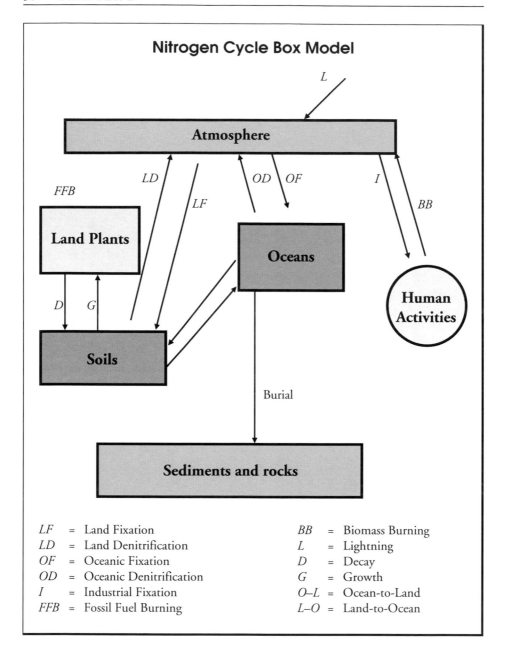

Nitrogen Cycle Box Model

LF	= Land Fixation	BB	= Biomass Burning
LD	= Land Denitrification	L	= Lightning
OF	= Oceanic Fixation	D	= Decay
OD	= Oceanic Denitrification	G	= Growth
I	= Industrial Fixation	O–L	= Ocean-to-Land
FFB	= Fossil Fuel Burning	L–O	= Land-to-Ocean

- Oceans: 23,000,000
- Sediments and rocks: 200,000,000,000

✔ *Important Point:* Buried rocks and sediments are the largest pool of nitrogen, but this reservoir is a minor part of the cycle.

In their voluminous and authoritative text *Wastewater Engineering*, Metcalf and Eddy, Inc. (1991) devotes several pages to describing the nitrogen cycle and its impact upon the wastewater treatment process. They point out that nitrogen is found in wastewater in the form of urea. During wastewater treatment, the urea is transformed into ammonia nitrogen. Since ammonia exerts a biochemical oxygen demand (BOD) and a chlorine demand, high quantities of ammonia in wastewater effluents are undesirable. The process of nitrification is utilized to convert ammonia to nitrates. Nitrification is a biological process that involves the addition of oxygen to the wastewater. If further treatment is necessary, another biological process called denitrification is used. In this process, nitrate is converted into nitrogen gas, which is lost to the atmosphere, as can be seen in figure 2.4.

✒ *Important Point:* Specialized bacteria and lightning are the only natural ways that nitrogen is fixed.

When attempting to address the important and complex factors that make up the topic of ecology, it is important to understand the impact that the nitrogen cycle can have on effluent that is dumped (outfalled) into the environment. At the same time, in regard to wastewater treatment, for example, one should remember that the nitrogen cycle that occurs in the wastewater stream is not the primary source of the nitrogen contamination of surface water bodies. As a case in point, Price (1984) points to the example of large inputs of nitrogen fertilizer from agricultural systems that "may result in considerable leaching and unidirectional flow of nitrogen into aquatic systems which become polluted with excessive nitrogen."

✒ *Important Note:* Spellman (1996) points out that nitrogen is a concern in stream quality when nitrogen in the soil is converted to nitrate (NO_3^-) form. It is a concern because nitrate is very mobile and easily moves with water in the soil. The concern of nitrates and water quality is generally directed at groundwater. However, nitrates can also enter surface waters such as ponds, streams, and rivers. Nitrate in drinking water can lead to a serious problem. Specifically, nitrate poisoning in infant humans or animals can cause serious problems and even death. This is the case because of a bacteria commonly found in the intestinal tract of infants that can convert nitrate to highly toxic nitrites (NO_2). Nitrite can replace oxygen in the bloodstream and result in oxygen starvation that causes a bluish discoloration of the infant ("blue baby" syndrome).

✒ *Important Point:* Lightning may have been necessary for life to begin:

No Lightning = → no bacteria = → no bacterial fixation = → no usable nitrogen = → no life . . .

Phosphorus Cycle

Phosphorus (P) is another element that is common in the structure of living organisms. However, of all the elements recycled in the biosphere, phosphorus is the scarcest and therefore the one most limiting in any given ecological system. It is indispensable to life, being intimately involved in energy transfer and in the passage of genetic information in the DNA of all cells.

The ultimate source of phosphorus is rock, from which it is released by weathering, leaching, and mining. Phosphorus has no stable gas phase, so the addition of P to the land is slow. Phosphorus occurs as phosphate or other minerals formed in past geological ages. These massive deposits are gradually eroding to provide phosphorus to ecosystems. A large amount of eroded phosphorus ends up in deep sediments in the oceans and lesser amounts in shallow sediments. Part of the phosphorus comes to land when marine animals are brought out. Birds also play a role in the recovery of phosphorus. The great guano deposit, bird excreta, of the Peruvian coast is an example. Man has hastened the rate of loss of phosphorus through mining activities and the subsequent production of fertilizers, which are washed away and lost. Even with the increase in human activities, however, there is no immediate cause for concern, since the known reserves of phosphate are quite large.

✔ *Important Point*: Humans have greatly accelerated P transfer from rocks to plants and soils (about five times faster than weathering).

Phosphorous has become very important in water quality studies, since it is often found to be a limiting factor (i.e., limiting plant nutrient). Control of phosphorus compounds that enter surface waters and contribute to growth of algal blooms is of much interest to stream ecologists. Phosphates, upon entering a stream, act as fertilizer, which promotes the growth of undesirable algae populations or algal blooms. As the organic matter decays, dissolved oxygen levels decrease, and fish and other aquatic species die. Figure 2.5 shows the phosphorus cycle. The Phosphorus Cycle Box Model is shown below.

While it is true that phosphorus discharged into streams is a contributing factor to stream pollution (and causes eutrophication), it is also true that phosphorus is not the lone factor. Odum (1975) warns against what he calls the one-factor control hypothesis, that is, the one-problem/one solution syndrome. He goes on to point out that environmentalists in the past have focused on one or two items, such as phosphorous contamination, and "have failed to understand that the strategy for pollution control must involve reducing the input of all enriching and toxic materials."

✔ *Important Point*: Because of its high reactivity, phosphorus exists in combined form with other elements. Microorganisms produce acids that form soluble phosphate from insoluble phosphorus compounds. The phosphates are utilized by algae and terrestrial green plants, which in turn pass into the bodies of animal consumers. Upon death and decay of an organism, phosphates are released for recycling (Spellman 1996).

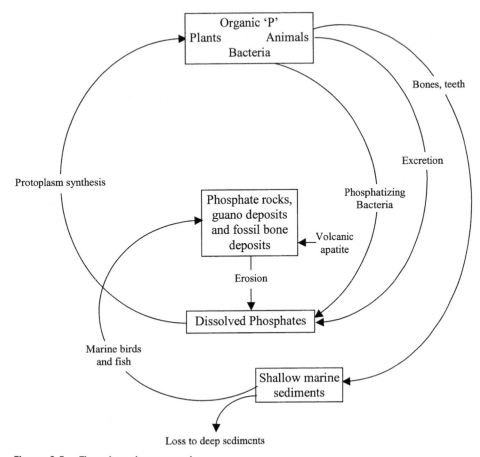

Figure 2.5.   The phosphorus cycle

Sulfur Cycle

Sulfur, like nitrogen and carbon, is characteristic of organic compounds. However, an important distinction between the cycling of sulfur and the cycling of nitrogen and carbon is that sulfur is "already fixed." That is, plenty of sulfate anions are available for living organisms to utilize; the largest reservoir is Earth's crust. By contrast, the major biological reservoirs of nitrogen atoms (N_2) and carbon atoms (CO_2) are gases that must be pulled out of the atmosphere. Sulfur is rarely a limiting nutrient for ecosystems or organisms.

✔   *Important Point*: In the sulfur cycle, elementary sulfur of the lithosphere is not available to plants and animals unless converted to sulfates.
   The sulfur cycle (see figure 2.6) is both sedimentary and gaseous. Tchobanoglous and Schroeder (1985) note that "the principal forms of sulfur that are of special significance in water quality management are organic sulfur, hydrogen sulfide, elemental sulfur and sulfate."

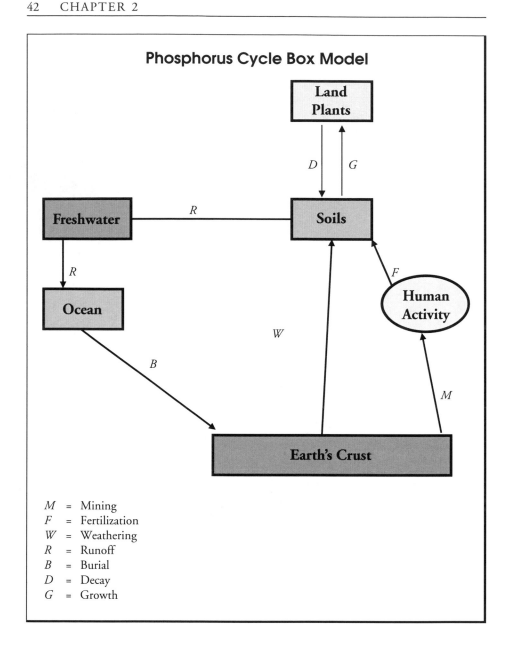

Phosphorus Cycle Box Model

Land Plants

Freshwater

Soils

Ocean

Human Activity

Earth's Crust

D
G
R
R
F
W
B
M

M = Mining
F = Fertilization
W = Weathering
R = Runoff
B = Burial
D = Decay
G = Growth

Bacteria play a major role in the conversion of sulfur from one form to another. In an anaerobic environment, bacteria break down organic matter, producing hydrogen sulfide with its characteristic rotten egg odor. A bacterium called *Beggiatoa* converts hydrogen sulfide into elemental sulfur. An aerobic sulfur bacterium, *Thiobacillus thiooxidans*, converts sulfur into sulfates. Other sulfates are contributed by the dissolving of rocks and some sulfur dioxide. Sulfur is incorporated by plants into proteins. Some of these

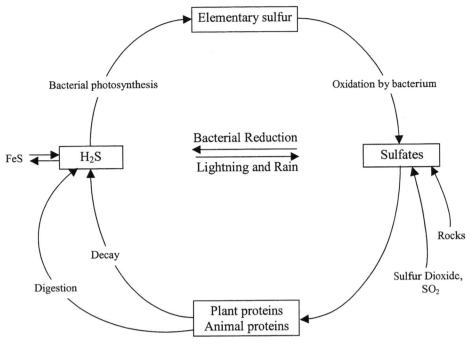

Figure 2.6. The sulfur cycle

plants are then consumed by organisms. Sulfur from proteins is liberated by many heterotrophic anaerobic bacteria as hydrogen sulfide.

Summary of Key Terms

Atmosphere—the gaseous mantle enveloping the hydrosphere and lithosphere, of which 78 percent is nitrogen, by volume.

Carbon—an essential part of all organic compounds; photosynthesis is a source of carbon. Photosynthesis is the chemical process by which solar energy is stored as chemical energy.

Hydrosphere—the water covering the Earth's surface, of which 80 percent is salt and 19 percent is groundwater.

Lithosphere—the solid components of the Earth's surface, such as rocks and weathered soil.

Nitrogen—required for construction of proteins and nucleic acids; the major source is the atmosphere.

Organisms—require 20 to 40 elements for survival.

Phosphorus cycle—very inefficient cycle; the greatest source is the lithosphere. Humans have greatly speeded up this cycle through mining.

Chapter Review Questions

2.1   Define biogeochemical cycle.

2.2   _____are those parts of the cycle where the chemical is held in large quantities for long periods of time.

2.3   The length of time a chemical is held in an exchange pool or a reservoir is termed its _____ time.

2.4   The average length of time a bioelement remains in a compartment is called the _____.

2.5   Name the three transport mechanisms.

2.6   Biochemical cycles can be _____ and/or _____.

2.7   Name the two types of biogeochemical cycles.

2.8   The three main subdivisions of the global environment are:

2.9   The carbon cycle is based on _____.

2.10   "The most important chemical process on Earth" is _____.

2.11   In the atmosphere, _____ acts as a _____ heat screen.

2.12   The atmosphere contains 78 percent by volume of _____.

2.13   Aerial nitrogen is converted into _____ mainly by microorganisms, bacteria, and blue-green algae.

2.14   Excretion by animals and anaerobic decomposition of dead organic matter by bacteria produce _____.

2.15   The process whereby ammonia converted by nitrification bacteria into nitrites and then into nitrates is known as _____.

2.16   The process whereby nitrate is converted into nitrogen gas is known as _____.

2.17   The ultimate source of phosphorus is _____.

2.18   Phosphate, upon entering a stream, acts as _____, which promotes the growth of _____.

2.19   Explain the "one-factor control hypothesis."

2.20   The _____ cycle is both sedimentary and gaseous.

Cited References and Recommended Reading

Abedon, S. T. (1997). *Ecosystems*. http//www.mansfield.ohio-state.edu (accessed February 15, 2007).

Abrahamson, D. E. (ed.). 1988. *The challenge of global warming*. Washington, DC: Island Press.

Asimov, I. 1989. *How did we find out about photosynthesis?* New York: Walker & Co.

Illinois State Water Survey. 2005. *Biogeochemical cycles II: The nitrogen cycle*. Illinois State Water Survey. www.sws.uiuc.edu (accessed February 15, 2007).

Killpack, S. C., and Buchholz, D. 1993. *Nitrogen in the environment: Nitrogen*. Columbia: University of Missouri.

Lowdermilk, W. C. 1953. *Conquest of the land through 7,000 years*. Washington, DC: U.S. Department of Agriculture, Soil Conservation Service.

Metcalf & Eddy, Inc. 1991. *Wastewater engineering: Treatment, disposal, reuse* (3rd ed.). New York: McGraw-Hill.

Miller, G. T. 1988. *Environmental science: An introduction*. Belmont, CA: Wadsworth.

Moran, J. M., Morgan, M. D., and Wiersma, J. H. 1986. *Introduction to environmental science*. New York: W. H. Freeman.

National Science Foundation (NSF). 2000. *Report of the workshop on the terrestrial carbon cycle*. http://www.carboncyclescience.gov (accessed February 18, 2007).

Odum, E. P. 1971. *Fundamentals of ecology*. Philadelphia: Saunders College.

Odum, E. P. 1975. *Ecology: The link between the natural and the social sciences*. New York: Holt, Rinehart & Winston.

Odum, E P. 1983. *Basic ecology*. Philadelphia: Saunders College.

Porteous, A. 1992. *Dictionary of environmental science and technology*. New York: John Wiley & Sons.

Price, P. W. 1984. *Insect ecology*. New York: John Wiley & Sons.

Smith, R. L. 1974. *Ecology and field biology*. New York: Harper & Row.

Spellman, F. R. 1996. *Stream ecology and self-purification*. Lancaster, PA: Technomic.

Tchobanoglous, G., and Schroeder, E. D. 1985. *Water supply*. Reading, MA: Addison-Wesley.

U.S. Geological Survey (USGS). 2005. The water cycle: Evapotranspiration. http//ga.water .usgs.Gov/edu/watercycleevapotranspiration.html (accessed February 16, 2007).

Spanish moss, Savannah, Georgia.
Photograph by Frank R. Spellman

CHAPTER 3

Energy Flow in the Ecosystem

All flesh is grass.

—Isaiah 40: 6–8

The original source of all energy going into food is the sun. This is because plants that have chlorophyll are able to combine water and carbon dioxide in the presence of light energy and produce sugar. This sugar can be converted to energy as the plant needs it. Of course, some of the sugar is converted into other complex chemicals that permit growth, reproduction, and other life processes.

—A. N. Tomera, American ecologist, 1989

Three hundred trout are needed to support one man for a year. The trout, in turn, must consume 90,000 frogs, that must consume 27 million grasshoppers that live off of 1,000 tons of grass.

—G. T. Miller, Jr., American chemist, 1988

Topics

Flow of Energy: The Basics
Food Chain Efficiency
Ecological Pyramids
Important Relationships in Living Communities
Productivity
Summary of Key Terms
Chapter Review Questions

As the chapter title suggests, the main concept covered in this chapter is how energy moves through an ecosystem. Again, it is important to point out that energy "moves" through an ecosystem and is not cycled through it. If you can understand this, you are in good shape, because then you have an idea of how ecosystems are balanced, how they may be affected by human activities, and how pollutants will move through an ecosystem.

Flow of Energy: The Basics

Simply defined, energy is the ability or capacity to do work. For an ecosystem to exist, it must have energy. All activities of living organisms involve work, which is the expenditure of energy—the degradation of a higher state of energy to a lower state. The flow of energy through an ecosystem is governed by two laws: the first and second laws of thermodynamics.

The first law, sometimes called the conservation law, states that energy cannot be created or destroyed. The second law states that no energy transformation is 100 percent efficient. That is, in every energy transformation, some energy is dissipated as heat. The term *entropy* is used as a measure of the nonavailability of energy to a system. Entropy increases with an increase in dissipation. Because of entropy, input of energy in any system is higher than the output or work done; thus, the resultant efficiency is less than 100 percent.

Odum (1975) explains that "the interaction of energy and materials in the ecosystem is of primary concern of ecologists." In chapter 2, we discussed the biogeochemical nutrient cycles. It is important to remember that it is the flow of energy that drives these cycles. Again, it should be noted that energy does not cycle as nutrients do in biogeochemical cycles. For example, when food passes from one organism to another, energy contained in the food is reduced step by step until all the energy in the system is dissipated as heat. Price (1984) refers to this process as "a unidirectional flow of energy through the system, with no possibility for recycling of energy." When water or nutrients are recycled, energy is required. The energy expended in this recycling is not recyclable. And, as Odum (1975) points out, this is a "fact not understood by those who think that artificial recycling of man's resources is somehow an instant and free solution to shortages."

While there is a slight input of geothermal energy, as pointed out earlier, the principal source of energy for any ecosystem is sunlight. Green plants, through the process of photosynthesis, transform the sun's light energy into chemical energy: carbohydrates that are consumed by animals. This transfer of energy, as stated previously, is unidirectional—from producers to consumers—and it is accomplished by cellular respiration, which is the process by which organisms (such as mammals) break the glucose back down into its constituents, water and carbon dioxide, thus regaining the stored energy the sun originally gave to the plants. Often this transfer of energy to different organisms is called a *food chain*. It is safe to say that food energy passes through a community in various ways—each separate way is called a food chain. Figure 3.1 shows a simple aquatic food chain.

All organisms, alive or dead, are potential sources of food for other organisms. All organisms that share the same general type of food in a food chain are

Figure 3.1. Aquatic food chain

said to be at the same trophic level (nourishment or feeding level; each level of consumption in a food chain is called a trophic level). Since green plants use sunlight to produce food for animals, they are called the producers, or the first trophic level. The herbivores, which eat plants directly, are called the second trophic level or the primary consumers. The carnivores are flesh-eating consumers; they include several trophic levels from the third on up. At each transfer, a large amount of energy (about 80 to 90 percent) is lost as heat and wastes. Thus, nature normally limits food chains to four or five links; however, in aquatic ecosystems, "food chains are commonly longer than those on land" (Dasmann 1984). The aquatic food chain is longer because several predatory fish may be feeding on the plant consumers. Even so, the built-in inefficiency of the energy transfer process prevents development of extremely long food chains.

Tomera describes a simple food chain that can be seen in a prairie dog community.

> The grass in the community manufactures food. The grass is called a food producer. The grass is eaten by a prairie dog. Because the prairie dog lives directly off the grass, it is termed a first-order consumer. A weasel [or other predator] may kill and eat the prairie dog. The weasel is, therefore, a predator and would be termed a second-order consumer. The second-order consumer is twice removed from the green grass. The weasel, in turn, may be eaten by a large hawk or eagle. The bird that kills and eats the weasel would therefore be a third-order consumer, three times removed from the grass. Of course, the hawk would give off waste materials and eventually die itself. Wastes and dead organisms are then acted on by decomposers. (1989, 50)

Only a few simple food chains are found in nature. Thus, when attempting to identify the complex food relationships among many animals and plants within a community, it is useful to create a food web. The fact is that most simple food chains are interlocked; this interlocking of food chains forms a food web. A *food web* can be characterized as a map that shows what eats what (Miller 1988). Most ecosystems support a complex food web. A food web involves animals that do not feed on one trophic level. For example, humans feed on both plants and animals. The point is, an organism in a food web may occupy one or more trophic levels. Trophic level is determined by an organism's role in its particular community, not by its species. Food chains and webs help to explain how energy moves through an ecosystem.

An important trophic level of the food web that has not been discussed thus far is comprised of the decomposers (bacteria, mushrooms, etc.). The decomposers feed on dead plants or animals and play an important role in recycling nutrients in the ecosystem. As Miller (1988) points out, "there is no waste in ecosystems. All organisms, dead or alive, are potential sources of food for other organisms." An example of an aquatic food web is shown in figure 3.2.

From the preceding discussion about food chains and food webs, the impor-

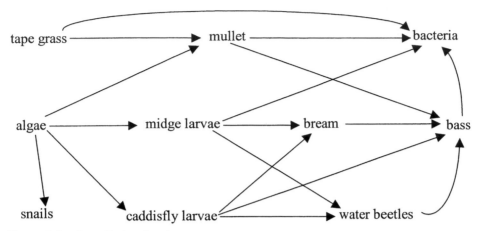

Figure 3.2. Aquatic food web

tant point to be gained is that there is a distinct difference between the two. A food chain, for example, is a simple straight-line process going from producer to first-, second-, and possibly third-order consumers, and ending with the decomposers. On the other hand, in a food web, there are a number of second- and third-order consumers.

Food Chain Efficiency

Earlier it was pointed out that energy from the sun is captured (via photosynthesis) by green plants and used to make food. Most of this energy is used to carry on the plant's life activities. The rest of the energy is passed on as food to the next level of the food chain.

✔ *Important Point*: A food chain is the path of food from a given final consumer back to a producer.

It is important to note that nature limits the amount of energy that is accessible to organisms within each food chain. Not all food energy is transferred from one trophic level to the next. For ease of calculation, "ecologists often assume an ecological efficiency of 10 percent (the 10 percent rule) to estimate the amount of energy transferred through a food chain" (Moran, Morgan, and Wiersma 1986). For example, if we apply the 10 percent rule to the diatoms-copepods-minnows-medium fish-large fish food chain shown in figure 3.3, we can predict that 1,000 grams of diatoms produce 100 grams of copepods, which will produce 10 grams of minnows, which will produce 1 gram of medium fish, which, in turn, will produce 0.1 gram of large fish. Thus, only about 10 percent of the chemical energy available at each trophic level is transferred and stored in usable form at the next level. What happens to the other 90 percent? The other 90 percent is lost to the

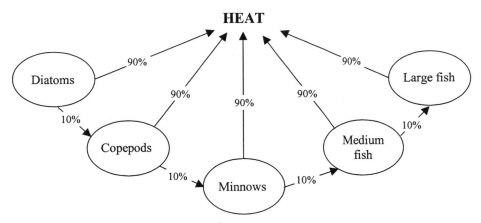

Figure 3.3.   Simple food chain

environment as low-quality heat in accordance with the second law of thermodynamics.

✔   *Important Point*: The ratio of net production at one level to net production at the next higher level is called the *conversion efficiency*.

✔   *Important Point*: When an organism loses heat, it represents a one-way flow of energy out of the ecosystem. Plants only absorb a small part of energy from the sun. Plants store half of the energy and lose the other half. The energy plants lose is metabolic heat. Energy from a primary source will flow in one direction through two different types of food chains. In a grazing food chain, the energy will flow from plants (producers) to herbivores, and then through some carnivores. In a detritus-based food chains, energy will flow from plants through detrivores and decomposers. In terms of the weight (or biomass) of animals in many ecosystems, more of their body mass can be traced back to detritus than to living producers. Most of the time the two food webs will intersect one another. For example, the Chesapeake Bay bass fish of the grazing food web will eat a crab of the detrital food web (Spellman 1996).

Ecological Pyramids

As we proceed in the food chain from the producer to the final consumer, it becomes clear that a particular community in nature often consists of several small organisms associated with a smaller and smaller number of larger organisms. A grassy field, for example, has a larger number of grass and other small plants, a smaller number of herbivores such as rabbits, and an even smaller number of carnivores such as fox. The practical significance of this is that we must have several more producers than consumers.

This pound-for-pound relationship, where we need more producers than consumers, can be demonstrated graphically by building an ecological pyramid. In an ecological pyramid, the number of organisms at various trophic levels in a food chain is represented by separate levels or bars placed one above the other with a base formed by producers and the apex formed by the final consumer. The pyramid shape is formed due to a great amount of energy loss at each trophic level. The same is true if numbers are substituted by the corresponding biomass or energy. Ecologists generally use three types of ecological pyramids: pyramids of number, biomass, and energy. Obviously, there will be differences among them. Following are some generalizations:

1. *Energy pyramids* must always be larger at the base than at the top (because of the second law of thermodynamics, which has to do with the dissipation of energy as it moves from one trophic level to another). Simply, energy pyramids depict the decrease in the total available energy at each higher trophic level.

2. Likewise, *biomass pyramids* (in which biomass is used as an indicator of production) are usually pyramid shaped. This is particularly true of terrestrial systems and aquatic ones dominated by large plants (marshes), in which consumption by heterotrophs is low and organic matter accumulates with time. A census of the population, multiplied by the weight of an average individual in it, gives an estimate of the weight of the population. This is called the *biomass* (or standing crop). However, it is important to point out that biomass pyramids can sometimes be inverted. This is especially common in aquatic ecosystems, in which the primary producers are microscopic planktonic organisms that multiply very rapidly and have very short life spans, and where there is heavy grazing by herbivores. At any single point in time, the amount of biomass in primary producers is less than that in larger, long-lived animals that consume primary producers.

3. *Numbers pyramids* can have various shapes (and not be pyramids at all, actually) depending on the sizes of the organisms that make up the trophic levels. In forests, the primary producers are large trees and the herbivore level usually consists of insects, so the base of the pyramid is smaller than the herbivore level above it; that is, the pyramid is inverted. In grasslands, the number of primary producers (grasses) is much larger than that of the herbivores above (large grazing animals) (Spellman 1996).

To get a better idea of how an ecological pyramid looks and how it provides its information, we need to look at an example. The example used here is the energy pyramid. According to Odum (1983), the energy pyramid is a fitting example because among the "three types of ecological pyramids, the energy pyramid gives by far the best overall picture of the functional nature of communities."

In an experiment conducted in Silver Springs, Florida, Odum (1983) measured the energy for each trophic level in terms of kilocalories. A kilocalorie is the amount of energy needed to raise 1 cubic centimeter of water 1 degree centigrade. When an energy pyramid is constructed to show Odum's findings, it takes on the typical upright form (as it must because of the second law of thermodynamics) as shown in figure 3.4.

Simply put, as reflected in figure 3.4 and according to the second law of thermodynamics, no energy transformation process is 100 percent efficient. This fact is demonstrated, for example, when a horse eats hay. The horse cannot obtain, for his own body, 100 percent of the energy available in the hay. For this reason, the energy productivity of the producers must be greater than the energy production of the primary consumers. When human beings are substituted for the horse, it is interesting to note that, according to the second law of thermodynamics, only a small population could be supported. But this is not the case. Humans also feed on plant matter, which allows a larger population. Therefore, if meat supplies become scarce, we must eat more plant matter. This is the situation we see today in countries where meat is scarce. Consider this: if we all ate soybeans, there would be at least enough food for ten times as many of us as compared to a world where we all eat beef (or pork, fish, chicken, etc.). Another way of looking at this: Every time we eat meat, we are taking food out of the mouths of nine other people, who could be fed with the plant material that was fed to the animal we are eating (Spellman 1996). Food-energy relationships are often referred to as eater-eaten relationships. It's not quite that simple, of course, but you probably get the general idea.

Important Relationships in Living Communities

In addition to the pyramid-shaped relationships, there are other important relationships in living communities. Some of these involve food energy and some do not. In this section, several of these relationships are described.

Symbiosis is a close (intimate) ecological relationship (organisms living together in close proximity) between the individuals of two or more different species. Sometimes a symbiotic relationship benefits both species, sometimes one species benefits at the other's expense, and in other cases neither species benefits. One thing is certain: The relationship is *obligate*, meaning at least one of the species must be involved in the relationship to survive.

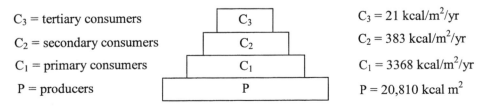

C_3 = tertiary consumers

C_2 = secondary consumers

C_1 = primary consumers

P = producers

C_3 = 21 kcal/m^2/yr

C_2 = 383 kcal/m^2/yr

C_1 = 3368 kcal/m^2/yr

P = 20,810 kcal m^2

Figure 3.4.   Energy flow pyramid

Adapted from Eugene P. Odum, *Fundamentals of Ecology* (Philadelphia: Saunders College, 1971), p. 80.

Ecologists use a different term for each type of symbiotic relationship:

- Mutualism   —   both organisms benefit
- Commensalism —   one organism benefits, the other is unaffected
- Parasitism   —   one organism benefits, the other is harmed
- Competition   —   neither organism benefits
- Neutralism   —   both organisms are unaffected

The effects of these interactions on population growth can be positive, negative, or neutral (see table 3.1).

Productivity

As mentioned previously, the flow of energy through an ecosystem starts with the fixation of sunlight by plants through photosynthesis. In evaluating an ecosystem, the measurement of photosynthesis is important. Ecosystems may be classified as highly productive or less productive. Therefore, the study of ecosystems must involve some measure of the productivity of that ecosystem.

Smith (1974) defines production (or, more specifically, primary production, because it is the basic form of energy storage in an ecosystem) as being "the energy accumulated by plants." Stated differently, primary production is the rate at which the ecosystem's primary producers capture and store a given amount of energy, in a specified time interval. In even simpler terms, primary productivity is a measure of the rate at which photosynthesis occurs. Odum (1971) lists four successive steps in the production process as follows:

1. *Gross primary productivity*—the total rate of photosynthesis in an ecosystem during a specified interval at a given trophic level
2. *Net primary productivity*—the rate of energy storage in plant tissues in excess of the rate of aerobic respiration by primary producers
3. *Net community productivity*—the rate of storage of organic matter not used
4. *Secondary productivity*—the rate of energy storage at consumer levels

Table 3.1. Population Interactions, Two-Species System

Type of Interaction	Response	
	A	B
Neutral	0	0
Mutualism	+	+
Commensalism	+	0
Parasitism	+	−
Competition	−	−

When attempting to comprehend the significance of the term *productivity* as it relates to ecosystems, it is wise to consider an example. Consider the productivity of an agricultural ecosystem such as a wheat field. Often its productivity is expressed as the number of bushels produced per acre. This is an example of the harvest method for measuring productivity. For a natural ecosystem, several one-square-meter plots are marked off, and the entire area is harvested and weighed to give an estimate of productivity as grams of biomass per square meter per given time interval. From this method, a measure of net primary production (net yield) can be measured.

Productivity, both in the natural and cultured ecosystem, may vary considerably, not only between types of ecosystems but also within the same ecosystem. Several factors influence year-to-year productivity within an ecosystem. Such factors as temperature, availability of nutrients, fire, animal grazing, and human cultivation activities are directly or indirectly related to the productivity of a particular ecosystem.

The following study of an aquatic ecosystem is used as an example of productivity. Productivity can be measured in several different ways in the aquatic ecosystem. For example, the production of oxygen may be used to determine productivity. Oxygen content may be measured in several ways. One way is to measure it in the water every few hours for a period of 24 hours. During daylight, when photosynthesis is occurring, the oxygen concentration should rise. At night, the oxygen level should drop. The oxygen level can be measured by using a simple *x-y* graph. The oxygen level can be plotted on the *y* axis with time plotted on the *x* axis, as shown in figure 3.5.

Another method of measuring oxygen production in aquatic ecosystems is to use light and dark bottles. Biochemical oxygen demand (BOD) bottles (300 ml) are filled with water to a particular height. One of the bottles is tested for the initial dissolved oxygen (DO), then the other two bottles (one clear, one dark) are suspended in the water at the depth they were taken from. After a 12-hour period,

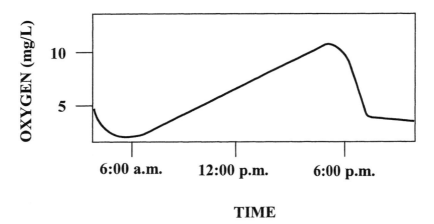

TIME

Figure 3.5.   The diurnal oxygen curve for an aquatic ecosystem

the bottles are collected and the DO values for each bottle are recorded. Once the oxygen production is known, the productivity in terms of grams/m/day can be calculated.

Table 3.2 shows representative values for the net productivity of a variety of ecosystems—both nature and managed. Keep in mind that these values are only approximations derived from Odum's (1971, 1983) work and are subject to marked fluctuations because of variations in temperature, fertility, and availability of water.

In the aquatic (and any other) ecosystem, pollution can have a profound impact upon the system's productivity. For example, certain kinds of pollution may increase the turbidity of the water. This increase in turbidity causes a decrease in energy delivered by photosynthesis to the ecosystem. Accordingly, this turbidity and its aggregate effects decrease net community productivity on a large scale (Laws 1993).

PRODUCTIVITY: THE BOTTOM LINE

The ecological trends paint a clear picture. Wherever we look, ecological productivity is limping behind human consumption. Since 1984, the global fish harvest has been dropping, and so has the per capita yield of grain crops (Brown 1994). Moreover, stratospheric ozone is being depleted—the release of greenhouse gases has changed the atmospheric chemistry and might lead to climate change; erosion and desertification are reducing nature's biological productivity; irrigation water tables are falling; contamination of soil and water is jeopardizing the quality of food; other natural resources are being consumed faster than they can regenerate; and biological diversity is being lost—to reiterate only a small part of a long list.

Table 3.2.   Estimated Net Productivity of Certain Ecosystems

Ecosystem	kilocalories/m²/year
Temperate deciduous forest	5,000
Tropical rain forest 15,000	
Tall-grass prairie	2,000
Desert	500
Coastal marsh	12,000
Ocean close to shore	2,500
Open ocean	800
Clear (oligotrophic) lake	800
Lake in advanced state of eutrophication	2,400
Silver Springs, Florida	8,800
Field of alfalfa (Lucerne)	15,000
Corn (maize) field, U.S.	4,500
Rice paddies, Japan	5,500
Lawn, Washington, D.C.	6,800
Sugar cane, Hawaii	25,000

These trends indicate a decline in the quantity and productivity of nature's assets (Wachernagel 1997).

Summary Key Terms

Ecological pyramids—three types of ecological pyramids are the pyramids of numbers, productivity, and energy. All of these pyramids are based on the fact that due to energy loss, fewer animals can be supported at each additional trophic level, which is the number of energy transfers an organism is from the rest of the pyramid indicates what happens to this energy.

Energy—the ability or capacity to do work. Energy is degraded from a higher to a lower state.

First law of thermodynamics—energy is transformed from one form to another, but is neither created nor destroyed. Given this principle, we should be able to account for all the energy in a system in an energy budget, a diagrammatic representation of the energy flows through an ecosystem.

Second law of thermodynamics—energy is only available due to degradation of energy from a concentrated to a dispersed form. This indicates that energy becomes more and more dissipated (randomly arranged) as it is transformed from one form to another or moved from one place to another. It also suggests that any transformation of energy will be less than 100 percent efficient (i.e., the transfers of energy from one trophic level to another are not perfect); some energy is dissipated during each transfer.

Chapter Review Questions

3.1   Define the conservation law.
3.2   The primary concern of ecologists is the interaction of _____ and _____ in the ecosystem.
3.3   Define unidirectional flow of energy in an ecosystem.
3.4   The transfer of energy to different organisms is called a _____.
3.5   Interlocked food chains are called a _____.
3.6   _____ feed on dead plants or animals and play an important role in recycling nutrients in the ecosystem.
3.7   Explain the 10-percent rule.
3.8   Explain an ecological pyramid.
3.9   What are the three types of ecological pyramids?
3.10 Define production.
3.11 The rate of storage of organic matter not used is known as _____.
3.12 Net yield is the same as _____.

Cited References and Recommended Reading

Brown, L. R. 1994. Facing food insecurity. In *State of the world*, ed. L. R. Brown et al. New York: W. W. Norton.

Dasmann, R. F. 1984. *Environmental conservation*. New York: John Wiley & Sons.

Laws, E. A. 1993. *Environmental science: An introductory text*. New York: John Wiley & Sons.

Miller, G. T. 1988. *Environmental science: An introduction*. Belmont, CA: Wadsworth.

Moran, J. M., Morgan, M. D., and Wiersma, H. H. 1986. *Introduction to environmental science*. New York: W. H. Freeman.

Odum, E. P. 1971. *Fundamentals of ecology*. Philadelphia: Saunders College.

Odum, E. P. 1975. *Ecology: The link between the natural and the social sciences*. New York: Holt, Rinehart & Winston.

Odum, E. P. 1983. *Basic ecology*. Philadelphia: Saunders College.

Price, P. W. 1984. *Insect ecology*. New York: John Wiley & Sons.

Smith, R. L. 1974. *Ecology and field biology*. New York: Harper & Row.

Spellman, F. R. 1996. *Stream ecology and self-purification*. Lancaster, Pa.: Technomic.

Tomera, A. N. 1989. *Understanding basic ecological concepts*. Portland, ME: J. Weston Walch.

Townsend, C. R., Harper, J. L., and Begon, M. 2000. *Essentials of ecology*. Malden, MA: Blackwell Science.

Wachernagel, M., 1997. *Framing the sustainability crisis: Getting from concerns to action*. http://www.sdri.ubc.ca/publications/wacherna.html (accessed February 26, 2007).

Wessells, N. K., and Hopson, J. L. 1988. *Biology*. New York: Random House.

Running Eagle Falls, Glacier National Park, Montana.
Photograph by Frank R. Spellman

CHAPTER 4

Population Ecology

The Earth is one but the world is not. We all depend on one biosphere for sustaining our lives. Yet each community, each country, strives for survival and prosperity with little regard for its impact on others. Some consume the Earth's resources at a rate that would leave little for future generations. Others, many more in number, consume far too little and live with the prospects of hunger, squalor, disease, and early death.

—World Commission on Environment
and Development 1987

Whether we look or whether we listen,
We hear life murmur or see it glisten.

—James Russell Lowell, 1819–1891

Topics

The 411 on Population Ecology
Laws of Population Ecology
Applied Population Ecology
Distribution
Population Growth
Population Response to Stress
Ecological Succession
Ecosystem Population Response to Stress
Summary of Key Terms
Chapter Review Questions

Population ecology owes its beginning to the contributions of Thomas Malthus, an English clergyman who, in 1798, published his *Essay on the Principle of Population*. Malthus introduced the concept that, at some point in time, an expanding population must exceed the supply of prerequisite natural resources—the "struggle for existence concept." Malthus's theories profoundly influenced Charles Darwin's 1859 *On the Origin of Species*—for example, his concept of natural selection.

The 411 on Population Ecology

Let's begin with the basics:

Population—defined by the word masters:
Webster's Third New International Dictionary defines population:
- "The total number or amount of things especially within a given area."
- "The organisms inhabiting a particular area or biotype."
- "A group of interbreeding biotypes that represents the level of organization at which speciation begins."

Population—defined by an ecologist (Abedon 2007):
- A population in an ecological sense is a group of organisms, of the same species, which roughly occupy the same geographical area at the same time.
- Individual members of the same population can either interact directly, or may interact with the dispersing progeny of the other members of the same population (e.g., pollen).
- Population members interact with a similar environment and experience similar environmental limitations.

Population System:
Population system or life-system (population system is definitely better, however) is a population with its effective environment (Clark, Gerier, Hughes, and Harris 1967; Berryman 1981; Sharov 1992).

Major Components of a Population System:
1. Population itself. Organisms in the population can be subdivided into groups according to their age, stage, sex, and other characteristics
2. Resources: food, shelters, nesting places, space, etc.
3. Enemies: predators, parasites, pathogens, etc.
4. Environment: air (water, soil) temperature, composition, variability of these characteristics in time and space (Sharov 1997)

Population Ecology (Sharov 1996):
Population ecology is the branch of ecology that studies the structure and dynamics of populations. Population ecology relative to other ecological disciplines is shown in figure 4.1.

The term *population* is interpreted differently in various sciences. For example, in human demography, a population is a set of humans in a given area. In genetics, a population is a group of interbreeding individuals of the same species, which is isolated from other groups. In population ecology, a population is a group of individuals of the same species inhabiting the same area.

✔   *Important Point*: Main axiom of population ecology—organisms in a population are ecologically equivalent. Ecological equivalency means:
   1. Organisms undergo the same life cycle.

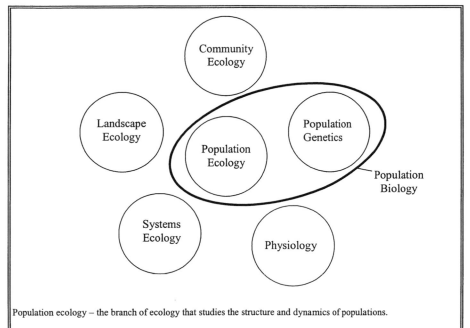

Population ecology – the branch of ecology that studies the structure and dynamics of populations.

Physiology – study of individual characteristics and individual processes. Used as a basis for prediction of processes at the population level.

Community ecology – study of the structure and dynamics of animal and plant communities. Population ecology provides modeling tools that can be used for predicting community structure and dynamics.

Population genetics – the study of gene frequencies and microevolution in populations. Selective advantages depend on the success of organisms in their survival, reproduction and competition. These processes are studied in population ecology. Population ecology and population genetics are often considered together and called 'population biology'. Evolutionary ecology is one of the major topics in population biology.

Systems ecology – a relatively new ecological discipline which studies interaction of human population with environment. One of the major concepts are optimization of ecosystem exploitation and sustainable ecosystem management.

Landscape ecology – another relatively new area in ecology. It studies regional large-scale ecosystems with the aid of computer-based geographic information systems. Population dynamics can be studied at the landscape level, and this is the link between landscape and population ecology.

Figure 4.1. Population ecology relative to other ecological disciplines
Adapted from Alexi Sharov, Dept. of Entomology, Virginia Tech, 1996, p. 1.

2. Organisms in a particular stage of the life cycle are involved in the same set of ecological processes.
3. The rates of these processes (or the probabilities of ecological events) are basically the same if organisms are put into the same environment (however, some individual variation may be allowed) (Sharov 1996).

Properties of Populations (Abedon 2007):
1. Population size (size)—depends on how the population is defined.
2. Population density (density)—the number of individual organisms per unit area.

3. Patterns of dispersion (dispersion)—individual members of populations may be distributed over a geographical area in a number of different ways, including clumped, uniform, and random distribution.
4. Demographics (demographics)—a population's vital statistics, including:
   - Education
   - Parental status
   - Work environment
   - Geographic location
   - Religious beliefs
   - Ethnicity
   - Age
   - Marital status
   - Income
   - Sex
   - Race
   - Gender
   - Sexual orientation
   - Physical ability
5. Population growth (growth)—simply, population growth occurs when there are no limitations on growth within the environment. When this happens, two situations occur: (1) the population displays its intrinsic rate of increase (i.e., the rate of growth of a population when that population is growing under ideal conditions and without limits); and (2) the population experiences exponential growth (i.e., exponential growth means that a population's size at a given time is equal to the population's size at an earlier time, times some greater-than-one number) (Abedon 2007).
6. Limits on population growth (limits)—exponential growth cannot go on forever; sooner or later any population will run into limits in their environment.

✔ *Important Point*: Note that all of these properties are not those of individual organisms but instead are properties that exist only if one considers more than one organism at any given time, or over a period of time.

Laws of Population Ecology

Note: The information in this section is based on and adapted from Haemig's (2006) *Laws of Population Ecology*.

According to Haemig (2006), the discovery of laws in ecology has lagged behind many of the other sciences (e.g., chemistry, physics, etc.) because ecology is a much younger science. However, as Colyvan and Ginzburg (2003) point out, misunderstandings and unrealistic expectations of what the laws are have also hindered the search, as have mistaken beliefs that ecology is just too complex a science to have laws. Nevertheless, over the years, researchers have been able to identify some of the laws that exist in ecology.

Ginzburg (1986) points out that while much remains to be learned, it now appears that laws of ecology resemble laws of physics. Colyvan and Ginsburg (2003) and Ginzburg and Colyvan (2004) point out that laws of ecology describe idealized situations, have many exceptions, and need not be explanatory or predictive. The laws of population ecology are listed and described below.

- *Malthusian law*—says that when birth and death rates are constant, a population will grow (or decline) at an exponential rate.
- *Allee's law*—says that there is a positive relationship between individual fitness and either the numbers or density of conspecifics (conspecifics are other individuals of the same species).
- *Verhulst's law*—this law deals with one factor: intraspecific competition (i.e., competition between members of the same species). Because the organisms limiting the population are also members of the population, this law is also called "population self-limitation" (Turchin 2001).
- *Lotka-Volterra's law*—says that "when populations are involved in negative feedback with other species, or even components of their environments," oscillatory (cyclical) dynamics are likely to be seen (Berryman 2002, 2003).
- *Liebig's law*—says that of all the biotic or abiotic factors that control a given population, one has to be limiting (i.e., active, controlling the dynamics) (Berryman 1993, 2003). Time delays produced by this limiting factor are usually one or two generations long (Berryman 1999). Krebs (2001) defines "a factor as limiting if a change in the factor produces a change in average or equilibrium density."
- *Fenchel's law*—says that species with larger body sizes generally have lower rates of population growth; the maximum rate of reproduction decrease with body size at a power of approximately one-quarter the body mass (Fenchel 1974). Fenchel's law is expressed by the following equation:

$$r = aW^{-1/4}$$

where

> r = the intrinsic rate of nature increase of the population
> a = constant (has three different values)
> W = average body weight (mass) of the organism (Fenchel 1974)

- *Calder's law*—says that species with larger body sizes generally have longer population cycles; the length of the population cycle increases with increasing body size at a power of approximately one-quarter the body mass (Calder 1983).

$$t = aW^{1/4}$$

where

> t = average time of the population cycle
> a = a constant
> W = average body weight (mass) of the organism

- *Damuth's law*—says that species with larger body sizes generally have lower average population densities; the average density of a population decreases with body size at a power of approximately three-quarters the body mass (Damuth 1981, 1987, 1991). Damuth's law is expressed by the following equation:

$$d = aW^{-3/4}$$

where

   d = the average density of the population
   a = a constant
   W = average body weight (mass) of the organism

- *Generation-Time law*—says that species with larger body sizes usually have longer generation-times; that the generation-time increases with increasing body size at a power of approximately one-quarter the body mass (Bonner 1965). Note: The body mass used in this law is the body mass of the organism at the time of reproduction. The generation-time law is expressed by the following equation:

$$g = aW^{1/4}$$

where

   g = average generation-time of the population
   a = a constant
   W = average body weight (mass) of the organism

- *Ginzburg's law*—says that the length of a population cycle (oscillation) is the result of the maternal effect and inertial populating growth. According to this law, the period lengths in the cycles of a population must be either two generations long or six or more generations long (Ginzburg and Colyvan 2004).

Applied Population Ecology

Note: In attempting to explain any concept, it is always best to do so with an example in mind—an illustrative example. In the following, a stream ecosystem is the illustrative example used to help explain population ecology.

   If stream ecology students wanted to study the organisms in a slow-moving stream or stream pond, they would have two options. They could study each fish, aquatic plant, crustacean, insect, and macroinvertebrate one by one. In that case, they would be studying individuals. It would be easy to do this if the subject were trout, but it would be difficult to separate and study each aquatic plant.

   The second option would be to study all of the trout, all of the insects of each specific kind, or all of a certain aquatic plant type in the stream or pond at the time of the study. When stream ecologists study a group of the same kinds of individuals in a given location at a given time, they are investigating a population. "Alternately, a population may be defined as a cluster of individuals with a high probability of mating with each other compared to their probability of mating with a member of some other population" (Pianka 1988). When attempting to determine the population of a particular species, it is important to remember that time is a factor. Whether it is at various times during the day, during the different seasons, or from year to year, time is important because populations change.

   When measuring populations, the level of species or density must be determined. Density (*D*) can be calculated by counting the number of individuals in the population (N) and dividing this number by the total units of space (S) the counted population occupies. Thus, the formula for calculating density becomes

$$D = N/S$$

When studying aquatic populations, the occupied space (S) is determined by using length, width, and depth measurements. The volumetric space is then measured in cubic units.

Population density may change dramatically. For example, if a dam is closed off in a river midway through spawning season, with no provision allowed for fish movement upstream (a fish ladder, for example), it would drastically decrease the density of spawning salmon upstream. Along with the swift and sometimes unpredictable consequences of change, it can be difficult to draw exact boundaries between various populations. Pianka (1988) makes this point in his comparison of European starlings that were introduced into Australia with starlings that were introduced into North America. He points out that these starlings are no longer exchanging genes with each other; thus, they are separate and distinct populations.

The population density or level of a species depends on natality, mortality, immigration, and emigration. Changes in population density are the result of both births and deaths. The birth rate of a population is called natality and the death rate, mortality. In aquatic populations, two factors besides natality and mortality can affect density. For example, in a run of returning salmon to their spawning grounds, the density could vary as more salmon migrated in or as others left the run for their own spawning grounds. The arrival of new salmon to a population from other places is termed immigration (ingress). The departure of salmon from a population is called emigration (egress). Thus, natality and immigration increase population density, whereas mortality and emigration decrease it. The net increase in population is the difference between these two sets of factors.

Population regulation is the control of the size of a population. Population is limited by various factors. There are basically two different types of population limiting factors—classified according to the types of factors that control the size of the population. The population limiting factors are (1) density-dependent control and (2) density-independent control (Winstead 2007).

1. *Density-dependent factors*—are ones where the effect of the factor on the size of the population depends upon the original density or size of the population. Density-dependent factors include (Abedon 2007) the following:
    *   Density-dependent limits on population growth are ones that stem from intraspecific competition.
    *   Typically, the organisms best suited to compete with another organism are those from the same species.
    *   Thus, the actions of conspecifics (an organism belonging to the same species as another) can very precisely serve to limit the environment (e.g., eat preferred food, obtain preferred shelter, etc.).
    *   Actions that serve to limit the environment for conspecifics—for example, eating, excreting wastes, using up nonfood resources, taking up space, defending territories—are those that determine carrying capacity (K).

- These limits are referred to as *density dependent* because the greater the density of the *population*, the greater their effects.
- Density-dependent factors may exert their effect by reducing birth rates, increasing death rates, extending generation times, or by forcing the migration of conspecifics to new regions.
- "The impact of disease on a population can be density dependent if the transmission rate of the disease depends on a certain level of crowding the population" (Campbell and Reece 2004).
- "A death rate that rises as population density rises is said to be density dependent, as in a birth rate that falls with rising density. Density-dependent rates are an example of negative feedback. In contrast, a birth rate or death rate that does change with population density is said to be density independent. . . . Negative feedback prevents unlimited population growth" (Campbell and Reece 2004).
- Predation can also be density dependent since predators often can switch prey preferences to match whatever prey organisms are more plentiful in a given environment.
- "Many predators, for example, exhibit switching behavior: They begin to concentrate on a particularly common species of prey when it becomes energetically efficient to do so" (Campbell and Reece 2004).

2. *Density-independent factors*—are ones "where the effect of the factor on the size of the population is independent and does not depend upon the original density or size of the population. The effect of weather is an example of a density-independent factor. A severe storm and flood coming through an area can just as easily wipe out a large population as small one. Another example would be a harmful pollutant put into the environment, e.g., a stream. The probability of that harmful substance at some concentration killing an individual would not change depending on the size of the population. For example, populations of small mammals are often regulated more by this type of regulation" (Winstead 2007).

- Density-independent effects on population sizes (or structures) occur to the same extent regardless of population size.
- These can be things like sudden changes in the weather.
- "Over the long term, many populations remain fairly stable in size and are presumably close to a carrying capacity that is determined by density-dependent factors. Superimposed on this general stability, however, are short-term fluctuations due to density-independent factors" (Campbell and Reece 2004).

Distribution

Each organism occupies only those areas that can provide for its requirements, resulting in an irregular distribution. How a particular population is distributed within a given area has considerable influence on density. As shown in figure 4.2,

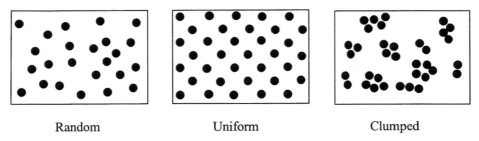

Random   Uniform   Clumped

Figure 4.2. Basic patterns of distribution

Adapted from Eugene P. Odum, *Fundamentals of Ecology* (Philadelphia: Saunders College, 1971), p. 205.

organisms in nature may be distributed in three ways, as a result of complex interactions among ecological variables.

In a random distribution, there is an equal probability of an organism occupying any point in space, and "each individual is independent of the others" (Smith 1974).

In a regular or uniform distribution, in turn, organisms are spaced more evenly; they are not distributed by chance. Animals compete with each other and effectively defend a specific territory, excluding other individuals of the same species. In regular or uniform distribution, the competition between individuals can be quite severe and antagonistic, to the point where the spacing generated is quite even (Odum 1983).

The most common distribution is the contagious or clumped distribution where organisms are found in groups; this may reflect the heterogencity of the habitat. Smith (1974) points out that contagious or clumped distribution "produce aggregations, the result of response by plants and animals to habitat differences."

Organisms that exhibit a contagious or clumped distribution may develop social hierarchies in order to live together more effectively. Animals within the same species have evolved many symbolic aggressive displays that carry meanings that are not only mutually understood but also prevent injury or death within the same species. For example, in some mountainous regions, dominant male bighorn sheep force the juvenile and subordinate males out of the territory during breeding season (Hickman, Roberts, and Hickman 1990). In this way, the dominant male gains control over the females and need not compete with other males.

As mentioned, distribution patterns are the result of complex interactions among ecological variables. For example, consider a study conducted by Hubbell and Johnson (1977) of five tropical bee colonies (the bees live in colonies in suitable trees) in the tropical dry forests of Costa Rica. The researchers set out to examine relationship between aggressiveness and patterns of colony distribution.

1. The researchers mapped locations of suitable nest trees. They found that the number of suitable trees was greater than number of colonies—thus, nest sites were not a limiting factor. Distribution of suitable trees was random.

2. The researchers next mapped locations of bee colonies. They found that colony sites for one species were dispersed randomly. Members of this species do not exhibit aggression toward one another. The colonies were sometimes quite close to one another.

On the other hand, colony sites for the other four species were dispersed in a regular fashion. Members of all four species were aggressive to members of other colonies of the same species. They also marked their colony sites with pheromones and engaged in ritualized battles for colony sites with conspecifics from other colonies.

Population Growth

The size of animal populations is constantly changing due to natality, mortality, emigration, and immigration. As mentioned, the population size will increase if the natality and immigration rates are high. On the other hand, it will decrease if the mortality and emigration rates are high. Each population has an upper limit on size, often called the carrying capacity. Carrying capacity can be defined as being the "optimum number of species' individuals that can survive in a specific area over time" (Enger, Kormelink, Smith, and Smith 1989). Stated differently, the carrying capacity is the maximum number of species that can be supported in a bioregion. A pond may be able to support only a dozen frogs depending on the food resources for the frogs in the pond. If there were thirty frogs in the same pond, at least half of them would probably die because the pond environment wouldn't have enough food for them to live. Carrying capacity, symbolized as K, is based on the quantity of food supplies, the physical space available, the degree of predation, and several other environmental factors.

The carrying capacity is of two types: ultimate and environmental. Ultimate carrying capacity is the theoretical maximum density; that is, it is the maximum number of individuals of a species in a place that can support itself without rendering the place uninhabitable. The environmental carrying capacity is the actual maximum population density that a species maintains in an area. Ultimate carrying capacity is always higher than environmental.

The population growth for a certain species may exhibit several types of growth. Smith (1974) states that "the rate at which the population grows can be expressed as a graph of the numbers in the population against time." Figure 4.3 shows one type of growth curve.

The J-shaped curve shown in figure 4.3 shows a rapid increase in size or exponential growth. Eventually, the population reaches an upper limit where exponential growth stops. The exponential growth rate is usually exhibited by organisms that are introduced into a new habitat, by organisms with a short life span such as insects, and by annual plants. A classic example of exponential growth by an introduced species is the reindeer transported to Saint Paul Island in the Pribilofs off Alaska in 1911. A total of twenty-five reindeer were released on the island and by 1938 there were over 2,000 animals on the small island. As time went by,

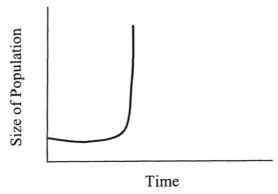

Figure 4.3.   J-shaped growth curve

however, the reindeer overgrazed their food supply and the population decreased rapidly. By 1950 only eight reindeer could be found (Pianka 1988).

Another example of exponential growth is demonstrated by the Lily Pond Parable. The "parable" (not really a parable) is an excellent example providing insight into long-term carrying capacity and population growth.

Another type of growth curve is shown in figure 4.4. This logistic or S-shaped (sigmoidal) curve is used for populations of larger organisms having a longer life span. This type of curve has been successfully used by ecologists and biologists to model populations of several different types of organisms, including water fleas, pond snails, and sheep, to name only a few (Masters 1991). The curve suggests an early exponential growth phase, while conditions for growth are optimal. As the numbers of individuals increase, the limits of the environment, or environmental resistance, begin to decrease the numbers of individuals, and the population size levels off near the carrying capacity, shown as K in figure 4.4. Usually there is some oscillation around K before the population reaches a stable size as indicated on the curve.

As pointed out in the Rule of 70 example, the S-shaped curve in figure 4.4 is derived from the following differential equation:

$$Dn/dt = Rn(1—N/K)$$

where N is population size, R is a growth rate, and K is the carrying capacity of the environment. The factor $(1—N/K)$ is the environmental resistance. As population grows, the resistance to further population growth continuously increases.

It is interesting to note that the S-shaped curve can also be used to find the maximum rate that organisms can be removed without reducing the population size. This concept in population biology is called the maximum sustainable yield value of an ecosystem. For example, imagine fishing steelhead fish from a stream. If the stream is at its carrying capacity, theoretically, there will be no population growth, so that any steelheads removed will reduce the population. Thus, the maximum sustainable yield will correspond to a population size less than the carrying capacity. If population growth is logistic or S-shaped, the maximum sustainable

Lily Pond Parable

1. If a pond lily doubles every day and it takes 30 days to completely cover a pond, on what day will the pond be one-quarter covered?
2. One-half covered?
3. Does the size of the pond make a difference?
4. What kind of environmental, social, and economic developments can be expected as the thirtieth day approaches?
5. What will begin to happen at one minute past the 30th day?
6. At what point (what day) would preventative action become necessary to prevent unpleasant events?

Answers:

1. Day 28. Growth will be barely visible until the final few days. (On the 25th day, the lilies cover 1/32 of the pond; on the 21st day, the lilies cover 1/512 of the pond).
2. The 29th day.
3. No. The doubling time is still the same. Even if you could magically double the size of the pond on day 30, it would still hold only one day's worth of growth!
4. The pond will become visibly more crowded each day, and this crowding will begin to exhaust the resources of the pond.
5. The pond will be completely covered. Even through the lilies will be reproducing, there will be no more room for additional lilies, and the excess population will die off. In fact, since the resources of the pond have been exhausted, a significant proportion of the original population may die off as well.
6. It depends on how long it takes to implement the action and how full you want the lily pond to be. If it takes two days to complete a project to reduce lily reproductive rates, that action must be started on day 28, when the pond is only 25 percent full—and that will still produce a completely full pond. Of course, if the action is started earlier, the results will be much more dramatic.

yield will be obtained when the population is half the carrying capacity. This can be seen in the following:

The slope of the logistic curve is given by

$$Dn/dt = Rn\ (1—N/K)$$

Setting the derivative to zero gives

$$d/dt\ (Dn/dt = r\ dn/dt—r/k\ (2N\ Dn/dt) = 0$$

yielding

$$1—2N/K = 0$$
$$N = K/2$$

The logistic growth curve is said to be density conditioned. As the density of individuals increases, the growth rate of the population declines.

Doubling Time and the Rule of 70

Population growing at a constant rate will have a constant doubling time . . . the time it takes for the population to double in size.

Population growing at a constant rate can be modeled with an exponential growth equation:

$$dN/dt = rN$$

The integral of the equation is

$$N_t = N_0 e^{rt}$$

How long will it take for the population to double growing at a constant rate r?

$$.69/r = T$$

The *Rule of 70* is useful for financial as well as demographic analysis. It states that to find the doubling time of a quantity growing at a given annual percentage rate, divide the percent at number into 70 to obtain the approximate number of years required to double. For example, at a 10 percent annual growth rate, doubling time is $70/10 = 7$ years.

Similarly, to get the annual growth rate, divide 70 by the doubling time. For example, 70/14 years doubling time = 5, or a 5 percent annual growth rate.

The following table shows some common doubling times:

Growth Rate (% per Year)	Doubling Time in Years
0.1	700
0.5	140
1	70
2	35
3	23
4	18
5	14
6	12
7	10
10	7

As stated previously, after reaching environmental carrying capacity, population normally oscillates around the fixed axis due to various factors that affect the size of population. These factors work against maintaining the level of population at the K level due to direct dependence on resource availability. The factors that affect the size of populations are known as population controlling factors. They are usually grouped into two classes: density dependent and density independent. Table 4.1 shows factors that affect population size.

Density-dependent factors are those that increase in importance as the size

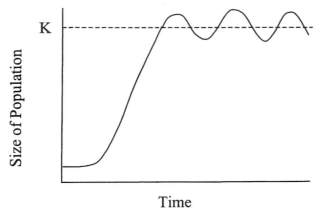

Figure 4.4.   S-shaped (sigmoidal) growth curve

of the population increases. For example, as the size of a population grows, food and space may become limited. The population has reached the carrying capacity. When food and space become limited, growth is suppressed by competition. Odum (1983) describes density-dependent factors as acting "like governors on an engine and for this reason are considered one of the chief agents in preventing overpopulation."

Density-independent factors are those that have the same effect on population regardless of size. Typical examples of density-independent factors are devastating forest fires, dried up streambeds, or the destruction of the organism's entire food supply by disease.

Thus, population growth is influenced by multiple factors. Some of these factors are generated within the population, others from without. Even so, usually no single factor can account fully for the curbing of growth in a given population. It should be noted, however, that humans are, by far, the most important factor; their activities can increase or exterminate whole populations.

Population Response to Stress

As mentioned earlier, population growth is influenced by multiple factors. When a population reaches its apex of growth (its carrying capacity), certain forces work

Table 4.1.   Some Factors Affecting Population Size

Density Independent	Density Dependent
drought	food
fire	pathogens
heavy rain	predators
pesticides	space
human destruction of habit	psychological disorders
	physiological disorders

to maintain population at a certain level. On the other hand, populations are exposed to small or moderate environmental stresses. These stresses work to affect the stability or persistence of the population. Ecologists have concluded that a major factor that affects population stability or persistence is species diversity.

Species diversity is a measure of the number of species and their relative abundance. There are several ways to measure species diversity. One way is to use the straight ratio, $D = S/N$. In this ratio, D = species diversity, N = number of individuals, and S = number of species. As an example, a community of 1,000 individuals is counted; these individuals are found to belong to 50 different species. The species diversity would be 50/1000 or 0.050. This calculation does not take into account the distribution of individuals of each species. For this reason, the more common calculation of species diversity is called the Shannon-Weiner index. The Shannon-Weiner index measures diversity by

$$\overset{s}{\underset{i=1}{H}} = -\Sigma \, (pi) \, (\log pi)$$

where

$H =$ the diversity index
$s =$ the number of species
$i =$ the species number
$pi =$ proportion of individuals of the total sample belonging to the
    ith species

The Shannon-Weiner index is not universally accepted by ecologists as being the best way to measure species diversity, but it is an example of a method that is available.

Species diversity is related to several important ecological principles. For example, under normal conditions, high species diversity—with a large variety of different species—tends to spread risk. This is to say that ecosystems that are in a fairly constant or stable environment, such as a tropical rain forest, usually have higher species diversity. However, as Odum (1983) points out, "diversity tends to be reduced in stressed biotic communities."

If the stress on an ecosystem is small, the ecosystem can usually adapt quite easily. Moreover, even when severe stress occurs, ecosystems have a way of adapting. Severe environmental change to an ecosystem can result from such natural occurrences as fires, earthquakes, or floods, and from people-induced changes such as land clearing, surface mining, and pollution.

One of the most important applications of species diversity is in the evaluation of pollution. As stated previously, it has been determined that stress of any kind will reduce the species diversity of an ecosystem to a significant degree. In the case of domestic sewage, for example, the stress is caused by a lack of dissolved oxygen (DO) for aquatic organisms. This effect is illustrated in figure 4.5. As illustrated in the graph, the species diversity of a stream exhibits a sharp decline after the addition of domestic sewage.

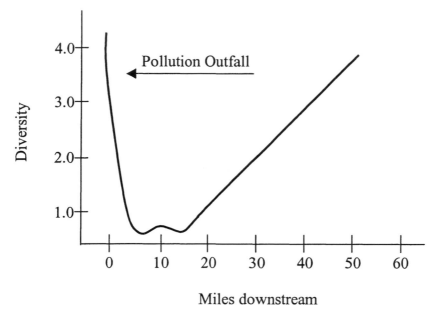

Figure 4.5. Effects of pollution on species diversity

Adapted from Eugene P. Odum, *Fundamentals of Ecology* (Philadelphia: Saunders College, 1971), p. 150.

Ecological Succession

Ecosystems can and do change. For example, if a forest is devastated by a fire, it will grow back, eventually, because of ecological succession. Ecological succession is the observed process of change (a normal occurrence in nature) in the species structure of an ecological community over time; that is, a gradual and orderly replacement of plant and animal species takes place in a particular area over time. The result of succession is evident in many places. For example, succession can be seen in an abandoned pasture. It can be seen in any lake and any pond. Succession can even be seen where weeds and grasses grow in the cracks in a tarmac, roadway, or sidewalk.

Additional specific examples of observable succession include:

1. Consider a red pine planting area where the growth of hardwood trees (including ash, poplar, and oak) occurs. The consequence of this hardwood tree growth is the increased shading and subsequent mortality of the sun-loving red pines by the shade-tolerant hardwood seedlings. The shaded forest floor conditions generated by the pines prohibit the growth of sun-loving pine seedlings and allow the growth of the hardwoods. The consequence of the growth of the hardwoods is the decline and senescence of the pine forest.

2. Consider raspberry thickets growing in the sunlit forest sections beneath the gaps in the canopy generated by wind-thrown trees. Raspberry plants require

sunlight to grow and thrive. Beneath the dense shade canopy, particularly of red pines but also dense stands of oak, there is insufficient sunlight for the raspberry's survival. However, in any place in which there has been a tree fall, the raspberry canes proliferate into dense thickets. Within these raspberry thickets, by the way, are dense growths of hardwood seedlings. The raspberry plants generate a protected "nursery" for these seedlings and prevent a major browser of tree seedlings (the white tail deer) from eating and destroying the trees. By providing these trees a shaded haven in which to grow, the raspberry plants are setting up the future tree canopy that will extensively shade the future forest floor and consequently prevent the future growth of more raspberry plants!

Succession usually occurs in an orderly, predictable manner. It involves the entire system. The science of ecology has developed to such a point that ecologists are now able to predict several years in advance what will occur in a given ecosystem. For example, scientists know that if a burned-out forest region receives light, water, nutrients, and an influx or immigration of animals and seeds, it will eventually develop into another forest through a sequence of steps or stages.

Two types of ecological succession are recognized by ecologists: primary and secondary. The particular type that takes place depends on the condition at a particular site at the beginning of the process.

Primary succession, sometimes called bare-rock succession, occurs on surfaces such as hardened volcanic lava, bare rock, and sand dunes, where no soil exists and where nothing has ever grown before (see figure 4.6). Obviously, in order to grow, plants need soil. Thus, soil must form on the bare rock before succession can begin. Usually this soil formation process results from weathering. Atmospheric exposure—weathering, wind, rain, and frost—forms tiny cracks and holes in rock surfaces. Water collects in the rock fissures and slowly dissolves the minerals out of the rock's surface. A pioneer soil layer is formed from the dissolved minerals and supports such plants as lichens. Lichens gradually cover the rock surface and secrete carbonic acid, which dissolves additional minerals from the rock. Eventually, the lichens are replaced by mosses. Organisms called decomposers move in and feed on dead lichen and moss. A few small animals such as mites and spiders arrive next. The result is what is known as a pioneer community. The pioneer community is defined as the first successful integration of plants, animals, and decomposers into a bare-rock community (Miller 1988).

After several years, the pioneer community builds up enough organic matter in its soil to be able to support rooted plants such as herbs and shrubs. Eventually, the pioneer community is crowded out and is replaced by a different environment. This, in turn, works to thicken the upper soil layers. The progression continues through several other stages until a mature or climax ecosystem is developed, several decades later. It is interesting to note that in bare-rock succession, each stage in the complex succession pattern dooms the stage that existed before it. According to Tomera (1990), "mosses provide a habitat most inhospitable to lichens, the

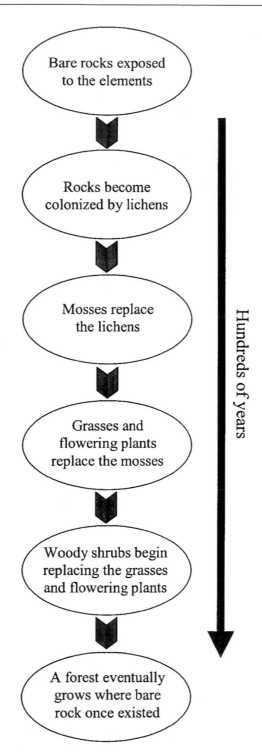

Figure 4.6.   Bare-rock succession

Adapted from A. N. Tomera, *Understanding Basic Ecological Concepts*
(Portland, ME: J. Weston Walch), p. 67.

herbs will eventually destroy the moss community, and so on until the climax stage is reached."

Secondary succession is the most common type of succession. Secondary succession occurs in an area where the natural vegetation has been removed or destroyed, but the soil is not destroyed. For example, succession that occurs in abandoned farm fields, known as old field succession, illustrates secondary succession. An example of secondary succession can be seen in the Piedmont region of North Carolina. Early settlers of the area cleared away the native oak-hickory forests and cultivated the land. In the ensuing years, the soil became depleted of nutrients, reducing the soil's fertility. As a result, farming ceased in the region a few generations later, and the fields were abandoned. Some 150 to 200 years after abandonment, the climax oak-hickory forest was restored.

Ecosystem Population Response to Stress

After reading the preceding sections in the chapter, the reader will have learned a few of the basic foundation blocks that support the science of ecology. However, the reader might also question if the preceding sections were necessary to gain a fundamental knowledge of ecology.

In answering this question, the short answer is: yes. The compound answer is, as you might expect, a bit more complicated.

To clear the air and explain the need for the information provided to this point, we must discuss the dynamics of a representative ecosystem. The focus of this discussion is on the stream ecosystem (but this information could be applied to any ecosystem).

STREAM ECOSYSTEM

It is important to understand that the dynamic balance of the stream ecosystem is between population growth and population reduction factors.

Factors that cause the population to increase in number are *growth factors*. Factors that cause the population to decrease in number are called *reduction factors*.

In a stream ecosystem, growth is enhanced by biotic and abiotic factors. These factors include:

1. ability to produce offspring
2. ability to adapt to new environments
3. ability to migrate to new territories
4. ability to compete with species for food and space to live
5. ability to blend into the environment so as not to be eaten
6. ability to find food

CASE STUDY 4.1
From Lava Flow to Forest: Primary Succession
(U.S. Geological Survey 1999)

Probably the best example of primary succession occurred (and is still occurring) on the Hawaiian Islands.

One of the most striking aspects of a newly formed lava flow is its barren and sterile nature. The process of colonization of new flows begins almost immediately as certain native organisms specially adapted to the harsh conditions begin to arrive from adjoining areas. A wolf spider and cricket may be the first to take up residence, consuming other invertebrates that venture onto the forbidding new environment. The succession process relies heavily on adjacent ecosystems. A steady rain of organic material, seeds, and spores slowly accumulates in cracks and pockets along with tiny fragments of the new lava surface. Some pockets of this infant soil retain enough moisture to support scattered 'ohi'a seedlings and a few hardy ferns and shrubs. Over time, the progeny of these colonizers, and additional species from nearby forests, form an open cover of vegetation, gradually changing the conditions to those more favorable to other organisms. The accumulation of fallen leaves, bark, and dead roots is converted by soil organisms into a thin but rich organic soil. A forest can develop in wet regions in less than 150 years.

On Hawaiian lava flows, primary succession proceeds rapidly on wet windward slopes, but more slowly in dry areas. The influence of moisture can be seen on the Kona side, where the same flow can support a forest along the Belt Highway but be nearly barren near the dry coast. Except for the newer flows and disturbed areas, the windward surfaces of Kilauea are heavily forested, but the leeward slope is barren or sparsely vegetated.

All the undisturbed flows on Kilauea, Mauna Loa, and Hualaiai volcanoes are young enough to be in some degree of primary succession, and the patterns and relative age of lava flows are reflected in the maturity of vegetation. Only a few of the newest flows on the dry upper slopes of dormant Mauna Kea are young enough to reflect primary succession. Extinct Kohala volcano is too old to find such flows, and vegetation differences reflect rainfall amounts and disturbance.

On wetter slopes of Hualaiai and Mauna Loa, younger flows stand out against a more uniform, older background, as the surfaces are recovered by lava at rates of only 20 to 40 percent a century. Small and more active, Kilauea renews about 90 percent of its surface in the same time period, and the resulting pattern is a patchwork of flows and vegetated remnants (*kipuka*). The many younger flows rely on the older *kipuka* to provide sources of plants and animals.

The native forest ecosystems have adapted to the overpowering nature of volcanic eruptions by being able to quickly recolonize from the many *kipuka* around new flows. However, the added losses due to forest clearing and alien invasion provide additional threats to which the native biota are not adapted. If too many of the native forest areas are cleared or taken over by introduced organisms, natural succession may not be able to provide a replacement native ecosystem on the younger flows. The continuing primary succession process may be already partially interrupted in low Puna, where so much of the native forest has been cleared for development and where colonizers from nearby areas are mostly introduced organisms.

7. ability to defend itself from enemies
8. favorable light
9. favorable temperature
10. favorable dissolved oxygen (DO) content
11. sufficient water level

The biotic and abiotic factors in a stream ecosystem that reduce growth include:

1. predators
2. disease
3. parasites
4. pollution
5. competition for space and food
6. unfavorable stream conditions (i.e., low water levels)
7. lack of food

✔   *Important Point*: When all populations within a stream ecosystem are in balance, the entire stream ecosystem is in balance.

In regard to stability in a stream ecosystem, the higher the species diversity, the greater the inertia and resilience of the ecosystem. At the same time, when the species diversity is high within a stream ecosystem, a population within the stream can be out of control because of an imbalance between growth and reduction factors, with the ecosystem at the same time still remaining stable.

In regard to instability in a stream ecosystem, recall that imbalance occurs when growth and reduction factors are out of balance. For example, when sewage is accidentally dumped into a stream, the stream ecosystem, via the self-purification process (discussed later), responds and returns to normal. This process is described as follows:

1. Raw sewage is dumped into the stream, which
2. decreases the oxygen available as the detritus food chain breaks down the sewage;
3. some fish die at the pollution site, and downstream
4. sewage is broken down and washes out to sea and is finally broken down in the ocean.
5. Oxygen levels return to normal;
6. fish populations that were deleted are restored as fish about the spill reproduce and the young occupy the real estate formerly occupied by the dead fish, and
7. populations all return to "normal."

A shift in balance in a stream's ecosystem (or in any ecosystem) similar to the one just described is a fairly common occurrence. In this particular case, the stream responded (on its own) to the imbalance the sewage caused and through the self-purification process returned to normal. Recall that we defined succession as being the method by which an ecosystem either forms itself or heals itself. Thus, we can say that a type of succession has occurred in the polluted stream described above, because, in the end, it healed itself. More importantly, it does not take a rocket scientist to determine that this healing process is a good thing; otherwise, long ago there would have been few streams on Earth suitable for much more than the dumping of garbage.

In summary, through research and observation, ecologists have found that the succession patterns in different ecosystems usually display common characteristics. First, succession brings about changes in the plant and animal members present. Second, organic matter increases from stage to stage. Finally, as each stage progresses, there is a tendency toward greater stability or persistence. Earlier it was stated that succession is usually predictable. This is the case unless humans interfere. Moreover, this illustrates Garrett Hardin's first law of ecology: We can never do merely one thing. Any intrusion into nature has numerous effects, many of which are unpredictable (Miller 1988).

Summary of Key Terms

Population density—the number of a particular species in an area. This is affected by natality (birth and reproduction), immigration (moving into), mortality (death), and emigration (moving out of).

Ultimate carrying capacity—the maximum number of a species an area can support; the environmental carrying capacity is the actual maximum capacity a species maintains in an area. Ultimate capacity is always greater than the environmental capacity.

Chapter Review Questions

4.1   Define population ecology.
4.2   What is the main axiom of population ecology?
4.3   When measuring populations, the level of _____ or _____ must be determined.
4.4   The arrival of new species to a population from other places is termed _____.
4.5   _____ studies the structure and dynamics of animal and plant communities.
4.6   _____ produce aggregation, the result of response by plants and animals to habitat differences.
4.7   _____ is the upper limit of population size.
4.8   The _____ is the actual maximum population density that a species maintains in an area.
4.9   _____ factors affect the size of populations.

4.10   _____ is the observed process of change in the species structure of an ecological community over time.

4.11   Describe bare-rock succession.

4.12   The first successful integration of plants, animals, and decomposers into a bare-rock community is called a _____.

4.13   In a stream ecosystem, growth is enhanced by _____ and _____ factors.

4.14   State Hardin's first law of ecology.

Cited References and Recommended Reading

Abedon, S. T. 2007. *Population ecology.* Email abdeon.1@osu.edu (accessed February 27, 2007).

Allee, W. C. 1932. *Animal aggregations: A study in general sociology.* Chicago: University of Chicago Press.

Berryman, A. A. 1981. *Population systems: A general introduction.* New York: Plenum Press.

Berryman, A. A. 1993. Food web connectance and feedback dominance, or does everything really depend on everything else? *Oikos* 68:13–185.

Berryman, A. A. 1999. *Principles of population dynamics and their application.* Cheltenham, UK: Stanley Thornes.

Berryman, A. A. 2002. *Population cycles: The case for trophic interactions.* New York: Oxford University Press.

Berryman, A. A. 2003. On principles, laws and theory in population ecology. *Oikos* 103:695–701.

Bonner, J. T. 1965. *Size and cycle: An essay on the structure of biology.* Princeton, NJ: Princeton University Press.

Calder, W. A. 1983. An allometric approach to population cycles of mammals. *Journal of Theoretical Biology* 100:275–82.

Calder, W. A. 1996. *Size, function and life history.* Mineola, NY: Dover.

Campbell, N. A., and Reece, J. B. 2004. *Biology* (7th ed.). San Francisco: Pearson, Benjamin Cummings.

Clark, L. R., Gerier, P. W., Hughes, R. D., and Harris, R. F. 1967. *The ecology of insect populations.* London: Methuen.

Colyvan, M., and Ginzburg, L. R. 2003. Laws of nature and laws of ecology. *Oikos* 101:649–53.

Damuth, J. 1981. Population density and body size in mammals. *Nature* 290:699–700.

Damuth, J. 1987. Interspecific allometry of population density in mammals and other animals: The independence of body mass and population energy-use. *Biological Journal of the Linnean Society* 31:193–246.

Damuth, J. 1991. Of size and abundance. *Nature* 351:268–69.

Enger, E., Kormelink, J. R., Smith, B. F., and Smith, R. J. 1989. *Environmental science: The study of interrelationships.* Dubuque, IA: William C. Brown.

Fenchel, T. 1974. Intrinsic rate of natural increase: The relationship with body size. *Oecologia* 14:317–26.

Ginzburg, L. R. 1986. The theory of population dynamics: 1. Back to first principles. *Journal of Theoretical Biology* 122:385–99

Ginzburg, L.R., and Colyvan, M. 2004. *Ecological orbits: How planets move and populations grow.* New York: Oxford University Press.

Ginzburg, L.R., and Jensen C. X. J. 2004. Rules of thumb for judging ecological theories. *Trends in Ecology and Evolution* 19:121–26.

Haemig, P. D. 2006. Laws of Population Ecology. *ECOLOGY.INFO* #23. http://www.ecolo gy.info/laws-population-ecology.htm (accessed September 2, 2007).

Hickman, C. P., Roberts, L. S., and Hickman, F. M. 1990. *Biology of animals*. St Louis: Time Mirror/Mosby College.

Hubbell, S. P., and Johnson, L. K. 1977. Competition and next spacing in a tropical stingless bee community. *Ecology* 58:949–63.

Krebs, R. E. 2001. Scientific laws, principles and theories. Westport, CT: Greenwood Press.

Liebig, J. 1840. *Chemistry and its application to agriculture and physiology*. London: Taylor & Walton.

Lotka, A. J. 1925. *Elements of physical biology*. Baltimore: Williams & Wilkens.

Malthus, T. R. (1798). *An essay on the principle of population*. London: J. Johnson.

Masters, G. M. 1991. *Introduction to environmental engineering and science*. Englewood Cliffs, NJ: Prentice Hall.

Miller, G. T. 1988. *Environmental science: An introduction*. Belmont, CA: Wadsworth.

Odum, E. P. 1983. *Basic ecology*. Philadelphia: Saunders College.

Pianka, E. R. 1988. *Evolutionary ecology*. New York: Harper Collins.

Sharov, A. 1992. Life-system approach: A system paradigm in Population ecology. *Oikos* 63:485–94.

Sharov, A. 1996. *What is population ecology?* Blacksburg: Department of Entomology, Virginia Technical University. http://www.ento.vt.edu/~sharov/PopEcol/lec1/whatis.html (accessed September 2, 2007).

Sharov, A. 1997. *Population ecology*. http://www.gypsymoth.ent.ut.edu/Sharov/population/ welcome (accessed February 28, 2007).

Smith, R. L. 1974. *Ecology and field biology*. New York: Harper & Row.

Spellman, F. R. 1996. *Stream ecology and self-purification*. Lancaster, PA: Technomic.

Tomera, A. N. 1990. *Understanding basic ecological concepts*. Portland, ME: J. Weston Walch.

Turchin, P. 2001. Does population ecology have general laws? *Oikos* 94:17–26.

Turchin, P. 2003. *Complex population dynamics: A theoretical/empirical synthesis*. Princeton, NJ: Princeton University Press.

U.S. Geological Survey (USGS). 1999. *Hawaiian Volcano Observatory*. http://hvo.wr.usgs.gov/ volcanowatch/1999/99_01_21.html (accessed March 1, 2007).

Verhulst, P. F. 1838. Notice sur la loi que la population suit dans son accrossement. *Corr. Math. Phys.* 10:113–21.

Volterra, V. 1926. Variazioni e fluttuazioni del numero d'indivudui in specie animali conviventi. *Mem. R. Accad. Naz. die Lincei Ser. VI* 2.

World Commission on Environment and Development (WCED). 1987. *Our common future*. New York: Oxford University Press.

Winstead, R. L. 2007. *Population regulation*. http://nsm1.nsm.iup.edu/rwinstea/popreg.shtm (accessed February 28, 2007).

Part II

BIODIVERSITY

Eastern gray squirrel, Tampa, Florida.
Photograph by Revonna Bieber

CHAPTER 5

Biodiversity

Nature is the incarnation of a thought, and turns to a thought again, as ice becomes water and gas. The world is mind precipitated, and the volatile essence is forever escaping again into the state of free thought. Hence the virtue and pungency of the influence on the mind, or natural objects, whether inorganic or organized. Man imprisoned, man crystallized, man vegetative, speaks to man impersonated. That power which does not respect quantity, which makes the whole and the particle its equal channel, delegates its smile to the morning, and distils its essence into every drop of rain. Every moment instructs, and every object: for wisdom is infused into every form. It has been poured into us as blood; it convulsed us as pain; it slid into us as pleasure; it enveloped us in dull, melancholy days, or in days of cheerful labor; we did not guess its essence, until after a long time.

—Ralph Waldo Emerson (1844)

Topics

Biodiversity
Loss of Biodiversity
Biodiversity and Stability
Biodiversity: Estimated Decline
Summary of Key Terms
Chapter Review Questions

Biodiversity

The Earth contains a diverse array of organisms whose species diversity, genetic diversity and ecosystems are together called biodiversity. The United Nations Environment Program (UNEP) defines biodiversity as "the variability among living organisms from all sources, including terrestrial, marine and other aquatic ecosystems and the ecological complexes of which they are a part; this includes diversity within species, between species and of ecosystems" (1995).

The U.S. Agency for International Development (USAID) defines biodiversity as the variety and variability of life on Earth (2007). This includes all of the plants and animals that live and grow on the Earth, all of the habitats that they call home, and all of the natural processes of which they are a part. The Earth supports an incredible array of biodiversity with plants and animals of all shapes and sizes. This fantastic variety of life is found in diverse habitats ranging from the hottest desert to tropical rainforests and the arctic tundra. Biodiversity is essential to every aspect of the way that humans live around the world. Plants and animals provide people with food and medicine, trees play an important role in absorbing greenhouse gases and cleaning the air we breathe, and rivers and watersheds provide the clean water that we drink.

Unfortunately, however we define it, the fact is the Earth's biodiversity is disappearing, with an estimated 1,000 species per year becoming extinct. Conversely, biodiversity is especially crucial in developing countries where people's livelihoods are directly dependent on natural resources such as forests, fisheries, and wildlife.

In its simplest terms, biodiversity is the variety of life at all levels; it includes the array of plants and animals; the genetic differences among individuals; the communities, ecosystems, and landscapes in which they occur; and the variety of process on which they depend (LaRoe 1995).

Biodiversity is important for several reasons. Its value is often reported in economic terms: for example, Keystone Center (1991) and Wilson (1992) report that about half of all medicinal drugs come from—or were first found in—natural plants and animals, and therefore these resources are critical for their existing and as-yet undiscovered medicinal benefits. Moreover, most foods were domesticated from wild stocks, and interbreeding of different, wild genetic stocks is often used to increase crop yield. LaRoe (1995) reports that today we use but a small fraction of the food crops used by native cultures; many of these underused plants may become critical new food sources for the expanding human population or in times of changing environmental conditions.

It should be noted that it is the great variety of life that makes existence on Earth possible—thus pointing out the greater importance of biodiversity. As a case in point, consider that plants convert carbon dioxide to oxygen during the photosynthetic process; animals breathe this fresh air, releasing energy and providing the second level of the food chain. In turn, animals convert oxygen back to carbon dioxide, providing the building blocks for the formation of sugars during photosynthesis by plants. Decomposers (microbes such as fungi, bacteria, and protozoans) break down the carcasses of dead organisms, recycling the minerals to make them available for new life; along with some algae and lichens, they create soils and improve soil fertility (LaRoe 1995).

Additionally, biodiversity provides the reservoir for change in our life-support systems, allowing life to adapt to changing conditions. This diversity is the basis not only for short-term adaptation to changing conditions but also for long-term evolution.

The Tragedy of the Commons

After the introduction of the term *biodiversity*, several researchers set out to define the term. During the defining stages, several buzzwords and interesting points of view on the subject area developed. One of the interesting concepts to come forth is known as the "tragedy of the commons." Simply, the tragedy of the commons, developed by Garrett Hardin, concerns how groups of people treat the common resources—air, freshwater, biodiversity, for example—they must share: Each individual person or nation will use as much of the common resource a possible to maximize their benefit from that resource (the "use it or lose it" mentality).

In the following, Hardin (1968) describes how the tragedy of the commons develops:

Picture a pasture pen open to all. It is to be expected that each herdsman will try to keep as many cattle as possible on the commons. Such an arrangement may work reasonably satisfactorily for centuries because tribal wars, poaching, and disease keep the numbers of both man and beast well below the carrying capacity of the land. Finally, however, comes the day of reckoning, that is, the day when the long-desired goal of social stability becomes a reality. At this point, the inherent logic of the commons remorselessly generates tragedy.

As a rational being, each herdsman seeks to maximize his gain. Explicitly or implicitly, more or less consciously, he asks, "What is the utility to me of adding one more animal to my herd?" This utility has one negative and one positive component:

1. The positive component is a function of the increment of one animal. Since the herdsman receives all the proceeds from the sale of the additional animal, the positive utility is nearly $+1$.
2. The negative component is a function of the additional overgrazing created by one more animal. Since, however, the effects of overgrazing are shared by all the herdsmen, the negative utility for any particular decision-making herdsman is only a fraction of -1.

Adding together the component partial utilities, the rational herdsman concludes that the only sensible course for him to pursue is to add another animal to his herd. And another, and another. . . . But this is the conclusion reached by each and every national herdsman sharing a commons. Therein is the tragedy. Each man is locked into a system that compels him to increase his herd without limit—in a world that is limited. Ruin is the destination toward which all men rush, each pursuing his own best interest in a society that believes in the freedom of the commons. Freedom in a commons brings ruin to all.

Humans cannot survive in the absence of nature. We depend on the diversity of life on Earth for about 25 percent of our fuel (wood and manure in Africa, India, and much of Asia, for example); more than 50 percent of our fiber (for clothes and construction, and so forth); almost 50 percent of our medicines; and, of course, for all our food (Miller et al. 1985).

Some people believe that because extinction is a natural process, we therefore should not worry about endangered species or the loss of biodiversity (LaRoe 1995).

✔   *Important Point:* NPG = Negative Population Growth. This means that the birth rate has fallen below replacement levels, as human longevity increases and the death rate falls—due to industrialized medicine, use of contraception, improved education, increased social opportunities for women, and economic stability—leading to a graying and shrinking population.

Loss of Biodiversity

According to the U.S. Geological Survey (USGS) in the Department of the Interior (1995), loss of biodiversity is real. Biologists have alerted each other and much of the general public to the contemporary mass extinction of species. Less recognized is loss of biodiversity at the ecosystem level, which occurs when distinct habitats, species assemblages, and natural processes are diminished or degraded in quality. Tropical forests, apparently the most species-rich terrestrial habitats on Earth, are the most widely appreciated, endangered ecosystems; they almost certainly are experiencing the highest rates of species extinction today (Myers 1984, 1988; Wilson 1988). However, biodiversity is being lost more widely than just in the tropics. Moyle and Williams (1990) point out that some temperate habitats, such as freshwaters in California and old-growth forests in the Pacific Northwest (Norse 1990) to name but two, are being destroyed faster than most tropical rainforests and stand to lose as great a proportion of their species. Because so much of the temperate zone has been settled and exploited by humans, losses of biodiversity at the ecosystem level have been greatest there so far.

Ecosystems can be lost or impoverished in basically two ways. The USGS (1995) reports that the most obvious kind of loss is quantitative—the conversion of a native prairie to a cornfield or to a parking lot. Quantitative losses, in principle, can be measured easily by a decline in a real extent of a discrete ecosystem type (i.e., one that can be mapped). The second kind of loss is qualitative and involves a change or degradation in the structure, function, or composition of an ecosystem (Franklin et al. 1981; Noss 1990a). At some level of degradation, an ecosystem ceases to be natural. For example, a ponderosa pine forest may be high-graded by removing the largest, healthiest, and frequently, the genetically superior trees; a sagebrush steppe may be grazed so heavily that native perennial grasses are replaced by exotic annuals; or a stream may become dominated by trophic generalist and exotic fishes. Qualitative changes may be expressed quantitatively—for instance, by reporting that 99 percent of the sagebrush steppe is affected by livestock grazing—but such estimates are usually less precise than estimates of habitat conversion. In some cases, as in the conversion of an old-growth forest to a tree farm, the qualitative changes in structure and function are sufficiently severe to qualify as outright habitat loss.

Several biologists (Ehrlich and Ehrlich 1981; Diamond 1984; Wilson 1985; Wilcox and Murphy 1985; Ehrlich and Wilson 1991; Soule 1991) agree that the major proximate causes of biotic impoverishment today are habitat loss, degradation, and fragmentation. Hence, modern conservation is strongly oriented toward habitat protection. The stated goal of the Endangered Species Act of 1973 is "to provide a means whereby the ecosystems upon which endangered species and threatened species depend may be conserved" (P.L. 94–325, as amended). The mission of the Nature Conservancy, the largest private land-protection organization in the United States, is to save "the last of the least and the best of the rest" (Jenkins 1985, 21) by protecting natural areas that harbor rare species and communities and high-quality samples of all natural communities.

The USGS (1995) reports that despite the many important accomplishments of natural-area programs in the United States, areas selected under conventional inventories tend to be small. As predicted by island biogeographic theory (MacArthur and Wilson 1967) and, more generally, by species-area relationships, smaller areas tend to have fewer species. All else being equal, smaller areas hold smaller populations, each of which is more vulnerable to extinction than larger populations (Soule 1987). Recognizing that small natural areas that are embedded in intensely used landscapes seldom maintain their diversities for long, scientists called for habitat protection and management at broad spatial scales such as landscapes and regions (Noss 1983, 1987, 1992; Harris 1984; Scott, Csuti, Smith, Estes, and Caicco 1991; Scott, Csuti, and Caicco 1991). In practice, however, most modern conservation continues to focus on local habitats of individual species and not directly on communities, ecosystems, or landscapes (Noss and Harris 1986).

🖢   *Important Point*: According to Homer-Dixon (1995), there are three types of scarcity: demand induced, supply induced, and structural scarcity. Demand-induced scarcity refers to the relative lack of a resource (for example, oil) due to its overuse by consumers. Supply-induced scarcity refers to the lack of a resource due to degradation or depletion, such as freshwater, as pollution and inefficient conservation have caused supplies to dwindle. Finally, structural scarcity refers to lack of resources due to a nature or human system: one country upstream (A) of another (B) may build a dam and cut off water flow to the downstream nation, resulting in a structural scarcity of water in country B.

Ecosystem conservation is a complement to—not a substitute for—species-level conservation. Protecting and restoring ecosystems serve to protect species about which little is known and to provide the opportunity to protect species while they are still common. Yet ecosystems remain less tangible than species (Noss 1991a). And, as the USGS (1995) points out, the logic behind habitat protection as a means of conserving biodiversity is difficult to refute, and conservationists face a major hurdle: convincing policymakers that significantly more and different kinds of habitats must be designated as reserves or otherwise managed for natural values. Scientists cannot yet say with accuracy how much land or what percentage of an ecosystem type must be kept in a natural condition to maintain viable populations of a given proportion of the native biota or the ecological processes of an ecosystem. However, few biologists doubt that the current level of protection is inadequate. Estimates of the fraction of major terrestrial ecosystem types that are not represented in protected areas in the United States range from 21 to 52 percent (Shen 1987). Probably a smaller percentage is adequately protected. For example, 60 percent of 261 major terrestrial ecosystems in the United States and in Puerto Rico, defined by the Bailey-Kuchler classification, were represented in designated wilderness areas in 1988 (Davis 1988). Only 19 percent of those ecosystem types, however, were represented in units of 100,000 hectares or more and only 2 percent in units of 1 million hectares or more—all of them in Alaska (Noss 1990b). Because the size of an area has a pronounced effect on the viability of species and on ecological processes, representation of ecosystem types in small units, in most cases, cannot be considered adequate protection.

Biodiversity and Stability

Biodiversity promotes stability. Meffe and Carroll (1997) purport a major benefit of biodiversity is that more diverse ecosystems may be more stable or more predictable through time when compared to species-poor ecosystems. Stability can be defined at the community level as fewer invasions and less extinction, meaning that a more stable community will contain a more stable composition of species. Stated differently, the stability of a system is an inherent property of its component populations and communities, and it is a measure of the ability of that system to accommodate environmental change (Jones 1997). Three main components of stability are:

- *persistence* (inertia): the ability of a community or ecosystem to resist disturbance or alteration
- *constancy:* the ability to maintain a certain size or maintain its number within limits—system remains unchanged
- *resilience:* the tendency of a system to return to a previous state after a perturbation

Biodiversity: Estimated Decline

In this section, the estimated decline (USGS 1995) of biodiversity, with emphasis on the United States, is presented. As noted below, estimated decline includes area loss and degradation.

50 United States
- 85 percent of original primary (virgin) forest destroyed by late 1980s (Postel and Ryan 1991)

- 90 percent loss of ancient (old-growth) forests (World Resources Institute 1992)
- 30 percent loss of wetlands from 1780s to 1980s (Dahl 1990)
- 12 percent loss of forested wetlands from 1940 to 1980 (Abernethy and Turner 1987)
- 81 percent of fish communities are adversely affected by anthropogenic limiting factors (Judy et al. 1982)

48 Conterminous States
- ca. 95–98 percent of virgin forests destroyed by 1990 (Postel and Ryan 1991)
- 99 percent loss of primary (virgin) astern deciduous forest (Allen and Jackson 1992)
- >70 percent loss of riparian forests since presettlement time (Brinson, Swift, Plantico, and Barclay 1981)
- 23 percent loss of riparian forest since the 1950s (Abernethy and Turner 1987)
- 53 percent loss of wetlands from 1780s to 1980s (Dahl 1990)
- 2.5 percent loss of wetlands between mid-1970s and mid-1980s (Dahl and Johnson 1991)
- 98 percent of an estimated 5.2 million kilometers of streams are degraded enough to be unworthy of federal designation as wild or scenic rivers (Benke 1990)

Summary of Key Terms

Stability—ability of a living system to withstand or recover from externally imposed changes or stresses.

Chapter Review Questions

1. Define biodiversity.
2. Freedom in a commons brings _____ to all.
3. Decomposers include _____.
4. Define NPG.
5. What region on Earth, at the ecosystem level, has suffered the greatest loss in biodiversity?
6. Ecosystem conservation is a complement to _____ conservation.
7. What are the three main components of stability?

Cited References and Recommended Reading

Abernethy, Y., and Turner, R. E. 1987. U.S. forested wetlands: 1940–1980. *Bioscience* 37:721–27.

Allen, E. G., and Jackson, L. L. 1992. The arid west. *Restoration Plans and Management Notes* 10(1):56–59.

Benke, A. C. 1990. A perspective on America's vanishing streams. *Journal of the North American Benthological Society* 91:77–88.

Brinson M. M., Swift, B. L., Plantico, R. C., and Barclay, J. S. 1981. *Riparian ecosystems: Their ecology and status*. FWS/OBS-83/17. Washington, DC: U.S. Fish and Wildlife Service, Biological Services Program.

Dahl, T. E. 1990. *Wetland losses in the United States 1780s to 1980s*. Washington, DC: U.S. Fish and Wildlife Service.

Dahl, T. E., and Johnson, C. E. 1991. *Wetlands: Status and trends in the conterminous United States mid-1970s to mid-1980s*. Washington, DC: U.S. Fish & Wildlife Service.

Davis, G. D. 1988. *Preservation of natural diversity: The role of ecosystem representation within wilderness*. Paper presented at National Wilderness Colloquium, Tampa, FL.

Diamond, J. M. 1984. Historic extinctions: A Rosetta stone for understanding prehistoric extinctions. In *Quaternary extinctions: A prehistoric revolution*, ed. P. S. Martin and R. G. Klein, 824–62. Tucson: University of Arizona Press.

Ehrlich, P. R., and Ehrlich, A. H. 1981. *Extinction: The causes and consequences of the disappearance of species*. New York: Random House.

Ehrlich, P. R., and Wilson, E. O. 1991. Biodiversity studies: Science and policy. *Science* 253:757–62.

Franklin, J. F., Cromack, K., Dension, W., McKee, A., Maser, C., Sedell, J., Swanson, F., and Juday, G. 1981. *Ecological characteristics of old-growth Douglas-fir forests*. Portland, OR: General Technical Report PNW-118, U.S. Forest Service, Pacific Northwest Forest and Range Experiment Station.

Hardin, G. 1968. The tragedy of the commons. *Science* 162 (3859):1243–48.

Harris, L. D. 1984. *Bottomland hardwoods: Valuable, vanishing, vulnerable*. Gainesville: Florida Cooperative Extension Service, University of Florida.

Homer-Dixon, T. F. 1995. *Environmental scarcity and violent conflict: The case of Gaza, Project on Environment, Population and Security*. Toronto, ON: American Association for the Advancement of Science and the University of Toronto.

Jenkins, R. E. 1985. Information methods: Why the heritage programs work. *Nature Conservancy News* 35 (6):21–23.

Jones, A. M. 1997. *Environmental biology*. New York: Routledge.

Judy, R. D., Seeley, P. N., Murray, T. M., Svirsky, S. C., Whitworth, M. R., and Ischinger, L. S. 1982. *National fisheries survey, Vol. I. Technical Report: Initial Findings*. FWS/OBS-84/06. Washington, DC: U.S. Environmental Protection Agency, U.S. Fish and Wildlife Service.

Keystone Center. 1991. *Biological diversity on federal lands: Report of a Keystone policy dialogue*. Keystone, CO: Keystone Center.

LaRoe, E. T. 1995. Biodiversity: A new challenge. In *Our living resources*. Washington, DC: U.S. Department of the Interior, National Biological Service.

MacArthur, R. H., and Wilson, E. O. 1967. *The theory of island biogeography*. Princeton, NJ: Princeton University Press.

Meffe, G. K., and Carroll, C. R. 1997. *Principles of conservation biology*. Sunderland, MA: Sinauer.

Miller, K. R., Fortado, J., De Klemm, C., McNeely, J. A., Myers, N., Soule, M. E., and Trexler, M. C. 1985. Issues on the preservation of biological diversity. In *The global possible*, ed. Robert Repetto. New Haven, CT: Yale University Press.

Moyle, P. B., and Williams, J. E. 1990. Biodiversity loss in the temperate zone: Decline of the native fish fauna of California. *Conservation Biology* 4:475–84.

Myers, N. 1984. *The primary source: Tropical forests and our future*. New York: W. W. Norton.

Myers, N. 1988. Tropical forests and their species. Going, going, . . . ? In *Biodiversity*, ed. E. O. Wilson. Washington, DC: National Academy Press.

Norse, E. A. 1990. *Ancient forests of the Pacific Northwest*. Washington, DC: Wilderness Society, Island Press.

Noss, R. F. 1983. A regional landscape approach to maintain diversity. *Bioscience* 33:700–706.

Noss, R. F. 1987. From plant communities to landscapes in conservation inventories: A look at the Nature Conservancy (USA). *Biological Conservation* 41:11–37.

Noss, R. F. 1990a. Indicators for monitoring biodiversity: A hierarchical approach. *Conservation Biology* 4:355–64.

Noss, R. F. 1990b. What can wilderness do for biodiversity? In *Preparing to manage wilderness in the 21st century*, ed. P. Reed, 49–61. Asheville, NC: U.S. Forest Service.

Noss, R. F. 1991a. From endangered species to biodiversity. In *Balancing on the brink of extinction: The Endangered Species Act and lessons for the future*, ed. K. A. Kohm, 227–46. Washington, DC: Island Press.

Noss, R. F. 1991b. Sustainability and wilderness. *Conservation Biology* 5:120–21.

Noss, R. F. 1992. The Wildlands Project: Land conservation strategy. *Wild Earth* (Special Issue): 10–25.

Noss, R. F., and Harris, L. D. 1986. Nodes, networks, and MUMs: Preserving diversity at all scales. *Environmental Management* 10:299–309.

Postel, S., and Ryan, J. C. 1991. Reforming forestry. In *State of the world 1991: A Worldwatch Institute report on progress toward a sustainable society*, ed. L. Starker, 74–92. New York: W. W. Norton.

Scott, J. M., Csuti, B., and Caicco, S. 1991. Gap analysis: Assessing protection needs. In *Landscape linkages and biodiversity*, ed. W. E. Hudson, 15–26. Washington, DC: Defenders of Wildlife, Island Press.

Scott, J. M., Csuti, B., Jacobi, J. D., and Estes, J. E. 1987. Species richness: A geographic approach to protecting future biological diversity. *Bioscience* 37:782–88.

Scott, J. M., Csuti, B., Smith, K., Estes, J. E., and Caicco, S. 1991. Gap analysis of species richness and vegetation cover: An integrated biodiversity conservation strategy. In *Balancing on the brink of extinction: The Endangered Species Act and lessons for the future*, ed. K. A. Kohm, 282–97. Washington, DC: Island Press.

Shen, S. 1987. Biological diversity and public policy. *Bioscience* 37:709–12.

Soule, M. E., ed. 1987. *Viable populations for conservation*. Cambridge: Cambridge University Press.

Soule, M. E. 1991. Conservation: Tactics for a constant crisis. *Science* 253:744–50.

United Nations Environmental Programme (UNEP). 1995. *Global biodiversity assessment*, ed. V. H. Heywood. Cambridge: Cambridge University Press.

U.S. Agency for International Development (USAID). 2007. *Environment*. http://www.usaid .gov/our_work/environment/biodiversity/index.html (accessed February 4, 2007).

U.S. Geological Survey (USGS). 1995. *Endangered ecosystem of the Unites State States: A preliminary assessment of loss and degradation*. Washington, DC: Author.

Wilcox, B. A., and Murphy, D. D. 1985. Conservation strategy: The effects of fragmentation on extinction. *American Naturalist* 125:879–87.

Wilson, E. O. 1985. The biological diversity crisis. *Bioscience* 35:700–706.

Wilson, E. O. 1988. *Biodiversity*. Washington, DC: National Academy Press.

Wilson, E. O. 1992. *The diversity of life*. Cambridge, MA: Belknap Press of Harvard University Press.

World Resources Institute. 1992. *The 1992 information please environmental almanac*. Boston: Houghton Mifflin.

Part III

DISTRIBUTION AND ABUNDANCE

Lion (Panthera leo), Tampa, Florida.
Photograph by Revonna Bieber

CHAPTER 6

Species

The fall of snowflakes in a still air, reserving to each crystal is perfect form;

The blowing of sleet over a wide sheet of water, and over plains, the waving rye—field, the mimic waving of acres of houstonia, whose innumerable floret whiten and ripple before the eye; the reflections of trees and flowers in glass lakes; the musical steaming odorous south wind, which converts all trees to windharps; the crackling and spurting of hemlock in the flames; or of pine logs, which yield glory to the walls and faces in the sitting-room—these are the music and pictures of the most ancient religion.

—Ralph Waldo Emerson (1844)

Topics

Birds
Mammals
Reptiles and Amphibians
Fishes
Invertebrates
Plants and Fungi
Summary of Key Terms
Chapter Review Questions
Note: Portions of this chapter are adapted from or based on information from United States Department of the Interior's *Our Living Environment* (1995).

Birds

For the novice, the word *animal* is often thought of as just the mammals. As a matter of fact, the fish, the insect, the snake, and the bird have as much right to be called animals as the raccoon and the bear. While in this study it is not important to precisely classify or memorize each individual animal organism by specific class, it is important to know the difference between one and the other. Well, this is rather easy to do, you say. And you would be correct, of course. For instance, we easily see that the fish differs in many ways from the dog and that the cat differs from the snake, and it is easy for us to grasp the fact that mammals differ from all other animals in that their young are nourished by milk from the breasts of the mother; when we learn to appreciate this fundamental fact, we will understand that such diverse forms as the whale, the horse, the cow, the bat, and man are members of one great class of animals.

Birds are animals (see figure 6.1) as well. A bird may be studied to ascertain what it is and what it does; however, there are those practitioners in the field who feel that it is necessary to only identify the bird and all the needed knowledge is thus attained. The professional ecologist knows better. As Comstock (1986) puts it, "the identification of birds is simply the alphabet to the real study, the alphabet by means of which we may spell out the life habits of the bird." Knowledge of birds adds a valuable tool to the ecologist's toolkit.

✔ *Interesting Point*: Each bird feather consists of three parts, the shaft or quill, which is the central stiff stem that gives the feather its strength. From this quill protrude the barbs that, toward the outer end, join together in a smooth web, making the thin, fanlike portion of the feather; at the base is the fluff, which is soft and downy and near to the body of the fowl.

Figure 6.1.   Flock of flamingos, Tampa, Florida
Photograph by Revonna Bieber

According to the National Biological Service (U.S. Department of the Interior 1995a), migratory bird populations are an international resource for which there is special federal responsibility. Moreover, birds are valued and highly visible components of natural ecosystems and may be indicators of environmental quality. Consequently, many efforts have been directed toward measuring and monitoring the condition of North America's migratory bird fauna. The task is not an easy one because the more than 700 U.S. species of migratory birds are highly mobile and may occur in the United States during only part of their annual cycle. One often cannot tell whether a bird observed at a given moment is a resident, a migrant, a visitor from another locality, or the same individual seen ten minutes earlier.

Determining status and trends is further complicated by the fact that each of these species has its own patterns of distribution and abundance, and each species has populations that respond to different combinations of environmental factors. Finally, the sheer abundance of birds—estimated at 20 billion individuals in North America at its annual late-summer peak (Robbins, Bruun, and Zim 1966)—may make it difficult to obtain accurate counts of common species, and the absolute abundance of some may mask important changes in their status.

Results from the nationwide Breeding Bird Survey (BBS; Peterjohn and Sauer 1993) and a portion of the large-scale Christmas Bird Count (CBC; Root and McDaniel 1995) show that some populations are declining, others increasing, and many show what appear to be normal fluctuations around a more or less stable average.

Of the 245 species considered in the BBS, 130 have negative trend estimates, 57 of which exhibit significant declines. Species with negative trend estimates are found in all families, but they are especially prevalent among the mimids (mockingbirds and thrashers) and sparrows. A total of 115 species exhibits positive trends, 44 of which are significant increases. Flycatchers and warblers have the largest proportions of species with increasing populations (Peterjohn and Sauer 1993). Some wading birds, such as the American white ibis (see figure 6.2), are holding their own with increases and offsetting decreases.

✔ *Interesting Point*: Among all the vocalists in the bird world, the mockingbird is seldom rivaled in the variety and richness of its repertoire.

SONGBIRDS

Of the 50 songbirds examined in the CBC, 27 (54 percent) exhibited a statistically and biologically significant trend in at least one state. Of these 27 species, 16 (59 percent) had populations declining in more states than states in which they were increasing; 12 exhibited only declines, and 4 had a population increase in at least one state. Ten (37 percent) of the 27 species had populations increasing in more states than states exhibiting declines, with 7 exhibiting only population increases.

Figure 6.2. American white ibis, Tampa, Florida
Photograph by Revonna Bieber

One (4 percent) species had populations increasing and decreasing in the same number of states (Root and McDaniel 1995).

✔ *Interesting Point:* Birds do most of their singing in the early morning and during the spring and early summer months.

Overall, approximately equal numbers of species appear to be increasing and decreasing over the past two to three decades. Groups of species with the most consistent declines are those characteristic of grassland habitats, apparently reflecting the conversion of these habitats to other types of vegetative cover.

✔ *Important Point*: Population health is a measure of a population's ability to sustain itself over time as determined by the balance between birth and death rates. Indexes of population size do not always provide an accurate measure of population health because population size can be maintained in unhealthy populations by immigration of recruits from healthy populations (Pulliam 1998). Poor population health across many populations in a species eventually results in the decline of that species. Early detection of population declines allows managers to correct problems before they are critical and widespread.

WATERFOWL POPULATIONS: GEESE

Waterfowl populations are monitored closely as a basis for regulating annual harvests at levels consistent with maintenance of populations. Goose populations (Rusch, Malecki, and Trost 1995; Hestback 1995a; Hupp, Schmutz, and Ely 2007) have shown some impressive gains over the past decades, but most gains have been registered by large-bodied geese, with several smaller species and smaller subspecies of the highly variable Canada goose having depressed populations.

> There is a sound, that, to the weather-wise farmer, means cold and snow, even though it is heard through the hazy atmosphere of an Indian summer day; and that is the honking of wild geese as they pass on their southward journey. And there is not a more interesting sight anywhere in the autumn landscape than the wedge-shaped flock of these long-necked birds with their leader at the front apex. (Comstock 1986)

According to Rusch et al. (1995), Canada geese are probably more abundant now than at any time in history. They rank first among wildlife watchers and second among harvests of waterfowl species in North America. Canada geese are also the most widely distributed and phenotypically (visible characteristics of the birds) variable species of bird in North America.

Breeding populations now exist in every province and territory of Canada and in 49 of the 50 United States. The size of the 12 recognized subspecies ranges from the 1.4 kg (3 lbs) cackling Canada goose to the 5.0 kg (11 lb) giant Canada goose (Delacour 1954; Bellrose 1976).

✔ *Important Point*: Market hunting and poor stewardship led to record low numbers of geese in the early 1900s, but regulated seasons including closures, refuges, and law enforcement led to restoration of most populations.

Large changes have occurred in the geographic wintering distribution and subspecies composition of the Atlantic flyway population of Canada geese over the past 40 years. The Atlantic flyway can be thought of as being partitioned into four regions: South, Chesapeake, Mid-Atlantic, and New England. Wintering numbers have declined in the southern states, increased then decreased in the Chesapeake region, and increased markedly in the mid-Atlantic region. In the New England region, wintering numbers increased from around 6,000 during 1948–1950 to between 20,000 and 30,000 today (Serie 1993).

North American populations of most goose species have remained stable or have increased in recent decades (U.S. Fish and Wildlife Service [USFWS] and Canadian Wildlife Service 1986). Some populations, however, have declined or historically have had small numbers of individuals, and thus are of special concern. Individual populations of geese should be maintained to ensure that they provide aesthetic, recreational, and ecological benefits to the nation. Monitoring and management efforts for geese should focus on individual populations to ensure that genetic diversity is maintained (Anderson, Rhymer, and Rohwer 1992).

The use of the census and determining the status of natural Canada goose populations are made more difficult by the widespread introduction and establishment of resident goose populations, which breed outside the traditional Arctic nesting areas and mix with migratory populations on the wintering grounds.

WATERFOWL POPULATIONS: DUCKS

Even though some species are stable or even increasing, many duck populations have declined in the past decade. Biologists attribute these declines to losses of breeding and wintering habitats and a long period of drought in breeding areas. Among species receiving special emphasis, canvasbacks showed a complex pattern with regional changes in distribution and abundance, and pintails showed a widespread and nearly consistent pattern of decline (U.S. Department of the Interior 1995).

Increased predation and habitat degradation and destruction coupled with drought, especially on breeding grounds, have caused the declines of some such populations. More than 30 species of ducks breed in North America, in areas as diverse as the arctic tundra and the subtropics of Florida and Mexico. For many of these species, however, the Prairie Pothole region of the north-central United States and south-central Canada is the most important breeding area, although migratory behavior and the life histories of different species lead them to use many wetland habitats (Caithamer and Smith 1995).

✔ *Important Point*: According to Ducks Unlimited (2007), the Prairie Pothole region is the core of what was once the largest expanse of grassland in the world, the Great Plains of North America. Its name comes from a geological phenomenon that left its mark 10,000 years ago. When the glaciers from the last ice age receded, they left behind millions of shallow depressions that are now wetland, known as prairie potholes. The potholes are rich in plant and aquatic life, and support globally significant populations of breeding waterfowl.

✔   *Interesting Point*: A duck has the same number of toes as a hen, but there is a membrane called the web that joins the second, third, and fourth toes, making a fan-shaped foot; the first toe or hind toe has a web of its own. A webbed foot is first of all a paddle for propelling its owner through the water; it is also a very useful foot on the shores of ponds and streams, since its breadth and flatness prevent it from sinking into the soft mud (Comstock 1986).

Duck population changes occur in breeding, staging, and wintering habitats, with the changes in breeding habitats having the greatest effect on populations. Degradation and destruction of wetlands over the past 200 years have diminished duck populations; wetland alteration and degradation continue. The rate of wetland loss has been greatest in prime agricultural area such as the Prairie Pothole region and lowest in northern boreal forests and tundra. Thus, species such as dabbling ducks that mostly nest in the severely altered Prairie Potholes have been harmed more than species such as sea ducks and mergansers that nest farther north (Bellrose 1980; Johnson and Grier 1988).

✔   *Interesting Point*: Dabbling ducks are so named because they feed mainly on vegetable matter by upending on the water surface, or grazing, and only rarely dive (Avianweb 2007).

Because most dabbling ducks need grassy cover for nesting (Kaminski and Weller 1992), conversion of native grasslands to agricultural production, including pastures, has reduced available nesting cover and contributed to a reduced nesting success for dabblers. This condition is especially true in the Prairie Pothole region of the United States and Canada. In addition, highly variable precipitation in the Prairie Potholes has changed the number of wetlands available for nesting. For example, in 1979 there were 6.3 million wetlands in the surveyed portion of the Prairie Pothole region, but by the next spring, wetlands in the same area had decreased 55 percent to 2.9 million. Two years later, they had increased more than 100 percent to 4.2 million. These annual changes can temporarily mask the long-term declining trend in wetland abundance across the Prairie Pothole region.

The changing availability of wetland habitats in the Prairie Pothole region causes substantial fluctuations in some duck populations. During periods of high precipitation, larger wetland basins are full or overflowing, and shallow wetlands are abundant. Species such as the northern pintail, which tend to use shallow or ephemeral wetlands for feeding, produce more young when wetland numbers increase (Smith 1970; Hochbaum and Bossenmaier 1972). Consequently, population numbers increase as they did during the 1970s.

Stewart and Kantrud (1973) report that during the driest periods, however, such as those in the 1980s, only the deepest and most permanent wetlands retain water, causing population declines in species such as pintails that rely primarily on shallow wetlands. Population numbers are more stable for species such as the canvasback, which rely on deeper marshes, and are therefore less affected by annual changes in wetland numbers because deeper marshes consistently retain water, providing ample habitat in most years.

Nest success in the Prairie Pothole region has declined in recent years largely because of increased nest predation caused by the range expansion of some predators and by reduced nesting habitat (Sargeant and Raveling 1992). Fewer and smaller areas of nesting habitats concentrate duck nests, enhancing the ability of predators to find nests. Predators such as raccoons have expanded their range northward, probably because they can den in buildings, rock piles, and other human-made sites during winter.

Although wetland drainage, urbanization, and other human-caused changes have resulted in wintering habitat losses, these loses have been offset, at least for dabbling ducks, by increased fall and winter food from waste grain left in stubble fields. In addition, the National Wildlife Refuge System has protected and managed many staging and wintering areas so that the ducks can benefit from the water protection (U.S. Department of the Interior 1995a).

Modern duck-hunting regulations are believed to keep recreational harvest at levels compatible with the long-term welfare of duck populations. The proportion of ducks harvested varies regionally and by species, age, and sex. In 1992, 2–12 percent of the adult mallards from the Prairie Pothole region were killed by hunters. Harvest rates of other species were generally lower. These conservative harvest rates are unlikely to cause population declines (Blohm 1989).

✔ *Interesting Point*: The U.S. Fish and Wildlife Service (2007c) points out that the northern pintail is one of several species of dabbling ducks that belong to the tribe Anatini in the family Anatidae. Pintails are medium-size ducks with slender, elegant lines and conservative plumage coloration. Males are larger than females. The male in breeding plumage is readily distinguished from other dabbling ducks by a combination of chocolate brown head, white neck stripes, black/gray body feathers, and very long black central tail feathers that give the species its name; the male has a bright green/purple iridescent speculum. The female is colored similarly to females of other dabbling ducks—basically, a dull brown with black markings. Pintails are graceful, acrobatic flyers capable of darting and wheeling routines especially during pursuit or courtship flights, which are fast and vigorous with sudden rapid dives from great heights to low ground-level flight. Common vocalizations by females include the traditional decrescendo "quack" call that consists of one loud quack followed by a softer quack; males issue a series of "chirps" and whistles. Pintails employ an elaborate series of displays during courtship, including "head-up-tail-up," "grunt-whistles," and "chin lifts." Pintails are very wary, long-lived ducks that tend to survive at higher rates than other ducks. The maximum longevity in the wild was recorded at 21 years 4 months based on a California-banded adult male record by a hunter in Idaho. Average life spans would be considerably shorter than this, but relative to other duck species, pintails are long-lived in the wild (Austin and Miller 1995; Bellrose 1980).

The size of the continental breeding population of northern pintail has greatly varied since 1955, with numbers in surveyed areas ranging from a high of 9.9 million in 1956 to a low of 1.8 million in 1991. This variation results primarily from differences in the numbers of breeding pintails in the prairie region of Canada and the United States; these numbers ranged from 8.6 million in 1956 to 9.5 million in 1991; numbers in the northern regions from Alaska to northern Alberta and northern Manitoba varied primarily between 1 and 2 million (Hestback 1995b).

Breeding pintails prefer seasonal shallow-water habitats without tall emergent aquatic vegetation (Smith 1968). The proportions and distribution of breeding pintails on the prairies vary annually depending on the amount of annual precipitation and the resulting increase or decrease in the availability of suitable breeding habitat (Smith 1970; Johnson and Grier 1988).

Changes in the size of the continental pintail population result from changes in production, survival, or both. Consequently, understanding population changes involves detecting variation in survival and production over time and relating that variation to changes in population size. Once the cause of the decline is determined, appropriate management strategies can be developed to reverse it.

✔   *Interesting Point*: The canvasback duck is a large diving duck that breeds in prairie potholes and winters on ocean bays. It sloping profile distinguishes it from other ducks.

Canvasbacks are unique to North America and are one of our most widely recognized waterfowl species. Unlike other ducks that nest and feed in uplands, diving ducks such as canvasbacks are totally dependent on aquatic habitats throughout their life cycle. Canvasbacks nest in prairie, parkland, subarctic, and Great Basin wetlands; stage during spring and fall on prairie marshes, northern lakes, and rivers; and winter in Atlantic, Pacific, and Gulf of Mexico bays, estuaries, and some inland lakes. They feed on plant and animal foods in wetland sediments. Availability of preferred foods, especially energy-rich subterranean plant parts, is probably the most important factor influencing geographic distribution and habitat use by canvasbacks (Hohman, Haramis, Jorde, Korschgen, and Takekawa 1995).

In spite of management efforts that have included restrictive harvest regulations and frequent hunting closures in all or some of the flyways (Anderson 1989), canvasback numbers declined from 1955 to 1993 and remain below the population goal (540,000) of the North American Waterfowl Management Plan (USFWS and Canadian Wildlife Service 1994). Causes for this apparent decline are not well understood, but habitat loss and degradation undoubtedly play a part.

Low rates of recruitment, a highly skewed sex ratio favoring males, and reduced survival of canvasbacks during their first year are considered important constraints on population growth.

SHOREBIRDS

Shorebirds are highly migratory, and status and trends of their populations are largely determined from observations made during periods in their life cycles in which birds congregate in limited breeding, staging, or migratory stopover areas. Populations of eastern and western species show general patterns of decline, although some species, including those using inland areas, are too poorly studied to detect trends. Apparent dependence on critical breeding and staging areas suggest that populations of many species are vulnerable to habitat loss and disturbance (U.S. Department of the Interior 1995a).

Shorebirds are a diverse group that includes oystercatchers, stilts, avocets, plovers, and sandpipers. They are familiar birds of seashores, mudflats, tundra, and other wetlands, and they also occur in deserts, high mountains, forests, and agricultural fields. Widespread loss and alteration of these habitats, coupled with unregulated shooting at the turn of the century, resulted in population declines and range contractions of several species throughout North America (Gill, Handel, and Page 1995). Most populations recovered after passage of the Migratory Bird Treaty Act of 1918, although some species never recovered and others have declined again. In the western portion of the continent, efforts to monitor the status and trends of shorebirds have been in effect for only the past 15–25 years and for only a few species (Harrington 1995).

SEABIRDS

Seabirds are birds that spend almost all their time on or near the sea (USFWS 2007d). They are medium-size to large birds; most are between the size of a robin and crow. They get all their food from the water. Some spend the winter at sea, several hundred miles from land. Seabirds come to land to raise the young birds each summer. They nest on protected cliffs or islands, often in dense groups called colonies.

✔ *Interesting Point*: Seabirds have special adaptations that allow them to live at sea and get all their food there. Some eat small fish or shrimplike invertebrates called zooplankton, which they catch from the sea. Seabirds such as kittiwakes pick their prey from the water's surface. Others, such as auks and cormorants, dive for their prey and chase it underwater (USFWS 2007d).

Seabirds in the Pacific region include many diverse species that respond differently to factors such as human proximity to nesting areas, oil spills, introduction of predators, depletion of fishery stocks, and availability of human refuse as food.

Some species, including certain gulls, brown pelicans, and double-crested cormorants, have responded positively to recent changes in some areas, whereas others, including murrelets and murres and kittiwakes, have shown declining trends. Populations of other species appear to fluctuate widely, and information for many species is insufficient to determine long-term trends (Carter, Gilmer, Takekawa, Lower, and Wilson 1995; Hatch and Platt 1995).

More than 2 million seabirds of 29 species nest along the coasts of California, Oregon, and Washington, including three species listed on the federal list of threatened and endangered species: the brown pelican, least tern, and marbled murrelet. The size and diversity of the breeding seabird community in this region reflect excellent near-shore prey conditions; subtropical waters within the southern California blight area; complex tidal waters of the Strait of Juan de Fuca and Puget Sound in Washington; large estuaries at San Francisco Bay, Columbia River, and Grays Harbor-Willapa bays; and the variety of nesting habitats used by seabirds throughout the region, including islands, mainland cliffs, old-growth forests, and artificial structures (U.S. Department of the Interior 1995a).

Breeding seabird populations along the West Coast have declined since European settlement began in the latter 1700s because of human occupation of, commercial use of, and introduction of mammalian predators to seabird nesting islands. In the 1900s, further declines occurred in association with rapid human population growth and intensive commercial use of natural resources in the Pacific region. In particular, severe adverse impacts have occurred from partial or complete nesting habitat destruction on islands or the mainland, human disturbance of nesting islands or areas, marine pollution, fisheries, and logging of old-growth forests (Ainley and Lewis 1974; Bartonek and Nettleship 1979; Hunt et al. 1979; Sowls, DeGange, Nelson, and Lester 1980; Nettleship, Sanger, and Springer 1984; Speich and Wahl 1989; Ainley and Boekelheide 1990; Sealy 1990; Ainley and Hunt 1991; Carter and Morrison 1992; Carter et al. 1992; Vermeer, Briggs, Morgan, and Siegel-Causey 1993).

> Paddling as quietly as I could, I slipped through the fog and stopped abruptly as a brown, robin-sized seabird—a marbled murrelet—dove almost immediately, reappearing again with a tiny fish in its bill. Loggers working along the Northwest coast in the early years gave this enigmatic little seabird the name "fog lark."
>
> —Audrey Benedict (2007)

About 100 million seabirds reside in marine waters of Alaska during some part of the year. Perhaps half this population is composed of 50 species of nonbreeding residents, visitors, and breeding species that use marine habitats only seasonally (Gould, Forsell, and Lensink 1982). Another 30 species include 40–60 million individuals that breed in Alaska and spend most of their lives in U.S. territorial waters (Sowls, Hatch, and Lensink 1978). Alaskan populations account for more than 95 percent of the breeding seabirds in the continental United States, and eight species nest nowhere else in North America (USFWS 1992a).

Seabird nest sites include rock ledges, open ground, underground burrows, and crevices in cliffs or talus. Seabirds take a variety of prey from the ocean, including krill, small fish, and squid. Suitable nest sites and oceanic prey are the most important factors controlling the natural distribution and abundance of seabirds (Hatch and Platt 1995).

The impetus for seabird monitoring is based partly on public concern for the welfare of these birds, which are affected by a variety of human activities such as oil pollution and commercial fishing. Equally important is the role seabirds serve as indicators of ecological change in the marine environment. Seabirds are long-lived and slow to mature, so parameters such as breeding success, diet, or survival rates often give earlier signals of changing environmental conditions than population size itself. Seabird survival data are of interest because they reflect conditions affecting seabirds in the nonbreeding season, when most annual mortality occurs (Hatch, Roberts, and Fadely 1993b).

Techniques for monitoring seabird populations vary according to habitat types and the breeding behavior of individual species (Hatch and Hatch 1978, 1989; Byrd, Day, and Knudson 1983). An affordable monitoring program can include but a few of the 1,300 seabird colonies identified in Alaska, and since the mid-1970s, monitoring efforts have emphasized a small selection of surface-feeding and diving species, primarily kittiwakes and murres. Little or no information on trends is available for other seabirds (Hatch 1993b). The exiting monitoring program occurs largely on sites within the Alaska Maritime National Wildlife Refuge, which was established primarily for the conservation of marine birds. Data are collected by refuge staff, other state and federal agencies, private organizations, university faculty, and students.

COLONIAL-NESTING WATERBIRDS

"Colonial-nesting waterbird" is a tongue-twister of a collective term used by bird biologists to refer to a large variety of different species that share to common characteristics: (1) they tend to gather in large assemblages, called colonies, during the nesting season, and (2) they obtain all or most of their food (fish and aquatic invertebrates) from the water. Colonial-nesting waterbirds can be further divided into two major groups depending on where they feed (USFWS 2002).

Seabirds (also called marine birds, oceanic birds, or pelagic birds) feed primarily in saltwater. Some seabirds are so marvelously adapted to marine environments that they spend virtually their entire lives at sea, returning to land only to nest; others (especially the gulls and terns) are confined to the narrow coastal interface between land and sea, feeding during the day and loafing and roosting on land. Included among the seabirds are such groups as the albatrosses, shearwaters, storm-petrels, tropic birds, boobies, pelicans, cormorants, frigate birds, gulls, terns, murres, guillemots, murrelets, auklets, and puffins. A few species of cormorants, gulls, and terns also occupy freshwater habitats.

Wading birds seek their prey in fresh or brackish waters. As the name implies, these birds feed principally by wading or standing still in the water, patently wait-ing for fish or other prey to swim within striking distance. The wading birds in-clude the bitterns, herons, egrets, night-herons, ibises (see figure 6.2), spoonbills, and storks.

✔   *Interesting Point*: Colonial-nesting waterbirds have attracted the attention of scientists, conservationists, and the public since the early 1900s when plume hunters nearly drove many species to extinction (Erwin 1995).

Colonial-nesting waterbirds of the continental and East Coast regions of the United States show trends related to many of the same factors operating in the Pacific region, with some species recovering from past losses from pesticides while some other species that exploit human refuse are increasing dramatically. Popula-tions of other species, especially certain terns, are declining, probably as a result of habitat loss and degradation or other kinds of human disturbance. Special efforts have been made to determine status and trends of the piping plover, a species listed as endangered in certain parts of its range and as threatened in others (Robbins et al. 1966).

RAPTORS

Raptors, or birds of prey, include the hawks, falcons, eagles, vultures, and owls and occur throughout North American ecosystems. As predators, most of them kill other vertebrates for their food. Compared to most other animal groups, birds of prey naturally exist at relatively low population levels and are widely dispersed within their habitats. The natural scarcity of raptors, combined with their ability to move quickly, the secretive behavior of many species, and the difficulties of detecting them in rugged terrain or vegetation, all make determining their popula-tion status difficult.

As top predators, raptors are key species for our understanding and conserva-tion of ecosystems. Changes in raptor status can reflect changes in the availability of their prey species, including population declines of mammals, birds, reptiles, amphibians, and insects. Changes in raptor status also can be indicators of more subtle detrimental environmental changes such as chemical contamination and the occurrence of toxic levels of heavy metals (e.g., mercury, lead). Consequently, determining and monitoring the population status of raptors are necessary steps in the wise management of our natural resources (Fuller, Henny, and Wood 1995).

The *California condor* is a member of the vulture family. With a wingspan of about nine feet and weighing about 20 pounds, it spends much of its time in soaring flight visually seeking dead animals as food. The California condor has always been rare (Wilbur 1978; Pattee and Wilbur 1989). Although probably numbering in the thousands during the Pleistocene epoch in North America, its numbers likely declined dramatically with the extinction of most of North Ameri-ca's large mammals 10,000 years ago. Condors probably numbered in the hundreds and were nesting residents in British Columbia, Washington, Oregon, California, and Baja California around 1800. In 1939 the condor population was estimated at 60–100 birds, and its home range was reduced to the mountains and foothills of California, south of San Francisco and north of Los Angeles (Pattee and Mesta 1995).

Franson, Sileo, and Thomas (1995) report that the U.S. Department of the Interior has investigated the deaths of more than 4,300 *bald* and *gold eagles* since the early1960s as part of an ongoing effort to monitor causes of wildlife mortality. The availability of dead eagles for study depends on finding carcasses in fair to good condition and transporting them to the laboratory. Such opportunistic collection and the fact that recent technological advances have enhanced our diagnostic capabilities, particularly for certain toxins, means that results reported here do not necessarily reflect actual proportional causes of death for all eagles in the United States throughout the 30-year period. This type of sampling does, however, identify major or frequent causes of death.

Most diagnosed deaths of eagles in the study (Franson et al. 1995) resulted from accidental trauma, gunshot, electrocution, and poisoning. Accidental trauma, such as impacts with vehicles, power lines, or other structures, was the most frequent cause of death in both eagle species (23 percent of bald and 27 percent of golden). Gunshot killed about 15 percent of each species. Electrocution was twice as frequent in golden (25 percent) than in bald eagles (12 percent), probably because of the preference of golden eagles for prairie habitats and their use of utility poles as perches.

Lead poisoning was diagnosed in 338 eagles from 34 states. Eagles become poisoned by lead after consuming lead shot and, occasionally, bullet fragments present in food items. Agricultural pesticides accounted for most remaining poisonings; organophosphorus and carbamate compounds killed 139 eagles in 25 states. Eagles are exposed to these chemicals in a variety of ways, often by consuming other animals that died of direct poisoning or from baits placed to deliberately kill wildlife.

Overall, poisonings were more frequent in bald eagles (16 percent) than golden eagles (6 percent). The reasons for this are unclear, but may be related to factors that influence submission of carcasses for examination or differences in species' preferences for agricultural, rangeland, and wetland habitats (Franson et al. 1995).

WILD TURKEYS

The wild turkey is a large gallinaceous (domestic fowl) bird characterized by strong feet and legs adapted for walking and scratching, short wings adapted for short rapid flight, a well-developed tail, and a stout beak useful for pecking. These birds probably originated some 2 to 3 million years ago in the Pliocene epoch (5.3 to 1.8 million years ago). Molecular data suggest this genetic line delivered from pheasant-like birds about 11 million years ago. There are two species in the genus, the wild turkey of the United States, portions of southern Canada, and northern Mexico; and the ocellated turkey in the Yucatan region of southern Mexico, Belize, and northern Guatemala. The wild turkey has shown dramatic increases in distribution and abundance in recent decades because of translocations, habitat restoration, and harvest control (Dickson 1995).

MOURNING DOVES

The mourning dove is one of the most widely distributed and abundant birds in North America (Droege and Sauer 1990). It is also the most important U.S. game bird in terms of numbers harvested. The U.S. fall population of mourning doves has been estimated to be about 475 million (Tomlinson, Dolton, Reeves, Nichols, and McKibben 1988; Tomlinson and Dunks 1993).

The breeding range of the mourning dove extends from the southern portions of the Canadian provinces throughout the continental United States into Mexico, the islands near Florida and Cuba, and scattered areas in Central America (Aldrich 1993). Although some mourning doves are nonmigratory, most migrate south to winter in the United States in areas from northern California and Connecticut, and south throughout most of Mexico and Central America to western Panama.

✔ *Interesting Point*: Mourning doves feed their nestlings crop milk or "pigeon milk," which is secreted by the crop lining. This is an extremely nutritious food with more protein and fat than is found in either cow or human milk. Crop milk, which is regurgitated by both adults, is the exclusive food of hatchlings for three days, after which it is gradually replaced by a diet of seeds (Cornell University 1999).

Within the United States, three areas contain breeding, migrating, and wintering mourning dove populations that are largely independent of each other (Kiel 1959). In 1960 three areas were established as separate management units: the Eastern, Central, and Western regions (Dolton 1995).

COMMON RAVENS

> But the raven, sitting lonely on the placid bust, spoke only
> That one word, as if his soul in that one word he did outpour.
> Nothing further then he uttered—not a feather then he
>   fluttered—
> Till I scarcely more than muttered, "other friends have flown
> before—
> On the morrow he will leave me, as my hopes have flown before."
> Then the bird said, "Nevermore."
>
> —Edgar Allan Poe, 1845

The common raven is a large black passerine (perching bird) bird found throughout the northern hemisphere, including western and northern North America. Long recognized as one of the most intelligent birds, the raven also has a less-than-savory image throughout history as a scavenger that does not discriminate between humans and animals (*Nature* 2007). Ravens are scavengers that frequently feed on road-killed animals, large dead mammals, and human refuse. They kill and eat prey, including rodents, lambs (Larsen and Dietrich 1970), birds, frogs, scorpions, beetles, lizards, and snakes. They also feed on nuts, grains, fruits, and other plant matter (Knight and Call 1980; Heinrich 1989). Their recent population increase is of concern because ravens eat agricultural crops and animals whose populations may be depleted.

Ravens are closely associated with human activities, frequently visiting solid-waste landfills and garbage containers at parks and food establishments, being pests of agricultural crops, and nesting on many human-made structures. In two recent surveys in the deserts of California (FaunaWest Wildlife Consultants 1989; Knight and Kawashima 1993), ravens were more numerous in areas with more human influences, and were often indicators of the degree to which humans affect an area (Boarman and Berry 1995).

MISSISSIPPI SANDHILL CRANES

Resident sandhill cranes formed a continuous population in Georgia and Florida and widely separated populations along the Gulf Coast plains of Texas, Louisiana, Mississippi, and Alabama. The Mississippi sandhill crane was one of the widely separated populations on the coastal plain, and bred in pine savannas in southeastern Mississippi, just east of the Pascagoula River to areas just west of the Jackson County line, south to Simmons Bayou, and north to an east-west line (5–10 miles) north of VanCleave (Gee and Hereford 1995).

✔ *Interesting Point:* Cranes are unique and are among the most spectacular of the bird families. In fact, they have captured the human imagination as few other birds have. Early naturalist and pioneering wildlife biologist Aldo Leopold called them "nobility in the midst of mediocrity." The Mississippi sandhill crane was described as a distinct subspecies in 1972 and there are physiological, morphological, behavioral, and other differences between them and other sandhill cranes. The Mississippi sandhill crane is a noticeably different darker shade of gray, resulting in a more distinct cheek patch (National Wildlife Refuge [NWR] 2007).

Mississippi sandhill cranes are a critically endangered subspecies found only on and adjacent to the Mississippi Sandhill Crane National Wildlife Refuge. There are only about 100 individuals remaining, including about 20 breeding pairs. Without intensive management from the U.S. Fish and Wildlife Service and its cooperators and partners, this unique bird may disappear from the wild (Gee and Hereford 1995).

PIPING PLOVERS

The piping plover is a wide-ranging, beach-nesting shorebird whose population viability continues to decline as a result of habitat loss from development and other human disturbance (Haig 1992). In 1985 the species was listed as endangered in the Great Lakes Basin and Canada and threatened in the northern Great Plains and along the U.S. Atlantic Coast (Haig and Plissner 1995).

Piping plovers are approximately seven inches long with sand-colored plumage on their backs and crown and white underparts. Breeding birds have a single black breastband, a black bar across the forehead, bright orange legs and bill, and a black tip on the bill. During winter, the birds lose their black bands, the legs fade to pale yellow, and the bill becomes mostly black (USFWS 2007a).

RED-COCKADED WOODPECKERS

The red-cockaded woodpecker, or RCW, is a territorial, nonmigratory, cooperative breed species (Lennartz, Hooper, and Harlow 1987). It is a small black-and-white bird about the size of a cardinal. The red patches, or "cockades," on either side of the head on males are rarely seen as they usually conceal the red until excited or agitated. This woodpecker usually does not frequent urban settings and is not a familiar backyard species. It is not likely to be observed at a bird feeder, unlike the downy and red-bellied woodpeckers, for example (Texas Parks and Wildlife Department [TPWD] 2007a).

Historically, the southern pine ecosystem, contiguous across a large area and kept open with recurring fire (Christensen 1981), provided ideal conditions for a nearly continuous distribution of RCWs throughout the South. Within this extensive ecosystem, red-cockaded woodpeckers were the only species to excavate cavities in living pine trees, thereby providing essential cavities for other cavity-nesting birds and mammals, as well as some reptiles, amphibians, and invertebrates (Kappes 1993). The loss of open pine habitat since European settlement precipitated dramatic declines in the bird's population and led to its being listed as endangered in 1970 (*Federal Register* 35:16047).

SOUTHWESTERN WILLOW FLYCATCHER

Sogge (1995) points out that the southwestern willow flycatcher occurs, as its name implies, throughout most of the southwestern United States. It is a Neotropical migrant songbird, that is, one of many birds that return to the United States and Canada to breed each spring after migrating south to the Neotropics (Mexico and Central America) to winter in milder climates. In recent years, there has been strong evidence of declines in many Neotropical migrant songbirds (e.g., Finch and Stangel 1993), including the southwestern willow flycatcher (*Federal Register* 1993). The flycatcher appears to have suffered significant declines throughout its range, including total loss from some areas where it historically occurred. These declines, as well as the potential for continued and additional threats, prompted the U.S. Fish and Wildlife Service to propose listing the southwestern willow flycatcher as an endangered species (*Federal Register* 1993).

✔ *Interesting Point*: The flycatcher reminds observers of a sentinel constantly at attention, whose flitty wing movements resemble salutes and constant tail motions signal a readiness for action. It feeds on insects in lush, multilayered riparian zones by snatching them on the wing or harvesting them from dense vegetation. Its mission to control insects in riparian areas is an essential function benefiting people as well as plant life (USFWS 2004).

BOTTOM LINE ON BIRDS

Even though the preceding information is representative of only a small sample of bird species, if any overall conclusion is possible on status and trends of bird populations it is this: apparent stability for many species; increase in some species, many of which are generalists adaptable to altered habitats; and decreases in other species, many of which are specialists most vulnerable to habitat loss and degradation (U.S. Department of the Interior 1995a).

Mammals

Mammals, like birds and in contrast to amphibians, fishes, and reptiles, are warm-blooded animals. In contrast to feather-covered birds, scale-covered fish, and skin-covered reptiles, the skin of mammals is more or less hairy. The young of most mammals are born alive, whereas the young of fish, birds, amphibians, and some species of reptiles hatch from eggs. After birth young mammals breathe by lungs rather than by gills as do fish; for a time they are nourished with milk produced by the mother (Comstock 1986). The small representative samples shown in figures 6.3–6.5 demonstrate that mammals are exceedingly diverse in size, shape, form, and function.

Tuggle (1995) reports many mammalian population studies have been initiated to determine a species' biological and/or ecological status. This studies were conducted because of the species' perceived economic importance, its abundance, its threatened or endangered state, or because it is viewed as our competitor. As a result, data on mammalian populations in North America have been amassed by researchers, naturalists, trappers, farmers, and land managers for years.

Inventory and monitoring programs that produce data about the status and

Figure 6.3. Spotted hyena, Tampa, Florida
Photograph by Revonna Bieber

trends of mammalian populations are significant for many reasons. One of the most important reasons, however, is that as fellow members of the most advanced class of organisms in the animal kingdom, the condition of mammal populations most closely reflects our condition. In essence, mammalian species are significant biological indicators for assessing the overall health of advanced organisms in an ecosystem.

Habitat changes, particularly those initiated by humans, have profoundly affected wildlife populations in North America. Although Native Americans used many wildlife species for food, clothing, and trade, their agricultural and land-use practices usually had minimal adverse effects on mammal populations during the pre-European settlement era. In general, during the post-Columbus era, most North American mammalian populations significantly declined, primarily because of their inability to adapt and compete with early European land-use practices and pressures.

Habitat modification and destruction during the settlement of North America occurred very slowly initially. Advances in agriculture and engineering accelerated the loss or modification of habitats that were critical to many species in climax communities. These landscapes transformations often occurred before we had any knowledge of how these environmental changes would affect native flora and fauna. Habitat alterations were almost always economically driven and

Figure 6.4. Silverback gorilla, Tampa, Florida
Photograph by Revonna Bieber

in the absence of land-use regulations and conservation measures many species were extirpated.

In addition to rapid and sustained habitat and landscape changes from agricultural practices, other factors such as unregulated hunting and trapping, indiscriminate predator and pest control, and urbanization also contributed significantly to the decline of once-bountiful mammalian populations. These practices, individually and collectively, have been directly correlated with the decline or extinction of many sensitive species.

The beginning of the twentieth century brought a new focus on conservation efforts in this country. Populations of some species, such as the white-tailed deer, showed marked recovery after regulatory and conservation strategies began. Ancient wildlife management and conservation programs, started primarily for game species, have increased our knowledge and understanding of species and habitat interactions. Conservation programs have also positively affected many species that share habitat with the target species the programs are designed to aid. To complement these efforts, however, integrated regulatory legislation and conservation policies that specifically help sustain nontarget species and their habitats are still imperative.

The increased emphasis on the importance of managing for biological diversity and adopting an ecosystem approach to management has enhanced our efforts

Figure 6.5.   Snow leopard, Tampa, Florida
Photograph by Revonna Bieber

to move from resource-management practices that are oriented to single species to strategies that focus on the long-term conservation of native populations and their natural habitats. Thus, an integrated and comprehensive inventory and monitoring program that coordinates data on the status and trends of our natural resources is critical to successfully manage habitats that support a diverse array of plant and animal species.

MARINE MAMMALS

At least 35 species of marine mammals are found along the U.S. Atlantic Coast and in the Gulf of Mexico: 2 seal species, 1 manatee, and 32 species of whales, dolphins (see figure 6.6), and porpoises. Seven of these species are listed as endangered under the Endangered Species Act (ESA). At least 50 species of marine mammals are found in U.S. Pacific waters: 11 species of seals and sea lions; walrus; polar bear; sea otter; and 36 species of whales, dolphins, and porpoises; 11 of these species are listed as endangered or threatened under the ESA (Kinsinger 1995).

Information on the size, distribution, and productivity of the California *sea otter* population is broadly relevant to two federally mandated goals: removing the population's listing as threatened under the Endangered Species Act and obtaining

Figure 6.6.   Bottlenose dolphin, Orlando, Florida
Photograph by Revonna Bieber

an "optimal sustainable population" under the Marine Mammal Protection Act. Except for the population in central California, sea otters were hunted to extinction between Prince William Sound, Alaska, and Baja California (Kenyon 1969). Wilson, Bogan, Brownell, Burdin, and Maminov (1991), based on variations in cranial morphology, recently assigned subspecific status to the California sea otter. Furthermore, mitochondrial DNA analysis has revealed genetic differences among populations in California, Alaska, and Asia (National Biological Service [NBS], unpublished data).

✔   *Interesting Point*: Sea otters have no blubber layer, unlike pinnipeds and cetaceans, and rely entirely on their thick fur's ability to trap air for insulation.

In 1977, the California sea otter was listed as threatened under the ESA, largely because of its small population size and perceived risks from such factors as human disturbance, competition with fisheries, and pollution. Because of unique threats and growth characteristics, the California population is treated separately from sea otter populations elsewhere in the North Pacific (Estes, Jameson, Bodkin, and Carlson 1995).

Whales, *dolphins*, and *porpoises* all belong to the same taxonomic order called Cetaceans. Cetaceans spend their whole lives in water and some live in family groups called "pods." Cetaceans are known for their seemingly playful behavior. Pinnipeds are carnivorous aquatic mammals that use flippers for movement on land and in the water. Seals, sea lions, and walruses all belong to the suborder called Pinnipedia or the "fin-footed." Pinnipeds spend the majority of their lives swimming and eating in water and have adapted their bodies to move easily through their aquatic habitat (National Oceanic and Atmospheric Administration [NOAA] 2007).

Manatees and *dugongs*, order Sirenia, are so ugly that they are really cute (NOAA 2007)!

Sirenians spend their whole lives in water. The word "Sirenia" came from the world "siren." Sirens are legendary Greek sea beauties who lured sailor in to the sea. It is thought that old-time mermaid sightings were actually Siernians rather than mythical half-women, half-fish.

INDIANA BATS

The Indiana bat is a medium-sized, dull gray, black, or chestnut bat listed as an endangered species and occurs throughout much of the eastern United States. Although bats are sometimes viewed with disdain, Drobney and Clawson (1995) write that they are of considerable ecological and economic importance. Bats consume a diet consisting largely of nocturnal insects and thereby are a natural control for both agricultural pests and insects that are annoying to humans. Furthermore, many forms of cave life depend upon nutrients brought into caves by bats in the form of guano or feces (Missouri Department of Conservation 1991).

✔ *Interesting Point*: Indiana bats live an average of 7.5 years, but some have reached 14 years of age.

Indiana bats use distinctly different habitats during summer and winter. In winter, bats congregate in a few large caves and mines for hibernation and have a more restricted distribution than at other times of the ear. Nearly 85 percent of the known population winters in only seven caves and mines in Missouri, Indiana, and Kentucky, and approximately one-half of the population uses only two of three hibernacula (i.e., hibernation location).

In spring, females migrate north from their hibernacula and form maternity colonies in predominately agricultural areas of Missouri, Iowa, Illinois, Indiana, and Michigan. Three colonies, consisting of 50 to 150 adults and their young, normally roost under the loose bark of dead, large-diameter trees throughout summer; however, living shagbark hickories and tree cavities are also used occasionally (Humphrey, Richter, and Cope 1977; Gardner, Garner, and Hofmann 1991; Callahan 1993; Kurta, King, Teramino, Stribley, and Williams 1993).

As a consequence of their limited distribution, specific summer and winter habitat requirements, and tendency to congregate in large numbers during winter, Indiana bats are particularly vulnerable to rapid population reductions resulting from habitat change, environmental contaminants, and other human disturbances (Brady et al. 1983). Additionally, because females produce only one young per year, recovery following a population reduction occurs slowly. Concerns arising from the high potential vulnerability and slow recovery rate have led to a long-term population monitoring effort for this species.

GRAY WOLVES

The USFWS (1998) points out that, historically, most Native Americans revered gray wolves, trying to emulate their cunning and hunting abilities. However, by 1960 the wolf was exterminated by federal and state governments from all of the United States except Alaska and northern Minnesota. Until recently, 24 subspecies of the gray wolf were recognized in North America, including eight in the contiguous 48 states. After the gray wolf was listed as an endangered species in 1967, recovery plans were developed for the eastern timber wolf, the northern Rocky Mountain wolf, and the Mexican wolf. The other subspecies in the contiguous United States were considered extinct (Mech, Pletscher, and Martinka 1995).

The Eastern Timber Wolf Recovery Plan (USFWS 1992b) set as criteria for recovery the following conditions: a viable wolf population in Minnesota consisting of at least 200 animals, and either a population of at least 100 wolves in the United States within 100 miles of the Minnesota population, or a population of at least 200 wolves if farther than 100 miles from the Minnesota population. The Northern Rocky Mountain Wolf Recovery Plan (USFWS 1987) defines recovery as when at least ten breeding pairs of wolves inhabit each of their specified areas in the northern Rockies for three successive years. The Mexican Wolf Recovery Plan (USFWS 1982) called for a self-sustaining population of at least 100 Mexican wolves in a 4,941-square-mile range.

A recent revision of wolf subspecies in North America (Novak 1994), however, reduced the number of subspecies originally occupying the contiguous 48 states from eight to four. It classified the wolf currently inhabiting northern Montana as being *C. I. occidentalis*, primarily a Canadian and Alaskan wolf. It considered *C. I. nubilus* to be the wolf remaining in most of the range of the former northern Rocky Mountain wolf and the present range of the eastern timber wolf; this leaves the eastern timber wolf extinct in its former U.S. range, surviving now only in southeastern Canada. The new classification may have implications for the recovery criteria propounded by the Eastern Timber Wolf and Northern Rocky Mountain Wolf recovery plans. The reclassification did not change the status of the Mexican wolf.

NORTH AMERICAN BLACK BEARS

Perhaps no other animals have so excited the human imagination as bears. References of bears are found in ancient and modern literature, folk songs, legends, mythology, children stories, and cartoons. Bears are among the first animals that children learn to recognize. Bear folklore is confusing because it is based on caricature, with Teddy Bears and the kindly Smoky on one hand and ferocious magazine cover drawings on the other. Dominant themes of our folklore are fear of the unknown and man against nature, and bears have traditionally been portrayed as the villains to support those themes, unfairly demonizing them to the public. A problem for black bears is that literature about bears often does not separate black bears from grizzly bears.

—Lynn L. Rogers, 2002

Habitat loss, habitat fragmentation, and unrestricted harvest have significantly changed the distribution and abundance of black bears in North America since colonial settlement. Even though bears have been more carefully managed in the past 50 years and harvest levels are limited, threats from habitat alteration and fragmentation still exist and are particularly acute in the southeastern United States. In addition, the increased efficiency in hunting techniques and the illegal trade in bear parts, especially gall bladders, have raised concerns about the effect of poaching on some bear populations. Because bears have lower reproductive rates, their populations recover more slowly from losses than do those of most other North American mammals (Vaughan and Pelton 1995).

Black bear populations are difficult to inventory and monitor because the animas occur in relatively low densities and are secretive by nature. Black bears are an important game species in many states and Canada and are an important component of their ecosystems. It is important that they be continuously and carefully monitored to ensure their continued existence.

GRIZZLY BEARS

Mattson, Wright, Kendall, and Martinka (1995) state that the grizzly bear, sometimes called the silvertip bear, is a powerful brownish-yellow bear that once roamed over most of the United States from the high plains of the Pacific Coast. In the Great Plains, they seem to have favored areas near rivers and streams, where conflict with humans was also likely. These grassland grizzlies also probably spent considerable time searching out and consuming bison that died from drowning, birthing, or winter starvation, and so were undoubtedly affected by the elimination of bison from most of the Great Plains in the late 1800s. They are potential competitors for most foods valued by humans, including domesticated livestock and agricultural crops, and under certain limited conditions are also a potential threat to human safety. For these and other reasons, grizzly bears in the United States were vigorously sought out and killed by European settlers in the 1800s and early 1900s.

✔ *Interesting Point*: Grizzly bears reach weights of 400–1,500 pounds; the male is on average 1.8 times as heavy as the female, an example of sexual dimorphism.

Between 1850 and 1920, grizzlies were eliminated from 95 percent of their original range, with extirpation occurring earliest on the Great Plains and later in remote mountainous areas. Unregulated killing of grizzlies continued in most places through the 1950s and resulted in a further 52 percent decline in their range between 1920 and 1970. Grizzlies survived this last period of slaughter only in remote wilderness areas larger than 10,000 square miles. Altogether, grizzly bears were eliminated from 98 percent of their original range in the contiguous United States during a 100-year period.

✔ *Interesting Point*: It is a common misconception that grizzly bears cannot climb trees. They will climb trees if they have a food incentive.

Because of the dramatic decline and the uncertain status of grizzlies in areas where they had survived, their populations in the contiguous United States were listed as threatened under the Endangered Species Act in 1975. High levels of grizzly bear mortality in the Yellowstone area during the early 1970s were also a major impetus for this listing.

BLACK-FOOTED FERRETS

Biggins and Godbey (1995) report that the black-footed ferret was a charter member of endangered species lists for North America, and was recognized as rare long before the passage of the Endangered Species Act of 1973. This member of the weasel family is closely associated with prairie dogs of three species, a specialization that contributed to its downfall. Prairie dogs make up 90 percent of the ferret diet; in addition, ferrets dwell in prairie dog burrows during daylight, venturing out mostly during darkness. Trappers captured black-footed ferrets during their quests for other species of furbearers. Although the species received increased attention as it because increasingly rare, the number of documented ferrets fell steadily after 1940, and little was learned about the animals before large habitat declines made studies of them difficult. These declines were brought about mainly by prairie dog control campaigns begun before 1900 and reaching high intensity by the 1920s and 1930s.

Much of what is known about black-footed ferret biology was learned from research during 1964–1974 on a remnant population in South Dakota (Linder, Dahlgren, and Hillman 1972; Hillman and Linder 1973), and from 1981 to the present on a population found at Meeteetse, Wyoming, and later transferred to captivity (Biggins, Schroeder, Forrest, and Richardson 1985; Forrest et al. 1988; Williams, Thorne, Appel, and Belitsky 1988). Nine ferrets from the sparse South Dakota population (only 11 ferret litters were located during 1964–1972) were taken into captivity from 1971 to 1973, and captive breeding was undertaken at the U.S. Fish and Wildlife Service's Patuxent Wildlife Research Center in Maryland (Carpenter and Hillman 1978). Although litters were born there, no young were successfully raised. The last of the Patuxent captive ferrets died in 1978, and no animals were located in South Dakota after 1979.

In 1981, at Meeteetse, black-footed ferrets were "rediscovered" in prairie dog complexes. This rediscovery gave conservationists what seemed a last chance to learn about the species and possibly save it from extinction. That population remained healthy (70 ferret litters were counted from 1982 to 1986) through 1984, a period when much was learned about ferret life history and behavior. In 1985, sylvatic plague, a disease deadly to prairie dogs, was confirmed in the prairie dogs at Meeteetse, creating fear that the prairie dog habitat vital for ferrets would be lost. In addition, field biologists were reporting a substantial decrease in the numbers of ferrets detected. The fear of plague was quickly overshadowed by the discovery of canine distemper in the ferrets themselves. It is a disease lethal to ferrets.

In 1985 six ferrets were captured to begin captive breeding, but two brought the distemper virus into captivity, and all six died (Williams et al. 1988). A plan was formulated to place more animals from Meeteetse into captivity to protect them from distemper and to start the breeding program. By December 1985, only ten ferrets were known to exist, six in captivity and four in Meeteetse. The following year, the surviving free-ranging ferrets at Meeteetse produced only two litters, a number thought too small to sustain the wild population. Because both the Meeteetse and captive populations were too small to sustain themselves, all remaining ferrets were removed from the wild, resulting in a captive population of 18 individuals by early 1987.

✔ *Interesting Point*: The ferret's large ears and eyes suggest it has acute hearing and sight, but smell is probably its most important sense for hunting prey underground in the dark (Black-footed Ferret Recovery Program [BFFRP] 2005).

Captive breeding of ferrets eventually became successful. Although the captive population is growing, researchers fear the consequences of low genetic diversity (already documented by O'Brien, Martenson, Eichelberger, Thorne, and Wright 1989) and of inbreeding depression (i.e., reduced fitness as a result of breeding of related individuals). A goal of the breeding program is to retain as much genetic diversity as possible, but the only practical way to increase diversity is to find more wild ferrets. In spite of intensive searches of the remaining good ferret habitat and investigations of sighting reports, no wild ferrets have been found.

Biggins and Godbey (1995) point out that the captive breeding program now is producing sufficient surplus ferrets for reintroduction into the wild; 187 ferrets were released into prairie dog colonies in Shirley Basin, Wyoming, during 1991–1993. Challenges facing the black-footed ferret reintroduction include low survivorship of released ferrets due to high dispersal and losses to other predators; unknown influence of low genetic diversity; canine distemper hazard; indirect effect of plague on prairie dogs and its possible direct effect on ferrets; and low availability of suitable habitat for reintroduction. The scarcity of habitat reflects a much larger problem with the prairie dog ecosystem and needs increased attention.

At the beginning of the 20th century, prairie dogs reportedly occupied more than 100 million acres of grasslands, but by 1960 that area had been reduced to about 1.5 million acres (Marsh 1984). Much reduction was attributed to prairie dog control programs, which continue. For example, in South Dakota in the late 1980s, millions were spent to apply toxicants to prairie dog colonies on Pine Ridge Indian Reservation (Sharp 1988). At least two states (Nebraska and South Dakota) have laws prohibiting landowners from allowing prairie dogs to flourish on their properties; if the land manager does not "control" the "infestation," the state can do so and bill expenses to the owner (Clarke 1988).

Several prairie dog complexes have been evaluated as sites for reintroduction of black-footed ferrets. The evaluation involves grouping clusters of colonies separated by fewer than 4.3 miles into complexes, based on movement capabilities of ferrets (Biggins et al. 1993); these areas include some of the best prairie dog complexes remaining in the states. Nevertheless, other extensive prairie dog complexes were not considered for ferret reintroduction.

Ramifications of a healthy prairie dog ecosystem extend well beyond black-footed ferrets. The prairie dog is a keystone species of the North American prairies. It is an important primary consumer, converting plants to animal biomass at a rate higher than that of other vertebrate herbivores of the short-grass prairies, and its burrowing provides homes for many other species of animals and increased nutrients in surface soil. This animal also provides food for many predators. We estimated it takes 700–800 prairie dogs to annually support a reproducing pair of black-footed ferrets and a similar biomass of associated predators (Biggins et al. 1993), suggesting that large complexes of prairie dog colonies are necessary to support self-sustaining populations of these second-order consumers.

The black-footed ferret cannot be reestablished on the grasslands of North America in viable self-sustaining populations without large complexes of prairie dog colonies. The importance of this system to other species is not completely understood, but large declines in some of its species should serve as a warning. The case of the black-footed ferret provides ample evidence that timely preventive action would be preferable to the inefficient "salvage" operations. Furthermore, there is considerable risk of irreversible damage (e.g., genetic impoverishment) with such rescue efforts (Biggins and Godbey 1995).

AMERICAN BADGERS

Primarily because of their burrows, which are dangerous for livestock and horsemen, the American badger has been considered "a malignant creature that had to be destroyed with all means possible, including trapping, shooting and other ways" (American-badgers.com 2007). In addition to the livestock problem with their burrows, the badger burrows also present a problem for farmers because the holes damage crops. Moreover, badgers prey on livestock. All of this led to significant species destruction and reduced the number of badgers in some regions to the point of extinction.

Steeg and Warner (1995) describe the American badger as a medium-sized carnivore found in treeless areas across North America, such as the tall-grass prairie (Lindzey 1982). Badgers rely primarily on small burrowing mammals as a prey source; availability of badger prey may be affected by changes in land-use practices that alter prey habitat. In the Midwestern United States, most native prairie was mowed for agricultural use in the mid-1800s (Burger 1978). In the past 100 years, Midwest agriculture has shifted from a diverse system of small farms with row crops, small grains, hay, and livestock pasture to larger agricultural operations employing a mechanized and chemical approach to cropping. The result is a more uniform agricultural landscape dominated by two primary row crops, corn and soybeans. The effects of such land-use alterations on badgers are unknown. In addition, other human activities such as hunting and trapping have no doubt had an impact on native vertebrates such as the badger.

Trends in carnivore abundance are difficult to evaluate because most species are secretive or visually cryptic. Trapping records, one of the earliest historical data sources for furbearers, are virtually nonexistent for badgers in the 1800s (Obbard et al. 1987).

✔ *Interesting Point*: Most research on badgers has been limited to the western United States. Although results have varied somewhat among these studies, average densities (estimated subjectively from mark-recapture and home-range data) have ranged from 0.98–12.95 badgers/square mile.

NORTHEASTERN WHITE-TAILED DEER

Storm and Palmer (1995) state that the populations of white-tailed deer have changed significantly during the past 100 years in the eastern United States (Halls 1984). After near extirpation in the eastern states by 1900, deer numbers increased during the first quarter of the 20th century. The effects of growing deer populations on forest regeneration and farm crops have been a concern to foresters and farmers for the past 50 years.

✔ *Interesting Point*: The fur of white-tail deer is a grayish color in the winter, and more red comes out during the summer. It has a band of white fur behind its nose, in circles around the eyes, and inside the ears. More white fur goes down the throat, on the upper insides of the legs, and under the tail.

In recent years, deer management plans have been designed to maintain deer populations at levels compatible with all land uses. Conflicts, however, between deer and forest management or agriculture still exist in the Northeast. Areas that were once exclusively forests are now a mixture of forest, farm, and urban environments that create increased interactions and conflicts between humans and deer, including deer-vehicle collisions. Management of deer near urban environments presents a unique challenge for local resource managers (Porter 1991).

NORTH AMERICAN ELK

North American elk or wapiti (the name "elk" was given by early explorers because they resembled the elk or moose of Europe, and the American Indian term "wapiti" is sometimes used to identify the animal) represent how a wildlife species can recover even after heavy exploitation of populations and habitats around the early 1900s. Elk is highly prized by wildlife enthusiasts and by the hunting public, which has provided the various state wildlife agencies with ample support to restore populations to previous occupied habitats and to manage populations effectively. Additionally, the Rocky Mountain Elk Foundation, founded in 1984, has promoted habitat management, acquisition, and proper hunting ethics among many segments of the hunting public (Peek 1995).

Current population size is estimated at 782,500 animals for the entire elk range (Rocky Mountain Elk Foundation 1989). Projections of population trends for the national forests and for the entire U.S. elk range are for continued increases through the year 2040 (Flather and Hoekstra 1989).

Indeed, the future of elk populations in North America seems secure. Demand for hunting as well as the nonconsumptive values of elk will ensure the success of substantial populations. Elk populations will benefit from improved habitat conditions on arid portions of the range, improved livestock management, more effective integrated management of forested habitats, and continued implementation of fire management policies in the major wilderness areas and national parks (Peek 1995).

Reptiles and Amphibians

There are more than 8,200 species of reptiles and more than 5,500 species of amphibians on Earth, including turtles, snakes, crocodiles, lizards (reptiles), frogs, toads, salamanders, and newts (amphibians). All reptiles have scales, but some are too small to be seen. Reptiles are ectothermic (i.e., they obtain heat from outside sources). Most lay eggs, but a few give birth to live young. For most amphibians, life begins in the water. They metamorphose, growing legs and changing in other ways to live on land. Like their "cold-blooded" reptile relatives, they depend on external energy sources (the sun) to maintain their body temperatures.

✔ *Important Point*: Amphibians became the first vertebrates to live on land and, like their "cold-blooded" reptile relatives, depend on external energy sources to maintain their body temperatures. Reptiles are ectotherms. They obtain heat from outside sources, such as the sun, and regulate their temperature through behaviors such as basking or seeking shade.

Amphibians and reptiles are crucial to the natural functioning of many ecological processes and key components for important ecosystems. In some areas, certain species are economically consequential; others are aesthetically pleasing to many people, and as a group they represent significant segments of the evolutionary history of North America. Knowledge gained from past study of amphibian development and metamorphosis has contributed immensely to our understanding of basic biological processes and has directly benefited humans (McDiarmid 1995).

Baseline information of the status and health of U.S. populations of amphibians and reptiles is remarkably sparse. No national program for monitoring populations of amphibians and reptiles is in operation. A recent publication (Heyer, Donnelly, McDiarmid, Hayek, and Foster 1994) recommended standard guidelines and techniques for monitoring amphibian populations and habitats; a similar volume on reptiles is planned.

Habitat degradation and loss seem to be the most important factors adversely affecting amphibian and reptile populations in North American. The drainage and loss of small aquatic habitat and their associated wetlands have had a major adverse effect on many amphibian species and some reptiles.

McDiarmid (1995) points out that many other factors in the decline of reptiles and amphibians have been implicated; most, perhaps all, are caused by humans. For example, non-native species of game fish introduced for sport have been implicated in the decline of frog populations in mountainous areas of some western states. Similarly, the introduction, accidental or intentional, of other non-native species (e.g., bullfrogs in western states, anoline lizards in south Florida, and snakes in Guam) has harmed native species in other parts of the country.

TURTLES

> A turtle is at heart a misanthrope; its shell is in itself proof of its owner's distrust of this world. But we need not wonder at this misanthropy, if we think for a moment of the creatures that lived on this earth at the time when turtles first appeared. Almost any of us would have been glad of a shell in which to retire if we had been contemporaries of the smilodon [saber-toothed cat] and other monsters of earlier geologic times.
>
> —Anna Botsford Comstock, 1986

As Comstock (1986) points out, turtles have existed for a very long time; virtually unchanged for the last 200 million years (see figure 6.7). Unfortunately, as Lovich (1995) points out, some of the same traits that allowed them to survive the ages often predispose them to endangerment. Delayed maturity and low and variable annual reproductive success make turtles unusually susceptible to increased mor-

Figure 6.7.   Tropical wildlife, Ocho Rios, Jamaica
Photograph by Revonna Bieber

tality through exploitation and habitat modifications (Brooks, Brown, and Galbraith 1991; Congdon, Dunham, and Van Loben Sels 1993).

✔   *Interesting Point*: A turtle eats grasses, mushrooms, berries, insects, flowers, worms, water plants, crayfish, sails, fish, frogs, and dead animals.

In general, turtles are overlooked by wildlife managers in spite of their ecological significance and importance to humans. Turtles are, however, important as scavengers, herbivores, and carnivores, and often contribute significant biomass to ecosystems. In addition, they are an important link in ecosystems, providing dispersal mechanisms for plants, contributing to environmental diversity, and fostering symbiotic associations with a diverse array of organisms. Adults and eggs of many turtles have been used as a food resource by humans for centuries (Brooks, Galbraith, Nancekivell, and Bishop 1988; Lovich 1995). As use pressures and habitat destruction increase, management that considers the life-history traits of turtles will be needed.

Marine Turtles

Marine turtles have outlived almost all of the prehistoric animals with which they once shared the planet. Five species of marine turtles frequent the beaches and offshore waters of the southeastern United States (Escambia 2007):

- *Loggerhead*—is the most common turtle to nest in Florida. Over 50,000 logger-head nests are recorded annually in Florida. This turtle is named for its dispro-portionately large head and feeds on crabs, mollusks, and jellyfish.
- *Green*—is the second-most common turtle in Florida waters. Green sea turtles are the only herbivorous sea turtles. They feed on seagrasses in shallow areas through the Gulf of Mexico. The lower jaw is serrated to help cut the seagrasses it eats.
- *Kemp's Ridley*—are the rarest sea turtle in the world. They primarily nest on one beach on the Gulf Coast of Mexico and are the smallest species of sea turtle. Scientists have been trying to transplant Kemp's Ridley eggs to Texas to establish a new nesting colony. They are the only species of sea turtle known to lay their eggs during the day.
- *Leatherback*—is the largest sea turtle in the world and can be over six feet long and weigh 1,400 pounds. It does not have a hard shell, but rather a leatherlike carapace with bony ridges underneath the skin. The leatherback makes long migrations to and from its nesting beaches in the tropics as far north as Canada. Jellyfish are the favored prey to these turtles.
- *Hawksbill*—is usually found feeding primarily on sponges in the southern Gulf of Mexico and Caribbean. The hawksbill sea turtle was hunted to near extinc-tion for its beautiful shell, which features overlapping scales.

All five are reported to nest, but only the loggerhead and green turtle do so in substantial numbers. Most nesting occurs from southern North Carolina to the middle west coast of Florida, but scattered nesting occurs from Virginia through southern Texas. The beaches of Florida, particularly in Brevard and Indian River counties, may host the world's largest population of loggerheads (Dodd 1995).

Marine turtles, especially juveniles and subadults, use lagoons, estuaries, and bays as feeding grounds. Areas of particular importance include Chesapeake Bay, Virginia (for loggerheads and Kemp's Ridleys); Pamlico Sound, North Carolina (for loggerheads); and Mosquito Lagoon, Florida, and Laguna Madre, Texas (for greens). Offshore waters also support important feeding grounds such as Florida Bay and the Cedar Keys, Florida (for green turtles), and the mouth of the Missis-sippi River and the northeast Gulf of Mexico (for Kemp's Ridleys).

Offshore reefs provide feeding and resting habitat (for loggerheads, greens, and hawksbills), and offshore currents, especially the Gulf Stream, are important migratory corridors (for all species, but especially leatherbacks).

✔   *Interesting Point*: Raccoons destroy thousands of sea turtle eggs each year and are the single greatest cause of sea turtle mortality in Florida.

Most marine turtles spend only part of their lives in U.S. waters. For exam-ple, hatchling loggerheads ride oceanic currents and gyres (giant circular oceanic surface currents) for many years before returning to feed as subadults in southeast-ern lagoons. They travel as far as Europe and the Azores, and even enter the Medi-terranean Sea, where they are susceptible to long-line fishing mortality. Adult loggerheads may leave U.S. waters after nesting and spend years in feeding grounds in the Bahamas and Cuba before returning. Nearly the entire world population of Kemp's Ridleys uses a single Mexican beach for nesting, although juveniles and subadults, in particular, spend much time in U.S. offshore waters (Dodd 1995).

The biological characteristics that make sea turtles difficult to conserve and manage include a long life span, delayed sexual maturity, differential use of habitats both among species and life stages, adult migratory travel, high egg and juvenile mortality, concentrated nesting, and vast areal dispersal of young and subadults. Genetic analyses have confirmed that females of most species return to their natal beaches to nest (Bowen, Meylan, et al. 1992; Bowen, Avise, et al. 1993). Nesting assemblages contain unique genetic markers showing a tendency toward isolation from other assemblages (Bowen et al. 1993); thus, Florida green turtles are genetically different from green turtles nesting in Costa Rica and Brazil (Bowen et al. 1992). Nesting on warm sandy beaches puts the turtles in direct conflict with human beach use and their use of rich off-shore waters subjects them to mortality from commercial fisheries (National Research Council 1990).

Marine turtles have suffered catastrophic declines since the European discovery of the New World (National Research Council 1990). In a relatively short time, the huge nesting assemblages in the Cayman Islands, Jamaica, and Bermuda were decimated. In the United States, commercial turtle fisheries once operated in south Texas (Doughty 1984), Cedar Keys, Florida Keys, and Mosquito Lagoon; these fisheries collapsed from overexploitation of the mostly juvenile green turtle populations. Today, marine turtle populations are threatened worldwide and are under intense pressure in the Caribbean Basin and Gulf of Mexico, including Cuba, Mexico, Hispaniola, the Bahamas, and Nicaragua. Marine turtles can be conserved only though international efforts and cooperation (Dodd 1995).

AMPHIBIANS

Bury, Corn, Dodd, McDiarmid, and Scott (1995) point out that amphibians are ecologically important in most freshwater and terrestrial habitats in the United States; they can be numerous, function as both predators and prey, and constitute great biomass. Amphibians have certain physiological (e.g., permeable skin) and ecological (e.g., complex life cycle) traits that could justify their use as bioindicators of environmental health. For example, local declines in adult amphibians may indicate losses of nearby wetlands. The aquatic breeding habits of many terrestrial species result in direct exposure of egg, larval, and adult stages to toxic pesticides, herbicides, acidification, and other human-induced stresses in both aquatic and terrestrial habitats. Reported declines of amphibian populations globally have drawn considerable attention (Bury, Dodd, and Fellers 1980; Bishop and Petit 1992; Richards, McDonald, and Alford 1993; Blaustein 1994; Pechmann and Wilbur 1994).

Approximately 230 species of amphibians, including about 140 salamanders and 90 anurans (frogs and toads) occur in the continental United States. Because of their functional importance in most ecosystems, declines of amphibians are of considerable conservation interest. If these declines are real, the number of listed or candidate species at federal, state, and local levels could increase significantly. Unfortunately, because much of the existing information on the status and trends of amphibians is anecdotal, coordinated monitoring programs are greatly needed (Bury et al. 1995).

North American amphibian species exhibit two major distributional patterns: endemic and widespread. Endemic species tend to have small ranges or are restricted to specific habitats (e.g., species that occur only in one cave or in rock talus on a single mountainside). Declines are documented best for endemic species, partly because their smaller range makes monitoring easier. Populations of endemics are most susceptible to loss or depletions because of localized activities (Bury et al. 1980; Dodd 1991). Examples of endemic species affected by different local impacts include the Santa Cruz long-toed salamander in California, the Texas blind salamander in Texas, and the Red Hills salamander in Alabama; these three species are listed as federally threatened or endangered.

The number of endemic species that have suffered losses or are suspected of having severe threats to their continued existence has increased in the past years. In part, the increase reflects descriptions of new species with restricted ranges, but the accelerating pace of habitat alteration is the primary threat.

The ranges of most endemics in the western states (26 species) are widely dispersed across the landscape. In contrast, endemics in the eastern and southeastern states (25 species) tend to be clustered in centers of endemism, such as in the Edwards Plateau (Texas), interior (Ozark) highlands (Arkansas, Oklahoma), Atlantic coastal plain (Texas to Virginia), and uplands or mountaintops in the (West Virginia to Georgia).

Widespread species often are habitat generalists. Many were previously common but have shown regional or range wide declines (Hine, Les, and Hellmich 1981; Corn and Fogelman 1984; Hayes and Jennings 1986). Reported declines of widespread species often lack explanation, perhaps because these observations have only recently received general attention or because temporal and spatial variations in population sizes of many amphibians are not well understood. Some reports are for amphibians in relatively pristine habitats where human impacts are not apparent (Bury et al. 1995).

No single factor has been identified as the cause of amphibian declines, and many unexpected declines likely result from multiple causes. Human-caused factors may intensify natural factors (Blaustein, Wake, and Sousa 1994) and exacerbate declines from which local populations cannot recover and thus they go extinct. Known or suspected factors in those declines include destruction and loss of wetlands (Bury et al. 1980); habitat alteration, such as the impacts from timber harvest and forest management (Corn and Bury 1989; Dodd 1991; Petranka, Eldridge, and Haley 1993); introduction of non-native predators, such as sport fish and bullfrogs, especially in western states (Hayes and Jennings 1986; Bradford 1989); increased variety and use of pesticides and herbicides (Hine et al. 1981); effects of acid precipitation, especially in eastern North America and Europe

(Freda 1986; Beebee et al. 1990; Dunson, Wyman, and Corbett 1992); increased ultraviolet radiation reaching the ground (Blaustein, Hoffman, et al. 1994); and diseases resulting from decreased immune system function (Bradford 1991; Carey 1993; Pounds and Crump 1994).

Amphibian populations also may vary in size because of natural factors, particularly extremes in the weather (Bradford 1983; Corn and Fogelman 1984). The size of amphibian populations may vary, sometimes dramatically, from year to year, and what is perceived as a decline may be part of long-term fluctuations (Pechmann et al. 1991). The effect of global climate change on amphibians is speculative, but it has the potential for causing the loss of many species.

American Alligators in Florida

Woodward and Moore (1995) point out that, as members of the crocodile family, alligators are living fossils that can be traced back 230 million years. The American alligator is an integral component of wetland ecosystems in Florida. Alligators also provide aesthetic, educational, recreational, and economic benefits to humans. Because of the commercial value of alligator hides for making high-quality leather products, alligator hunting was a major economic and recreational pursuit of many Floridians from the mid-1800s to 1970. The Florida alligator population varied considerably during the 1900s in response to fluctuating hunting pressure caused by unstable markets for luxury leather products.

The declining abundance of alligators during the late 1950s and early 1960s led to the 1967 classification of the Florida alligator population as endangered throughout its range. Federal and international regulations imposed during the 1970s and 1980 have helped control trade of alligator hides, and illegal hunting of alligators was checked. The Florida alligator population responded immediately to protection and was reclassified as threatened in 1977 and as threatened because of its similarity in appearance to the American crocodile in 1985 (Neal 1985).

Native Ranid Frogs

Many recent declines and extinctions of native amphibians have occurred in certain parts of the world (Wake and Morowitz 1991). All species of native true frogs have declined in the western United States over the past decade (Hayes and Jennings 1986). Most of these native amphibian declines can be directly attributed to habitat loss or modification, which is often exacerbated by natural events such as droughts or floods (Wake 1991). A growing body of research, however, indicates that certain native frogs are particularly susceptible to population declines and extinctions in habitats that are relatively unmodified by humans (e.g., wilderness areas and national parks in California; Bradford 1991; Fellers and Drost 1993; Kagarise Sherman and Morton 1993). To understand these declines, we must document the current distribution of these species over their entire historical range to learn where they have disappeared (Jennings 1995).

In 1988 the California Department of Fish and Game commissioned the

California Academy of Sciences to conduct a six-year study on the status of the state's amphibians and reptiles not currently protected by the Endangered Species Act. The study's purpose was to determine amphibians and reptiles most vulnerable to extinction and to provide suggestions for future research, management, and protection by state, federal, and local agencies (Jennings and Hayes 1993).

Desert Tortoises

Desert tortoises are any of the land-dwelling turtles that are widespread throughout the southwestern United States and Mexico. Berry and Medica (1995) point out that within the United States, desert tortoises live in the Mojave, Colorado, and Sonoran deserts of southeastern California, southern Nevada, southwestern Utah, and western Arizona. A substantial portion of the habitat is on lands administered by the U.S. Department of the Interior.

✔ *Interesting Point*: Although 95 percent of the desert tortoise's life is spent in underground burrows, it is able to live where ground temperature may exceed 140 F.

The U.S. government treats the desert tortoise as an indicator or umbrella species to measure the health and well-being of the ecosystems it inhabits. The tortoise functions well as an indicator because it is long-lived, takes 12–20 years to reach reproductive maturity, and is sensitive to changes in the environment. In 1990 the U.S. Fish and Wildlife Service listed the species as threatened in the northern and western parts of its geographic range because of widespread population declines and overall habitat loss, deterioration, and fragmentation.

✔ *Interesting Point*: Ravens have caused more than 50 percent of juvenile desert tortoise deaths in some areas of the Mojave Desert.

Because some populations exhibit significant genetic, morphologic, and behavioral differences, the Desert Tortoise Recovery Team identified various population segments for critical habitat protection and long-term conservation within the Mojave and Colorado deserts (e.g., Lamb, Avise, and Gibbons 1989; USFWS 1994).

Fringe-Toed Lizards

> From tiny to gigantic, from drab to remarkably beautiful, from harmless to venomous, lizards are spectacular products of natural selection.
>
> —Pianka and Vitt, 2006

The fringe-toed lizard (*Uma* spp.; order Squamata) is a medium-sized, whitish lizard that inhabits many of the scattered windblown sand deposits of southeastern

California and southwestern Mexico. These lizards have several specialized adaptations: elongated scale on their hind feet ("fringes") for added traction in loose sand, a shovel-shaped head and a lower jaw adapted to aid diving into and moving short distances beneath the sand, elongated scales covering their ears to keep sand out, and unique morphology (form or structure) or internal nostrils that allows them to breathe below the sand without inhaling sand particles (Barrows, Muth, Fisher, and Lovich 1995).

While these adaptations enable fringe-toed lizards to successfully occupy sand dune habitats, the same characteristics have restricted them to isolated sand "islands." Three fringe-toed lizard species live in the United States: the Mojave (*U. scoparia*), the Colorado Desert (*U. notata*), and the Coachella Valley (*U. inornata*). Of the three, the Coachella Valley fringe-toed lizard has the most restricted range and has been most affected by human activities. In 1980 this lizard was listed as a threatened species by the federal government.

In 1986 the Coachella Valley Preserve system was established to protect habitat for the Coachella Valley fringe-toed lizard. This action set several precedents: It was the first Habitat Conservation Plan established under the revised (1982) Endangered Species Act and the newly adopted Section 10 of the act, it established perhaps the only protected area in the set aside for a lizard, and its design was based on a model of sand dune ecosystem processes, the sole habitat for this lizard. Three disjunctive sites in California, each with a discrete source of windblown sand, were set aside to protect fringe-toed lizard populations: Thousand Palms, Willow Hole, and Whitewater River. Collectively, the preserves protect about 2 percent of the lizards' original range.

Barrows (unpublished data) points out that eight years after the establishment of the preserve system, few Coachella valley fringe-toed lizards exist outside the boundaries of the three protected sites. Barrows recently identified scattered pockets of windblown sand occupied by fringe-toed lizards in the hill along the northern fringe of the valley, but only at low densities. Fringe-toed lizard populations within the protected sites have been monitored yearly since 1986. During this period, California experienced one of its most severe droughts, which ended in spring 1991. Numbers of fringe-toed lizards within the Thousand Palms and Willow Hole sites declined during the drought, but rebounded after 1991. By 1993, after three wet springs, lizard numbers had increased substantially.

✔ *Interesting Point*: A limbless lizard has eyelids and ear openings.

Barrows et al. (1995) report that the lizards at the Whitewater River site have been intensively monitored since 1985 by using mark-recapture methods to count the population on 2.25-hectare (5.56-acre) plots. In 1986 this site had the highest population density of the three protected sites. As with the other two sites, the Whitewater River population declined throughout the drought, but only increased slightly after the drought broke in the 1991. Compounding the drought effect, much of the fine sand preferred by fringe-toed lizards was blown off the site during the dry years. This condition was unique to the Whitewater River site; the other two protected sites have much deeper sand deposits and are less susceptible to wind erosion. New windblown sand was deposited on the Whitewater River site in 1993 after a period of high rainfall. The population appears to be increasing in response to these favorable conditions.

Tarahumara Frog

The Tarahumara frog is a medium-sized, drab green-brown frog with small brown to black spots on the body and dark crossbars on the legs. The hind feet are extensively webbed (USFWS 2007e).

Hale et al. (1995) point out that in the spring of 1983 the last known Tarahumara frog in the United States was found dead. Overall, the species seems to be doing well in Mexico, although the decline of more northern populations is of concern. The Tarahumara frog inhabits seasonal and permanent bedrock and boulder streams in the foothills and main mountain mass of the Sierra Madre Occidental of northwestern Mexico. It ranges from northern Sinaloa, through western Chihuahua and eastern and northern Sonora, and until recently into extreme south-central Arizona.

USFWS (2007e) points out that the reasons why the Tarahumara frog disappeared from Arizona are not clear. However, the following hypotheses have been presented: (1) winter cold; (2) flooding or severe drought; (3) competition; (4) predation; (5) disease; and (6) heavy metal acid precipitation. Metals occur naturally in streamside deposits and may be mobilized by acid precipitation events.

Fishes

> I have laid aside business, and gone afishing.
>
> —Izaac Walton, 1593–1683

Anna Comstock points out that Izaac Walton "discovered that nature-study, fishing, and philosophy were akin and as inevitably related as the three angles of a triangle." One thing is certain: old Izaac Walton loved to fish. This point is made clear by Walton in his *The Compleat Angler* (1653):

> It remains yet unresolved whether the happiness of a man in this World doth consist more in contemplation or action . . . Concerning which two opinions I shall forebear to add a third by declaring my own, and rest myself contented in telling you that both of these meet together, and do most properly belong to the most honest, ingenious, quiet and harmless art of angling.
>
> And first I tell you what some have observed, and I have found to be a real truth, that every sitting by the riverside is not only the quietest and the fittest place for contemplation, but will invite an angler to it.

From the material presented in this section, the inescapable conclusion must be that within historical time, native fish communities have undergone significant and adverse changes (Maughan 1995). These changes generally tend toward reduced distributions, lowered diversity, and increased numbers of species considered rare. These changes have been more inclusive and more dramatic in the arid western regions where there are primarily endemic (native) species, but similar though more subtle changes have occurred throughout the country. These trends are the same whether one focuses on faunas or on populations of genetic variation within a single species. Changes in fish communities may be indicative of the overall health of an aquatic system; some species have narrow habitat requirements.

It should not come as any great surprise that fish populations have changed over time; all things change with time. We have, however, massively modified fish habitat through the very water demands that define our society (domestic, agricultural, and industrial water supplies; waste disposal; power generation; transportation; and flood protection). All of these activities have resulted in controlling or modifying the flow or degrading the quality of natural waters. In addition, almost all contaminants ultimately find their way into the aquatic system. Species of fishes that have evolved under the selection pressures imposed by natural cycles have often been unable to adapt to the changes imposed on them as a result of human activities.

Physical and chemical changes in their habitats are not the only stresses that fishes have encountered over time. Through fish management programs, the aquarium trade, and accidental releases, many aquatic species have been introduced to new areas far beyond their native ranges. Although these introductions were often done with the best of intentions, they have sometimes subjected native fish species to new competitors, predators, and disease agents that they were ill-equipped to withstand.

✔   *Interesting Point:* Theoretically, the smaller the gene pool, the less likely a species is able to adapt to changing environmental conditions.

It appears unlikely that the forces that have led to these changes in our fish fauna will lessen significantly in the immediate future. Therefore, if we are to preserve the diversity and adaptive potential of our fishes, we must understand much more of their ecology. Vague generalizations about habitat requirements or the results of biotic interactions are no longer enough. We must know quantitatively and exactly how fishes use habitat and how that use changes in the face of biotic pressures. Only when armed with such information are we likely to reduce the current trends among our native fishes (Maughan 1995).

FRESHWATER FISHES OF THE CONTIGUOUS UNITED STATES

Up to 800 species of freshwater fishes are native to the United States (Lee et al. 1980; Moyle and Cech 1988; Warren and Burr 1994). These fishes range from old, primitive forms such as paddlefish, bowfin, gar, and sturgeon, to younger, more advanced fishes, such as minnows, darters, and sunfishes. They are not equally distributed across the nation, but tend to concentrate in larger, more diverse environments such as the Mississippi River drainage (375 species; Robison 1986; Warren and Burr 1994). Drainages that have not undergone recent geological change, such as the Tennessee and Cumberland rivers, are also rich in native freshwater fishes (250 species; Starnes and Etnier 1986). Fewer native fishes are found in isolated drainages such as the Colorado River (36 species; Carlson and Muth 1989). More arid states west of the 100th meridian average about 44 native fish species per state, while states east of that boundary average more than three times that amount (138 native species per state; Johnson 1995).

Extinction, dispersal, and evolution are naturally occurring processes that influence the kinds and numbers of fishes inhabiting our streams and lakes. More recent human-related impacts to aquatic ecosystems, such as damming of rivers, pumping of aquifers, addition of pollutants, and introductions of non-native species, also affect native fishes, but at a more rapid rate than natural processes. Some fishes are better able to withstand these rapid changes to their environments or are able to find temporary refuge in adjacent habitats; fishes that lack tolerance or are unable to retreat face extinction (Johnson 1995).

In 1979 the Endangered Species Committee of the American Fisheries Society (AFS) developed a list of 251 freshwater fishers of North America judged in danger of disappearing (Deacon et al. 1979), 198 of which are found in the United States. A decade later, AFS updated the list (Williams et al. 1989), noting 364 taxa of fish are in some degree of danger, 254 of which are native to the United States. Both AFS lists used the same endangered and threatened categories defined in the endangered Species Act of 1973, and added a special concern category to include fishes that could become threatened or endangered with relatively minor disturbances to their habitat. These imperiled native fishes are the first to indicate changes in our surface waters; thus, their status provides us with a method of judging the health of our streams and lakes.

MANAGED POPULATIONS: LOSS OF GENETIC INTEGRITY THROUGH STOCKING

Philipp and Claussen (1995) point out that species are composed of genetically divergent units usually interconnected by some (albeit low) level of gene flow (Soule 1987). Because of this restriction in gene flow, natural selection can genetically tailor populations to their environments through the process of local adaptation (Wright 1931).

Because freshwater and anadromous (i.e., adults travel upriver from the sea to spawn) fishes are restricted by the boundaries of their aquatic habitats, genetic subdivisions may be more pronounced for these vertebrates than for others. Consequently, managers of programs for these species must realize that the stock (i.e., local discrete populations), and not the species as a whole, must be the units of primary management concern (Kutkuhn 1981).

Genetic variability in a species occurs both among individuals *within* population as well as among populations (Wright 1978). Variation *within* populations is lost through genetic drift (Allendorf, Ryman, and Utter 1987), a process increased when population size becomes small. Variation *among* populations is lost when previously restricted gene flow between populations is increased for some reason (e.g., stocking, removal of natural barriers such as waterfalls); differentiation between populations is lost as a result of the homogenization of two previously distinct entities (Altukhov and Salmenkova 1987; Campton 1987).

Beyond this loss of genetic variation, mixing two groups can result in outbreeding depression, which is the loss of fitness in offspring that results from the mating of two individuals that are too distantly related (Templeton 1987). This loss in fitness is caused by the disruption of the process that produced advantageous local adaptations through natural selection. Inbreeding depression, on the other hand, is the loss of fitness produced by the repeated crossing of related organisms. The area of optimal relatedness occurs between inbreeding depression and outbreeding depression.

Many sport-fish populations are managed by using a combination of harvest regulation, habitat manipulation, and stocking. Jurisdiction for these activities falls to federal, state, tribal, and local governments, as well as private citizens. Many resource managers in the past were unaware of the long-term consequences that stocking efforts would have on the genetic integrity of local populations (Philipp, Epifanio, and Jennings 1993).

Fish populations can be classified into three types: non-native introductions, in which a given species of fish is introduced into a body of water outside its native range (regardless of any political boundaries); stock transfers, in which fish from one stock are introduced into a water body in a different geographic region inhabited by a different stock of that same species, yet are still within their native range; and genetically compatible introductions, in which fish are removed from a given water body and they—or, more often, their offspring—are introduced back into that water body or another water body that is still within the boundaries of the genetic stock serving as the hatchery brood source (Philipp et al. 1993).

Although non-native introduction may often cause ecological problems for the environments in which they are introduced, they can also cause genetic problems if they hybridize with closely related native species. Examples of this are the hybridization of introduced small-mouth bass and spotted bass with native Guadalupe bass in Texas, and the hybridization of introduced rainbow trout with native Apache trout (Carmichael, Hanson, Schmidt, and Morizot 1993). The greatest degree of genetic damage—that is, the loss of genetic variation among populations—is caused by stock transfers, a common practice among fisheries management agencies and the private sector.

✔   *Interesting Point*: According to the TPWD (2007b), stocking can be a useful fisheries management tool but is not a cure-all for poor fishing. The numbers of fish spawned by wild populations usually outweigh the numbers produced and stocked by hatcheries. However, stocking can be helpful for:

- starting populations in new or renovated waters
- supplementing populations that have insufficient natural reproduction
- increasing species diversity by introducing fishes such as striped bass
- restoring populations that have been reduced or eliminated by natural or man-made catastrophes
- providing catchable-size fish for educational activities and community fishing lakes
- enhancing the genetic makeup of a population (e.g., Florida largemouth bass)
- taking advantage of improved habitat resulting from increased water level or new vegetation

The genetic integrity of many other managed fish species is eroding as a result of management programs that inadvertently permit or deliberately promote stock transfers. This causes not only the loss of genetic variation among populations, but through outbreeding depression it is also probably negatively affecting the fitness of many native stocks involved. We need to address genetic integrity when restoring native populations (Philipp and Claussen 1995).

COLORADO RIVER BASIN FISHES

Starnes (1995) explains that the Colorado River and its tributaries have undergone drastic alterations from their natural states over the past 125 years. These alterations include both physical change or elimination of aquatic habitats and the introductions of numerous non-native species, particularly fish. Ironically, several more species occur at most localities today than were historically present before these alterations. This situation complicates the use of biodiversity as a litmus test for monitoring trends of either the deterioration or the health of an aquatic ecosystem.

Over its entire basin, the Colorado River has been changed from its natural state perhaps as much as any river system in the world. The demands for water and power in the arid West have drastically altered the system by impoundments, irrigation diversions, diking, channelization, pollutants, and destruction of bank habitats by cattle grazing and other practices. Some reaches, ranging from desert spring runs to main rivers, have been completely dewatered or, seasonally, their flows consist almost entirely of irrigation return laden with silt and chemical pollutants. The Gila River of Arizona, one of the Colorado's largest tributaries, has not flowed over its lower 248 miles since the early 1900s. These alterations and their effects on the fish fauna have been discussed by several authors (Miller 1961; Minckley and Deacon 1968, 1991; Stalmaker and Holden 1973; Carlson and Muth 1989). Only a few small tributaries, mostly at higher elevations, retain most of their natural characteristics.

✔ *Interesting Point*: Despite the expansive drainage basin (243,937 square miles) of the Colorado River, the system supported only a relatively small number of native fish species compared with basins of much smaller size east of the Continental Divide. The Colorado Basin's native fauna, however, was nearly unique (Starnes 1995).

CUTTHROAT TROUT: GLACIER NATIONAL PARK

Marnell (1995) points out that the indigenous fishery of Glacier National Park has been radically altered from its pristine condition during the past half-century through introductions of non-native fishes and the entry of non-native species from waters outside the park. These introductions have adversely affected the native westslope cutthroat trout (also known Clark's trout, red-throated trout, and short-tailed trout) throughout much of its park range.

The effects of non-native fishes on indigenous fisheries have been reviewed by Taylor, Courtenay, and McCann (1984), Marnell (1986), and Moyle, Li, and Baron (1986). Effects of fish introductions in Glacier National Park include establishment of non-native trout populations in historically fishless waters, genetic contamination (i.e., hybridization) of some native westslope cutthroat trout stocks, and ecological interferences with various life-history stages of native trout.

Research conducted in the park during the 1980s addressed the genetic effects of fish introductions on native trout. Of 47 lakes known or suspected to contain cutthroat trout or trout hybrids, 32 lakes contained viable populations of cutthroat trout, rainbow trout, or hybrids. Trout introduced in the other waters were evidently unable to sustain themselves through natural reproduction. Research conducted includes the following:

- About 30 trout sampled from each lake underwent laboratory genetic analyses. Close agreement of the results from two analytical procedures yielded a high degree of confidence in the conclusions (Marnell, Behnke, and Allendorf 1987).
- Fourteen pure-strain populations of westslope cutthroat trout persist in 15 lakes (i.e., some interconnected lakes contain a single trout population) in the North and Middle Fork drainages of the Flathead River; the species was historically present in these waters.
- Pure-strain native trout also inhabit four other Middle Fork lakes (i.e., Avalanche, Snyder, and Upper and Lower Howe lakes), but it is unclear whether they are indigenous or were transplanted from other park waters. Recent findings from sediment paleolimnology studies suggest that trout have been present in at least one other lake for more than 300 years (Verschuren, University of Minnesota, and Spellman, unpublished data). Hence, trout populations in these four lakes are tentatively classified as indigenous.
- Introduced populations of Yellowstone cutthroat trout and trout hybrids including cutthroat-rainbow trout occur in 13 lakes distributed among the three continental drainages that form their headwaters in Glacier National Park. Native

cutthroat trout were not found east of the Continental Divide in the Missouri River on South Saskatchewan River drainages within the park.

In addition to genetic concerns, ecological disturbances associated with the presence of introduced fishes have compromised the native westslope cutthroat fishery. Fish are no longer stocked in park waters; however, several waters, including some that contain undisturbed native fisheries, remain vulnerable to invasion by non-native migratory species. Introduced kokanee salmon, a specialized planktivore, are believed to be competing with juvenile stages of native trout in some waters, especially during periods of winter ice cover when plankton may be limited. Predation by introduced lake trout has also been implicated in the decline of native cutthroat trout in several large glacial lakes in North and Middle Fork drainages (Marnell 1988). Native cutthroat trout have been compromised by fish introductions and invasions throughout about 84 percent of their historic range in Glacier National Park (Marnell 1988).

Although native cutthroat trout have been adversely affected throughout a large portion of their park range, the species has not been lost from any water where it was historically present. Glacier National Park remains one of the last strongholds of genetically pure strains of lacustrine (i.e., lake-adapted) westslope cutthroat trout. This fact could have important implications for reestablishment of this unique subspecies throughout the central Rocky Mountains, where this trout has disappeared from most of its original range (Marnell 1995).

WHITE STURGEON

Scott and Crossman (1973) explain that white sturgeon, the largest freshwater fish in North America, live along the West Coast from the Aleutian Islands to central California. Genetically similar reproducing populations inhabit three major river basins: Sacramento-San Joaquin, Columbia, and Fraser. The greatest numbers of white sturgeon are in the Columbia River Basin.

✔   *Interesting Point*: Instead of scales, the white sturgeon is covered with patches of miniscule dermal denticles and isolated rows of large bony plates (Interactive Broadcasting Corporation, IBC 2007).

Historically, white sturgeon inhabited the Columbia River from the mouth upstream into Canada, the Snake River upstream to Shoshone Falls, and the Kootenai River upstream in Kootenai Falls (Scott and Crossman 1973). White sturgeon also used the extreme lower reaches of other tributaries but not extensively. Current populations in the Columbia River Basin can be divided into three groups: fish below the lowest dam, with access to the ocean (the lower Columbia River); fish isolated (functionally but not genetically) between dams; and fish in several large tributaries.

The Columbia River has supported important commercial, treaty, and recreational white sturgeon fisheries. A commercial fishery that began in the 1880s peaked in 1892 when 2.5 million kilograms (5.5 million pounds) were harvested (Craig and Hacker 1940). By 1899 the population had been severely depleted, and annual harvest was very low until the early 1940s, but the population recovered enough by the later 1940s that the commercial industry expanded. A six-foot maximum size restriction was enacted to prevent another population collapse. Total harvest doubled in the1970s and again in the 1980s because of increased treaty and recreational fisheries. From 1983 to 1994, 15 substantial regulatory changes were implemented on the main-stem Columbia River downstream from McNary Dam as a result of increased fishing. Columbia River white sturgeon are still economically important. Recreational, commercial, and treaty fisheries into the Columbia River downstream from McNary Dam were valued at $10.1 million in 1992 (Tracy 1993).

Several factors make white sturgeon relatively vulnerable to overexploitation and changes in their environment. The fish may live more than 100 years (Rieman and Beamesderfer 1990), and overexploitation is well documented for long-lived, slow-growing fish (Ricker 1963). Female white sturgeon are slow to reach sexual maturity; in the Snake River they mature at age 15–32 (Cochnauer 1981). Mature females in the Columbia River Basin only spawn every 2–11 years (Stockley 1981; Cochnauer 1983; Welch and Beamesderfer 1993). Sustainable harvest levels vary for impoundments in the Columbia River. Several impoundments are managed as groups, making overexploitation more likely in impoundments with low sustainable harvest levels.

White sturgeon populations in free-flowing and inundated reaches of the Columbia River Basin have been negatively affected by the abundant hydropower dams in most of the main-stem Columbia and Snake rivers (Rieman and Beamesderfer 1990). These dams have altered the magnitude and timing of discharge, water depths, velocities, temperatures, turbidities, and substrates, and have restricted sturgeon movement within the basin. Sturgeons in other river basins have declined in response to dam-induced habitat alterations (Artyukhin, Sukhoparova, and Fimukhira 1978).

Invertebrates

Invertebrates are among the most interesting and available of all living creatures for study. They are impressive in abundance and diversity, living on land, water, and air. Many species are borne to distant places on air and water currents, and via modern transportation (Mason 1995).

Of the millions of species of animals worldwide, about 90 percent are invertebrates, that is, animals without backbones. The arthropods, or jointed-legged invertebrates such as beetles, account for 75 percent of this total. More than 90,000 described insect species inhabit North America; the Lepidoptera (butterflies and moths) alone account for about 11,500.

Mason (1995) points out that within an acre of land and water, hundreds of different invertebrates form an ecological web of builders, gatherers, collectors, predators, and grazers, all interacting with each other and each a necessary component of a healthy ecosystem. The large macroscopic invertebrates—like bees, beetles, butterflies, grasshoppers, snails, and earthworms—are well known, but other invertebrates are almost invisible because they are extremely tiny or camouflaged for protection. We have just begun to understand the ecology of some commercially important species, but we understand very little about the behavior, communication, and function of many other invertebrates within various ecosystems.

Each individual invertebrate is a highly complex, specialized animal. Some morph (change or metamorphose) into several distinct life stages. For example, some insects transform from egg to larva, then to pupa, and finally emerge as a terrestrial winged adult. Some aquatic invertebrates do not have pupal stages, and the larvae (nymphs or naiads) grow progressively larger by molts. Earthworms bear cocoons, and each contains about six miniature juveniles; they also reproduce by fragmentation (architomy). The summary of structure of an insect is shown in figure 6.8.

Changes to the environment can disrupt basic interactions of invertebrate species, thereby affecting other organisms in the food chain. Disruptions of natural food cycles may cause drastic changes in the community structure and ecological web of life. This is especially true of the fauna that dwell in fragile ecosystems such as caves and springs. Eventually even humans are affected by changes to food webs and by the destruction of beneficial habitat for wildlife.

Most invertebrates can survive extreme natural events such as severe storms, blizzards, and flooding. When confronted by unnatural disturbances, however, such as excessive siltation from urban and highway developments, eutrophication (excessive nutrients) by runoff from agricultural lands, and contamination of aquatic habitats by toxic substances and acids, invertebrate populations can be severely damaged. Airborne toxicants such as acid rain are harmful to the long-term well-being of insects. If disturbances are sufficient, natural fauna may be extirpated (removed or lost) and replaced by more tolerant kinds. This "unbalanced" situation usually results in a population explosion of a few species. Such a biological reaction makes these aquatic invertebrates excellent bioindicators of overall environmental conditions (Bartsch and Ingram 1959). The use of aquatic invertebrates for bioassay (testing the toxicity of substances to "standard" test organisms) has greatly helped to minimize adverse effects of contaminants on aquatic life.

Butterflies and moths are particularly susceptible to environmental disturbances, although their responses to mild disturbances and changes may be slow, lasting decades (Otte 1995; Swengel and Swengel 1995). McCabe (1995) concludes that some of the flux in biodiversity is likely due to the "edge effect" at the interface from one habitat to another, and not necessarily to anthropogenic (human-caused) disturbances.

✔ *Important Point*: In regard to "edge effect," when habitat areas are fragmented, the result is more edge area where patches interface with the sur-

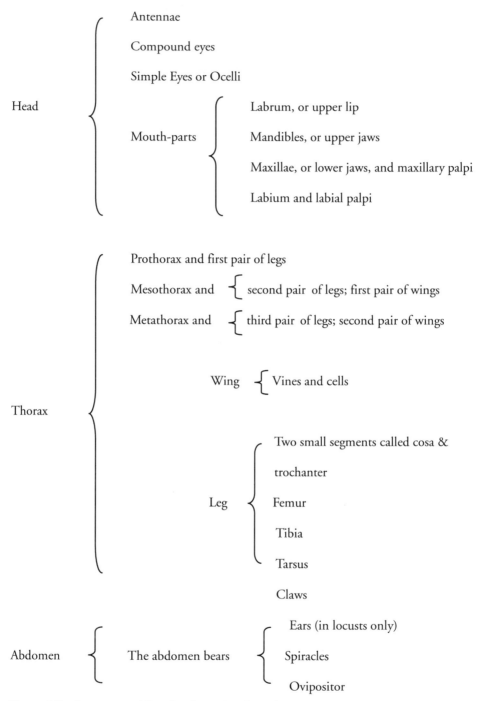

Figure 6.8. Summary of the structure of an insect

rounding environment. According to Cunningham and Cunningham (2002), these patch areas "with a relatively large ratio of edge to interior have some unique characteristics. They are often distinguished by increased predation when predators are able to hunt or forage along this edge more easily."

✔ *Interesting Point*: In regard to butterflies and moths being particularly susceptible to environmental disturbances, it is also interesting to note that the all-time predictor of the winter to come is the woolly worm, also known as fuzzy bear, hedgehog caterpillar, or woolly bear (Sault Ste. Marie Horticultural Society [SSMHS] 2007; see figure 6.9).

In the aquatic realm, organic chemicals and other toxic substances, acids and alkalis, and mine drainage can quickly decimate populations of mussels, mayflies, and stoneflies, whereas reduced water flow and introduction of pollutants such as silt and excessive nutrients (Mason, Fremling, and Nebeker 1995; Webb 1995) cause a slow, relentless destruction of the indigenous fauna.

In the past 50 years, nearly 72 percent of the United States' 297 native mussel species have become endangered, threatened, or of special concern (Williams and Neves 1995). Their populations have been damaged because of siltation, point and nonpoint source pollution, and outright habitat destruction.

Figure 6.9.   Woolly bear caterpillar, George Washington National Forest, Virginia
Photograph by Revonna Bieber

This zebra mussel and some other nonindigenous species represent "biological pollution" (Schloesser and Nalepa 1995), and should be considered much like toxic pollution for control and treatment. Nonnative zebra mussels lack predators and have invaded nearly the full length of the Mississippi River and its major tributaries, threatening the native mussel fauna of the eastern United States (Williams and Neves 1995).

It is important to point out that historical databases have traditionally focused on commercially important invertebrate species such as clams and oysters (Otte 1995). In contrast, little information exists on the status and trends of nonconsumptive, indigenous invertebrate life, and existing data are often not in formats for use in modern decision-making tools (Messer, Linthurst, and Overton 1991).

An important, often-overlooked problem with providing scientifically credible data involves the taxonomy and systematics (identification and classification) of organisms. Today, our museum collections of invertebrates are often old and worn out, and there are few trained taxonomists to renew archival materials. In fact, many "type" specimens used for original species' descriptions in the early 1900s are unusable, making comparisons of recently collected specimens impossible.

Canada has been doing continuous biomonitoring for several decades, which has now resulted in status and trend analyses of subtle perturbations such as acidification (Chmielewski and Hall 1993). It is clear that the success of future assessments in the United States will greatly depend on availability of and access to high-quality data; stop-gap measures are unlikely to prove successful because of inconsistencies caused by differing collection methods, taxonomy, and reporting units.

DIVERSITY OF INSECTS

> The mute insect, fix't upon the plant
> On whose soft leaves it hangs, and from
>   whose cup
> Drains imperceptibly its nourishment,
> Endear'd my wanderings.
>
> —William Wordsworth

Insects are the most diverse group of organisms (Wheeler 1990); potentially, they are highly indicative of environmental change through close adaptation to their environment; they represent the majority of links in the community food chain; and they have likely the largest biomass of the terrestrial animals (Holden 1989). Thus, knowledge about them is fundamental to studying the environment (Hodges 1995).

GRASSHOPPERS

> The grasshoppers weave their autumn song by the golden railing
> of the well.
>
> —Chinese, *Long Yearning*

A grasshopper is an amazing insect that can leap at least 20 times the length of its own body. If we could do that, we would be able to jump almost 120 feet! While it would be great to be able to jump so far, we might not be able to make as graceful a landing as the grasshopper (*Grasshopper Facts* 2007).

✔   *Interesting point*: Grasshoppers consume green forage roughly eight times as fast in proportion to their weight as beef animals on a good range (Pohly 2007).

According to Otte (1995), grasshoppers are perhaps the most important grazing herbivores in the nation's grasslands, which from a human standpoint are the most important food-producing areas. The damage that grasshoppers do to plants varies with the species. A few dozen species at most are highly injurious to crops, while those that feed on economically unimportant plants may have no measurable impact, and those that feed on detrimental plants are highly beneficial. Given such differences, it becomes important to distinguish properly between harmful and beneficial species. Grasshopper abundance in all kinds of grasslands means they are an important factor in the ecological equation. Their economic importance—positive and negative—means that they must be included in all studies of grassland and desert-grassland communities.

LEPIDOPTERA: BUTTERFLIES AND MOTHS

> WINGED BEAUTY
> When attention is drawn to magnificence resting on a leaf,
> We remain attentive prior to that pre-flight opening of wings;
> The dense mosaic of tiny individually colored scales forms a sight
>   beyond belief;
> The ethereal beauty of form and design captivate us;
> Upon gaining flight and exposing colors to light, a faint breeze
>   sings.
>
> —Frank R. Spellman

Whether it be the fritillaries, coppers, monarchs, painted ladies, sulphurs, or owl butterflies, Lepidoptera (Latin: scale wing; butterflies and moths) make up about 13 percent of the described and named 90,000 insect species of North American (11,500 named); they are among the better known large orders, although no complete inventory of the Lepidoptera species exists for any state, county, or locality in North America (Powell 1995; see figure 6.10).

➤ *Interesting Point*: What is the difference between moths and butterflies? Moths fly at night, have feathered antennae, and rest with their wings open. In contrast, butterflies fly during the day, have knob-ended antennae, and rest with their wings closed.

The inventorying of butterflies, both official (scientific) and unofficial (unscientific), has been going on for some time. For example, the Xerces Society started the Fourth of July Butterfly Count (FJC) in 1975, sponsoring it annually until 1993, when the North American Butterfly Association (NABA) assumed administration (Swengel 1995). The general methods of the butterfly count are patterned after the highly successful Christmas Bird Count (CBC), founded in 1900 and sponsored by the National Audubon Society (Swengel 1990).

➤ *Interesting Point*: An interesting observation about wing patterns is referred to as Oudemans' principle. As you study the ventral wing pattern of a resting butterfly, you'll notice the pattern often smoothly translates from the hindwing to the visible tip portion of the forewing. In contrast, the covered portion of the forewing lacks the patterning and is often more brightly colored making for a disorienting flash of color as the butterfly launches into flight. Oudemans' principle can also be observed on the forewing patterns where design elements align between the fore and hind wings when the butterfly is displaying its dorsal surfaces (Bugbios 2007).

The results of the FJC, including butterfly data, count-site descriptions, and weather information on count day, are published annually. The count was designed as an informal program for butterfly enthusiasts and the general public. These counts can never substitute for more formal scientific censuses because data sets from the counts have flaws that impair scientific analysis. Nevertheless, the FJC program does provide data that, with considerable caution, can be useful for science and conservation (Swengel 1990). FJC data have been used to study the biology, status, and trends of both rare and widely distributed species (Swengel 1990; Nagel, Nightengale, and Dankert 1991; Nagel 1992; Swengel, unpublished data).

In regard to insect sampling (Opler 1995), butterflies and large moths are among the best-sampled insects and as such are excellent indicators of ecological conditions or environmental change. Because the caterpillars of most Lepidoptera are herbivorous, their species richness is most often a reflection of plant diversity (Brown and Opler 1990).

Butterflies and moths are particularly susceptible to environmental disturbances, although their responses to mild disturbances and changes may be slow, lasting decades (Swengel and Swengel 1995). This is no better demonstrated than by the butterfly community of the tall-grass prairie region of the United States.

Figure 6.10.   Butterfly, Victoria, Canada
Photograph by Revonna Bieber

The prairie biome, between the Missouri River and the Rocky Mountains, is a plant community dominated by grasses and nongrassy herbs (wildflowers or "forbs"), with some woody shrubs and occasional trees. Prairies are classified into three major types by rainfall and consequent grass composition. The easternmost and moistest division is the tall-grass prairie (Risser et al. 1981). Although tall-grass prairie once broadly covered the middle of the United States, this biome is now estimated to be at least 99 percent destroyed from presettlement by pioneers, who converted it for agricultural uses. Prairie loss continues through plowing, extreme overgrazing, and development, but at varying degrees. Prairie is also lost passively because the near-total disruption of previous ecological processes causes shifts in floristic composition and structure (Swengel and Swengel 1995).

✔   *Interesting Point*: The tallgrass prairie is a complex ecosystem, including flowers, trees, birds, mammals, insects, and microorganisms. But grass dominates. Like other grasses, tallgrasses do not form woody tissue or increase in girth. Their stems are hollow except where the leaves join, leaves are narrow with parallel veins, and flowers are small and inconspicuous. Tallgrass prairie is so-named because the component grasses—big bluestem, little bluestem, Indiangrass. and switchgrass—can reach eight or nine feet (National Park Service [NPS] 1995).

As a result of this habitat destruction, butterflies and other plants and animals that are obligate to the prairie ecosystem are rare and primarily restricted to prairie preserves. The Dakota skipper and the regal fritillary are federal candidates for listing under the Endangered Species Act, and additional prairie butterfly species are on state lists as officially threatened or endangered. Patches of original prairie vegetation remain in preserves, parks, unintensively used farmlands such as hayfields and pastures, and in unused land. These remnants of prairie, however, are isolated and often in some state of ecological degradation (Swengel and Swengel 1995).

The existence of the prairie depends on the occurrence of certain climatic conditions and disturbance processes such as animal herbivory and fire. These natural process, however, are severely disrupted today because of the destruction and fragmentation of the prairie biome. Without management intervention, the vegetational composition and structure of prairie sites are altered through invasion of woody species and smothering under dead plant matter. Prairies usually require active management to maintain the ecosystem and its biodiversity, but it is difficult to know exactly which processes once naturally maintained the prairie ecosystem. Frequent fire, whether caused by lightning or intentionally set, is usually considered the dominant prehistoric process that maintained prairies; thus, management for tallgrass prairie in most states relies primarily or solely on frequent fire (e.g., Sauer 1950; Hulbert 1973; Vogl 1974). Other researchers, however, assert that prairies were the result of grazing by large herds of ungulates (England and DeVos 1969).

Despite this scientific conflict, it appears certain that successful management for maintaining the prairie landscape and its native species should be based on these natural processes, whatever they were. The vast diversity and specificity of insects to certain plants and habitat features make them fine-tuned ecological indicators. Thus, butterfly conservation is useful not only for maintaining these unique species but also for helping us monitor and learn about the soundness of our general ecosystem management (Swengel and Swengel 1995).

AQUATIC INSECTS AND BIOTIC INDEXES

Mason et al. (1995) state that aquatic insects are among the most prolific animals on Earth but are highly specialized and represent less than 1 percent of the total animal diversity (Pennak 1978). Most people know the 12 orders and about 11,000 species of North American aquatic insects (Merritt and Cummins 1984) only by the large adults that fly around or near wetlands.

Aquatic insects are excellent overall indicators of both recent and long-term environmental conditions (Patrick and Palavage 1994). The immature stages of aquatic insects have a short life cycle, often several generations a year, and remain in the general area of propagation. Thus, when environmental changes occur, the species must endure the disturbance, adapt quickly, or die and be replaced by more tolerant species. These changes often result in an overabundance of a few tolerant species, and the communities become destabilized or "unbalanced" (Mason et al. 1995).

Spellman (1996) points out that environmental scientists find of interest and often use four different indicators of water quality: coliform bacteria count, concentration of dissolved oxygen (DO), biochemical oxygen demand (BOD), and the *biotic index*. The biota that exist at or near a stream, for example, are, as pointed out earlier, direct indicators (a biotic index) of the condition of the water. This biotic index is often more reliable than many of the laboratory chemical tests that environmental scientist/toxicologists use in attempting to determine the pollutant level in a stream. Indicator species help determine when pollutant levels are unsafe.

How does the biotic index actually work? How does it indicate pollution? Certain common aquatic organisms, by indicating the extent of oxygenation of a stream, may be regarded as indicators of the intensity of pollution from organic waste. The responses of aquatic organisms in streams to large quantities of organic wastes are well documented. They occur in a predictable cyclical manner. For example, upstream from an industrial waste discharge point (an end-of-pipe, point-source polluter), a stream can support a wide variety of algae, fish, and other organisms, but in the section of the stream where oxygen levels are low (below five parts per million [5 ppm]), only a few types of worms survive. As the stream flow courses downstream, oxygen levels recover and those species that can tolerate low rates of oxygen (such as gar, catfish, and carp) begin to appear. Eventually, at some further point downstream, a clean water zone reestablishes itself and a more diverse and desirable community of organisms returns (Spellman 1996).

During this characteristic pattern of alternating levels of dissolved oxygen (in response to the dumping of large amounts of biodegradable organic material), a stream, as stated above, goes through a cycle. This cycle is called an *oxygen sag curve*. Its state can be determined using the biotic index as an indicator of oxygen content.

The biotic index is a systematic survey of invertebrate organisms. Since the diversity of species in a stream is often a good indicator of the presence of pollution, the biotic index can be used to correlate with water quality. A knowledgeable person (an environmental scientist or ecologist, for example) can easily determine the state of water quality of any stream simply through observation—observation of the types of species present or missing used as an indicator of stream pollution. The biotic index, used in the determination of the types, species, and numbers of biological organisms present in a stream, is commonly used as an auxiliary to BOD determination in determining stream pollution. The disappearance of particular organisms tends to indicate the water quality of the stream.

The biotic index is based on two principles:

1. A large dumping of organic waste into a stream tends to restrict the variety of organisms at a certain point in the stream.
2. As the degree of pollution in a stream increases, key organisms tend to disappear in a predictable order.

Several different forms of the biotic index are commonly used. In Great Britain, for example, the Trent Biotic Index (TBI), the Chandler score, the Biological Monitoring Working Party (BMWP) score, and the Lincoln Quality Index (LQI) are widely used. Most forms use a biotic index that ranges from 0 to 10. The most polluted stream, which contains the smallest variety of organisms, is at the lowest end of the scale (0); the clean streams are at the higher end (10). A stream with a biotic index greater than 5 will support game fish; a stream with a biotic index less that 4 will not support game fish (Spellman 1996).

Because they are easy to sample, macroinvertebrates have predominated in biological monitoring. Macroinvertebrates are a diverse group. They demonstrate tolerances that vary between species. Discrete differences tend to show up, and often contain both tolerant and sensitive indicators. In addition, comparison with identification keys, which are portable and conveniently used in field settings, can easily identify invertebrates. Present knowledge of invertebrate tolerances and responses to stream pollution is well documented. In the United States, for example, the Environmental Protection Agency (EPA) has required states to incorporate narrative biological criteria into its water quality standards since 1993.

The biotic index provides a valuable measure of pollution, especially for species very sensitive to lack of oxygen. Consider the stonefly. Stonefly larvae live underwater and survive best in well-aerated, unpolluted waters with clean gravel bottoms. When the stream deteriorates from organic pollution, stonefly larvae cannot survive. The degradation of stonefly larvae has an exponential effect upon other insects and fish that feed off the larvae; when the stonefly larvae disappear, so in turn do many insects and fish (Spellman 1996).

✔ *Important Point*: Webb (1995) points out that preliminary analysis of the recent collections of Illinois stoneflies indicates a reduction in the species' richness in Illinois, a reduction in the spatial distribution of many species, the dominance of more generalist species more tolerant to environmental perturbations, and the extirpation of several species.

These general trends can be expanded for all of the central United States. The reduction in stream flow through the construction of locks and dams and the resulting effect of increased sedimentation have severely affected the habitat and niche selection available to species such as stoneflies that require rapidly flowing streams. This situation has been compounded by the erosional effects of deforestation and agricultural practices, which are maximizing the amount of land put into cultivation, as well as the increased problems related to nonpoint pollution from agricultural pesticides and fertilizers. To properly delineate these trends, the status of stoneflies and most other groups of aquatic organisms in the central United States needs to be evaluated.

Table 6.1 shows a modified version of the BMWP biotic index. Since the BMWP biotic index indicates ideal stream conditions, this index takes into account the sensitivities of different macroinvertebrate species to stream contamination. Aquatic macroinvertebrate species are represented by diverse populations and are excellent indicators of pollution. These organisms are large enough to be seen by the unaided eye. Most aquatic macroinvertebrates live for at least a year. They are sensitive to stream water quality, on both a short-term and a long-term basis. Mayflies, stoneflies, and caddisflies are aquatic macroinvertebrates considered clean-water organisms; they are generally the first to disappear from a stream if water quality declines and are therefore given a high score. Tubicid worms (tolerant to pollution) are given a low score.

Table 6.1. The BMWP Score System (modified for illustrative purposes)

Families	Common-Name Examples	Score
Hepatagenidae	Mayflies	
Leuctridae	Stoneflies	10
Aeshnidae	Dragonflies	8
Polycentropidae	Caddisflies	7
Hydrometridae	Water strider	
Gyrinidae	Whirligig beetle	5
Chironomidae	Mosquitoes	2
Oligochaera	Worms	1

Note: Organisms with high scores, especially mayflies and stoneflies (the most sensitive), and others (dragonflies and caddis flies) are very sensitive to any pollution (deoxygenation) of their aquatic environment.

In table 6.1, a score from 1 to 10 is given for each family present. A site score is calculated by adding the individual family scores. The site score (total score) is then divided by the number of families recorded to derive the average score per taxon (ASPT). High ASPT scores result from such taxa as stoneflies, mayflies, and caddisflies present in the stream. A low ASPT score is obtained from heavily polluted streams dominated by tubicid worms and other pollution-tolerant organisms (Spellman 1996).

In using the biotic index, environmental scientists/ecologists make use of the fact that unpolluted streams normally support a wide variety of macroinvertebrates and other aquatic organisms with relatively few of one kind in making determinations about water quality in the field. While some aquatic species, such as mayflies and stoneflies, are more sensitive than others to certain pollutants, and succumb more readily to the effects of pollution, other species, such as mussels and clams, accumulate toxic materials in their tissues at sublethal levels. These species can be monitored (*must* be monitored to protect public health) to track pollution movement and buildup in aquatic systems.

Using a biotic index to determine the level of pollution in a water body demonstrates only one application. Levine, Hall, Barret, and Taylor (1989) state that similar determinations regarding soil quality can be made by observing and analyzing organisms (such as earthworms) in the soil. Studies conducted to assess the impact of sewage biosolids (sludge) treatments on old-field communities revealed that earthworms concentrate cadmium, copper, and zinc in their tissues at levels that exceed those found in the soil. Cadmium levels even exceed the concentrations found in the biosolids. Thus, earthworms may provide an "index" to monitor the effects of biosolids disposal on terrestrial communities.

INSECT MACROINVERTEBRATES

If the biotic index is to be used as an indicator of environmental pollution in water bodies (lakes and streams), for example, then one must have some understanding of insect macroinvertebrates. Insect macroinvertebrates are ubiquitous in streams and are often represented by many species. Although the numbers refer to aquatic species, a majority is to be found in streams (Spellman 1996).

The most important macroinvertebrate insect groups in streams are Ephemeroptera (mayflies), Plecoptera (stoneflies), Trichoptera (caddisflies), Diptera (true flies), Coleoptera (beetles), Hemiptera (bugs), Megaloptera (alderflies and dobsonflies), and Odonata (dragonflies and damselflies). The identification of these different orders is usually easy and there are many keys and specialized references (e.g., Merritt and Cummins, *An Introduction to the Aquatic Insects of North America*, 1984) available to help in the identification to species. In contrast, some genera and species, particularly the Diptera, can often only be diagnosed by specialist taxonomists (Spellman 2003).

Mayflies (Order: Ephemeroptera)

Streams and rivers are generally inhabited by many species of mayflies and, in fact, most species are restricted to streams. For the experienced freshwater ecologist who looks upon a mayfly nymph, recognition is obtained through trained observation: abdomen with leaflike or featherlike gills, legs with a single tarsal claw, generally (but not always) with three cerci (three "tails"; two cerci, and between them usually a terminal filament; see figure 6.11). The experienced ecologist knows that mayflies are hemimetabolous insects (i.e., where larvae or nymphs resemble wingless adults) that go through many postembryonic molts, often in the range between 20 and 30. For some species, body length increases about 15 percent for each instar (developmental stage).

Mayfly nymphs are mainly grazers or collector-gatherers feeding on algae and fine detritus, although a few genera are predatory. Some members filter particles from the water using hair-fringed legs or maxillary palps. Shredders are rare among mayflies. In general, mayfly nymphs tend to live mostly in unpolluted streams, where with densities of up to 10,000 per square meter they contribute substantially to secondary producers.

Adult mayflies resemble nymphs, but usually possess two pair of long, lacy wings folded upright; adults usually have only two cerci. The adult lifespan is short, ranging from a few hours to a few days, rarely up to two weeks, and the adults do not feed. Mayflies are unique among insects in having two winged stages, the subimago (winged and capable of flight but not sexually mature) and the imago (adult). The emergence of adults tends to be synchronous, thus ensuring the survival of enough adults to continue the species (Spellman 2003).

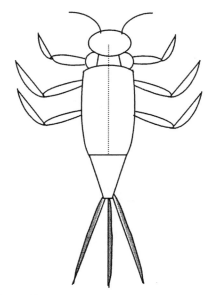

Figure 6.11. Mayfly (Order: Emphemeroptera)

Stoneflies (Order: Plecoptera)

Although many freshwater ecologists would maintain that the stonefly is a well-studied group of insects, this is not exactly the case. Despite their importance, less than 5–10 percent of stonefly species are well known with respect to life history, trophic interactions, growth, development, spatial distribution, and nymphal behavior.

Notwithstanding our lacking of extensive knowledge in regard to stoneflies, enough is known to provide an accurate characterization of these aquatic insects. We know, for example, that stonefly larvae are characteristic inhabitants of cool, clean streams (i.e., most nymphs occur under stones in well-aerated streams). While they are sensitive to organic pollution—or, more precisely, to low oxygen concentrations accompanying organic breakdown processes—stoneflies seem rather tolerant to acidic conditions. Lack of extensive gills at least partly explains their relative intolerance of low oxygen levels.

Stoneflies are drab-colored, small- to medium-sized (1/6 to 2-1/4 inches [4 to 60 mm]), rather flattened insects. Stoneflies have long, slender, many-segmented antennae and two long narrow antenna-like structures (cerci) on the tip of the abdomen (see figure 6.12). The cerci may be long or short. At rest, the wings are held flat over the abdomen, giving a "square-shouldered" look compared to the rooflike position of most caddisflies and vertical position of the mayflies. Stoneflies have two pair of wings. The hind wings are slightly shorter than the forewings and much wider, having a large anal lobe that is folded fanwise when the wings are at rest. This fanlike folding of the wings gives the order its name: "*pleco*" (folded or plaited) and "*-ptera*" (wings). The aquatic nymphs are generally very similar to mayfly nymphs except that they have only two cerci at the tip of the abdomen. The stoneflies have chewing mouthparts. They may be found any-

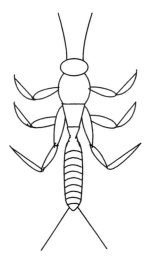

Figure 6.12. Stonefly (Order: Plecoptera)

where in a nonpolluted stream where food is available. Many adults, however, do not feed and have reduced or vestigial mouthparts.

Stoneflies have a specific niche in high-quality streams where they are very important as a fish food source at specific times of the year (winter to spring, especially) and of the day. They complement other important food sources, such as caddisflies, mayflies, and midges (Spellman 2003).

Caddisflies (Order: Trichoptera)

Trichoptera (Greek: *trichos*, a hair; *ptera*, wing), is one of the most diverse insect orders living in the stream environment, and caddisflies have nearly a worldwide distribution (the exception: Antarctica). Caddisflies may be categorized broadly into free-living (roving and net spinning) and case-building species.

Caddisflies are described as medium-sized insects with bristlelike and often long antennae. They have membranous hairy wings (which explains the Latin name "trichos"), which are held tentlike over the body when at rest; most are weak fliers. They have greatly reduced mouthparts and five tarsi (end segments of legs). The larvae are mostly caterpillar-like and have a strongly sclerotized (hardened) head with very short antennae and biting mouthparts. They have well-developed legs with a single tarsi. The abdomen is usually ten-segmented; in case-bearing species, the first segment bears three papillae, one dorsally and the other two laterally, which helps hold the insect centrally in its case and allows a good flow of water past the cuticle and gills, and the last or anal segment bears a pair of grappling hooks.

In addition to being aquatic insects, caddisflies are superb architects. Most caddisfly larvae (see figure 6.13) live in self-designed, self-built houses, called *cases*. They spin out silk, and either live in silk nets or use the silk to stick together bits of whatever is lying on the stream bottom. These houses are so specialized that you can usually identify a caddisfly larva to genus if you can see its house (case). With nearly 1,400 caddisfly species in North America (north of Mexico), this is a good thing!

Caddisflies are closely related to butterflies and moths (order: Lepidoptera). They live in most stream habitats and that is why they are so diverse (have so many species). Each species has special adaptations that allow it to live in the environment it is found in.

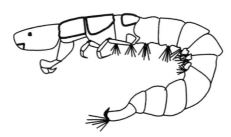

Figure 6.13.   Caddis larvae, *Hydropsyche spp.*

Mostly herbivorous, most caddisflies feed on decaying plant tissue and algae. Their favorite algae are diatoms, which they scrape off rocks. Some of them, though, are predacious.

Caddisfly larvae can take a year or two to change into adults. They then change into *pupae* (the inactive stage in the metamorphosis of many insects, following the larval stage and preceding the adult form) while still inside their cases for their metamorphosis. It is interesting to note that caddisflies, unlike stoneflies and mayflies, go through a "complete" metamorphosis.

Caddisflies remain as pupae for two to three weeks and then emerge as adults. When they leave their pupae, splitting their case, they must swim to the surface of the water to escape it. The winged adults fly during the evening and night, and some are known to feed on plant nectar. Most of them will live less than a month; like many other winged stream insects, their adult lives are brief compared to the time they spend in the water as larvae.

Caddisflies are sometimes grouped by the kinds of cases they make into five main groups: free-living forms that do not make cases, saddle-case makers, purse-case makers, net-spinners and retreat makers, and tube-case makers.

Caddisflies demonstrate their "architectural" talents in the cases they design and make. For example, a caddisfly might make a perfect, four-sided box case of bits of leaves and bark or tiny bits of twigs. It may make a clumsy dome of large pebbles. Others make rounded tubes out of twigs or very small pebbles. In our experience in gathering caddisflies, we have come to appreciate not only their architectural ability but also their flare in the selection of construction materials. For example, we have found many caddisfly cases constructed of silk, emitted through an opening at the tip of the labium, used together with bits of ordinary rock mixed with sparkling quartz and red garnet, green peridot, and bright fool's gold.

Besides the protection their cases provide them, the cases provide another advantage. The cases actually help caddisflies breathe. They move their bodies up and down, back and forth inside their cases, and this makes a current that brings them fresh oxygen. The less oxygen there is in the water, the faster they have to move. It has been seen that caddisflies inside their cases get more oxygen than those that are outside of their cases—and this is why stream ecologists think that caddisflies can often be found even in still waters, where dissolved oxygen is low, in contrast to stoneflies and mayflies (Spellman 2003).

True Flies (Order: Diptera)

True or two- (*Di-*) winged (*ptera*) flies not only include the flies that we are most familiar with, like fruitflies and houseflies, they also include midges (see figure 6.14), mosquitoes, craneflies (see figure 6.15), and others. Houseflies and fruitflies live only on land, and we do not concern ourselves with them. Some, however, spend nearly their whole lives in water; they contribute to the ecology of streams.

True flies are in the order Diptera, and are one of the most diverse orders of the class Insecta, with about 120,000 species worldwide. Dipteran larvae occur almost everywhere except Antarctica and deserts where there is no running water.

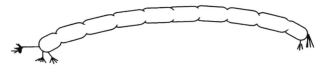

Figure 6.14.   Midge larvae

They live in a variety of places within a stream: buried in sediments, attached to rocks, beneath stones, in saturated wood or moss, or in silken tubes, attached to the stream bottom. Some even live below the stream bottom.

True fly larvae eat almost anything, depending on their species. Those with brushes on their heads use them to strain food out of the water that passes through. Others eat algae, detritus, plants, and even other fly larvae.

The longest part of the true fly's life cycle, like that of mayflies, stoneflies, and caddisflies, is the larval stage. It may remain an underwater larva anywhere from a few hours to five years. The colder the environment, the longer it takes to mature. It pupates and emerges, and then becomes a winged adult. The adult might live four months—or it might only live for a few days. While reproducing, it will often eat plant nectar for the energy it needs to make its eggs. Mating sometimes takes place in aerial swarms. The eggs are deposited back in the stream; some females will crawl along the stream bottom, losing their wings, to search for the perfect place to put their eggs. Once they lay them, they die.

Diptera are especially good "bioindicators" of aquatic environmental conditions because, in addition to the attributes of other aquatic insects, they occupy the full spectrum of habitats and conditions (Paine and Gaufin 1956; Roback 1957; Mason 1975; Hudson, Lenat, Caldwell, and Smith 1990; Spellman 2003).

Moreover, Diptera serve an important role in cleaning water and breaking down decaying material, and they are a vital food source (i.e., they play pivotal roles in the processing of food energy) for many of the animals living in and around streams. However, the true flies most familiar to us are the midges, mosquitoes, and the craneflies because they are pests. Some midge flies and mosquitoes bite; the cranefly, however, does not bite but looks like a giant mosquito.

Like mayflies, stoneflies, and caddisflies, true flies are mostly in larval form. Like caddisflies, you can also find their pupae because they are holometabolous insects (go through complete metamorphosis). Most of them are free-living; that is, they can travel around. Although none of the true fly larvae have the six jointed legs we see on the other insects in the stream, they sometimes have strange little almost-legs—prolegs—to move around with.

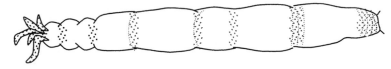

Figure 6.15.   Cranefly larvae

Others may move somewhat like worms do, and some—the ones who live in waterfalls and rapids—have a row of six suction discs that they use to move much like a caterpillar does. Many use silk pads and hooks at the ends of their abdomens to hold them fast to smooth rock surfaces (Spellman 2003).

Beetles (Order: Coleoptera) (Hutchinson 1981)

Of the more than 1 million described species of insect, at least one-third are beetles, making the Coleoptera not only the largest order of insects but also the most diverse order of living organisms. Even though the most speciose order of terrestrial insects, surprisingly their diversity is not so apparent in running waters. Coleoptera belongs to the infraclass Neoptera, division Endpterygota. Members of this order have an anterior pair of wings (the *elytra*) that are hard and leathery and not used in flight; the membranous hind wings, which are used for flight, are concealed under the elytra when the organisms are at rest. Only 10 percent of the 350,000 described species of beetles are aquatic.

Beetles are holometabolous. Eggs of aquatic Coleopterans hatch in one or two weeks, with diapause occurring rarely. Larvae undergo from three to eight molts. The pupal phase of all Coleopterans is technically terrestrial, making this life stage of beetles the only one that has not successfully invaded the aquatic habitat. A few species have diapausing prepupae, but most complete transformation to adults in two to three weeks. Terrestrial adults of aquatic beetles are typically short-lived and sometimes nonfeeding, like those of the other orders of aquatic insects. The larvae of Coleoptera are morphologically and behaviorally different from the adults, and their diversity is high.

Aquatic species occur in two major suborders, the Adephaga and the Polyphaga. Both larvae and adults of six beetle families are aquatic: Dytiscidae (predaceous diving beetles), Elmidae (riffle beetles), Gyrinidae (whirligig beetles), Halipidae (crawling water beetles), Hydrophilidae (water scavenger beetles), and Noteridae (burrowing water beetles). Five families, Chrysomelidae (leaf beetles), Limnichidae (marsh-loving beetles), Psephenidae (water pennies), Ptilodactylidae (toe-winged beetles), and Scirtidae (marsh beetles) have aquatic larvae and terrestrial adults, as do most of the other orders of aquatic insects; adult limnichids, however, readily submerge when disturbed. Three families have species that are terrestrial as larvae and aquatic as adults: Curculionidae (weevils), Dryopidae (long-toed water beetles), and Hydraenidae (moss beetles), a highly unusual combination among insects. (Note: Because they provide a greater understanding of a freshwater body's condition—that is, they are useful indicators of water quality—we focus our discussion on the riffle beetle, water penny, and whirligig beetle).

Riffle beetle larvae (most commonly found in running waters, hence the name riffle beetle) are up to 3/4" long (see figure 6.16). Their body is not only long but also hard, stiff, and segmented. They have six long segmented legs on the upper middle section of body, and the back end has two tiny hooks and short hairs.

Figure 6.16.   Riffle beetle larvae

Larvae may take three years to mature before they leave the water to form a pupa; adults return to the stream.

Riffle beetle adults are considered better indicators of water quality than larvae because they have been subjected to water quality conditions over a longer period. They walk very slowly under the water (on the stream bottom) and do not swim on the surface. They have small oval-shaped bodies (see figure 6.17) and are typically about 1/4″ in length.

Both adults and larvae of most species feed on fine detritus with associated microorganisms that are scraped from the substrate, although others may be xylophagous, that is, wood eating (e.g., *Lara*, Elmidae). Predators do not seem to include riffle beetles in their diet, except perhaps for eggs, which are sometimes attacked by flatworms.

The adult *water penny* is inconspicuous and often found clinging tightly in a suckerlike fashion to the undersides of submerged rocks, where they feed on attached algae. The body is broad, slightly oval and flat in shape, ranging from 4 to 6 mm (1/4″) in length. The body is covered with segmented plates and looks like a tiny round leaf (see figure 6.18). It has six tiny jointed legs (underneath). The color ranges from light brown to almost black.

There are 14 water penny species in the United States. They live predominately in clean, fast-moving streams. Aquatic larvae live one year or more (they are aquatic); adults (they are terrestrial) live on land for only a few days. They scrape algae and plants from surfaces.

Whirligig beetles are common inhabitants of streams and normally are found on the surface of quiet pools. The body has pincherlike mouthparts, and six seg-

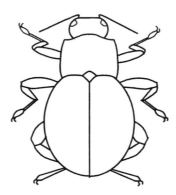

Figure 6.17.   Riffle beetle adult

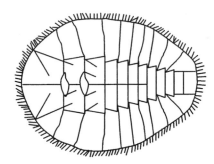

Figure 6.18.   Water Penny larva

mented legs on the middle of the body; the legs end in tiny claws. Many filaments extend from the sides of the abdomen. They have four hooks at the end of the body and no tail (see figure 6.19).

✔   *Note*: When disturbed, whirligig beetles swim erratically or dive while emitting defensive secretions.

As larvae, they are benthic predators, whereas the adults live on the water surface, attacking dead and living organisms trapped in the surface film. They occur on the surface in aggregations of up to thousands of individuals. Unlike the mating swarms of mayflies, these aggregations serve primarily to confuse predators. Whirligig beetles have other interesting defensive adaptations. For example, the Johnston's organ at the base of the antennae enables them to echolocate using surface wave signals; their compound eyes are divided into two pairs, one above and one below the water surface, enabling them to detect both aerial and aquatic predators; and they produce noxious chemicals that are highly effective at deterring predatory fish (Spellman 2003).

Water Strider ("Jesus Bugs"; Order: Hemiptera)

It is fascinating to sit on a log at the edge of a stream pool and watch the drama that unfolds among the small water animals. Among the star performers in small streams are the water bugs. These are aquatic members of that large group of

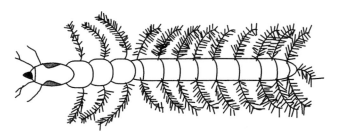

Figure 6.19.   Whirligig beetle larva

insects called the "true bugs," most of which live on land. Moreover, unlike many other types of water insects, they do not have gills but get their oxygen directly from the air.

Most conspicuous and commonly known are the water striders or water skaters. These ride the top of the water, with only their feet making dimples in the surface film. Like all insects, the water striders have a three-part body (head, thorax, and abdomen), six jointed legs, and two antennae. They have a long, dark, narrow body (see figure 6.20). The underside of the body is covered with water-repellent hair. Some water striders have wings, others do not. Most water striders are over 0.2″ (5 mm) long.

Water striders eat small insects that fall on the water's surface, and larvae. Water striders are very sensitive to motion and vibrations on the water's surface. They use this ability in order to locate prey. They push their mouths into their prey, paralyze it, and suck the insect dry. Predators of the water strider, such as birds, fish, water beetles, backswimmers, dragonflies, and spiders, take advantage of the fact that water striders cannot detect motion above or below the water's surface (Spellman 2003).

Alderflies and Dobsonflies (Order: Megaloptera)

Larvae of all species of Megaloptera ("large wing") are aquatic and attain the largest size of all aquatic insects. Megaloptera is a medium-sized order with fewer

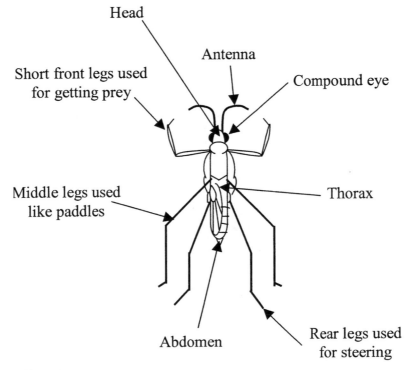

Figure 6.20.   Water strider

than 5,000 species worldwide. Most species are terrestrial; in North America, 64 aquatic species occur.

In running waters, alderflies (family: Sialidae) and dobsonflies (family: Corydalidae; sometimes called hellgrammites or toe biters) are particularly important, as they are voracious predators, having large mandibles with sharp teeth.

Alderfly larvae are brownish-colored and possess a single tail filament with distinct hairs. The body is thick-skinned with six to eight filaments on each side of the abdomen; gills are located near the base of each filament. Mature body size is 0.5″ to 1.25″ (see figure 6.21). Larvae are aggressive predators, feeding on other adult aquatic macroinvertebrates (they swallow their prey without chewing); as secondary consumers, other larger predators eat them. Female alderflies deposit eggs on vegetation that overhangs water, larvae hatch and fall directly into water (i.e., into quiet but moving water). Adult alderflies are dark with long wings folded back over the body; they only live a few days.

Dobsonfly larvae are extremely ugly (thus, they are rather easy to identify) and can be rather large, anywhere from 25 to 90 mm (13″) in length. The body is stout, with eight pairs of appendages on the abdomen. Brushlike gills at the base of each appendage look like "hairy armpits" (see figure 6.22). The elongated body has spiracles (spines), and has three pairs of walking legs near the upper body and one pair of hooked legs at the rear. The head bears four segmented antennae, small compound eyes, and strong mouth parts (large chewing pinchers). Coloration varies from yellowish, brown, gray, and black, often mottled. Dobsonfly larvae, commonly known as hellgrammites, are customarily found along stream banks under and between stones. As indicated by the mouthparts, they are predators and feed on all kinds of aquatic organisms.

Dragonflies and Damselflies (Order: Odonata)

The Odonata (dragonflies, suborder Anisoptera; and damselflies, suborder Zygoptera) is a small order of conspicuous, hemimetabolous insects (lack a pupal stage) of about 5,000 named species and 23 families worldwide. Odonata is a Greek word meaning "toothed one." It refers to the serrated teeth located on the insect's chewing mouthparts (mandibles).

Characteristics of dragonfly and damselfly larvae include

• Large eyes
• Three pairs of long segmented legs on upper middle section (thorax) of body

Figure 6.21. Alderfly larva

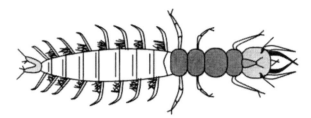

Figure 6.22.   Dobsonfly larva

- Large scooplike lower lip that covers bottom of mouth
- No gills on sides or underneath abdomen

✔   *Note*: Dragonflies and damselflies are unable to fold their four elongated wings back over the abdomen when at rest.

Dragonflies and damselflies are medium to large insects with two pairs of long equal-sized wings. The body is long and slender, with short antennae. Immature stages are aquatic and development occurs in three stages (egg, nymph, adult).

Dragonflies are also known as darning needles. (Note: Myths about dragonflies warned children to keep quiet or less the dragonfly's "darning needles" would sew the child's mouth shut.) The nymphal stage of dragonflies is a grotesque creature, robust and stoutly elongated. They do not have long "tails" (see figure 6.23). They are commonly gray, greenish, or brown to black in color. They are medium to large aquatic insects with size ranging from 15 to 45 mm, and the legs are short and used for perching. They are often found on submerged vegetation and at the bottom of streams in the shallows. They are rarely found in polluted waters. Food consists of other aquatic insects, annelids, small crustacea, and mollusks. Transformation occurs when the nymph crawls out of the water, usually onto vegetation. There it splits its skin and emerges prepared for flight. The adult dragonfly is a strong flier, capable of great speed (>60 mph) and maneuverability (fly backward,

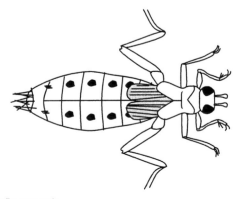

Figure 6.23.   Dragonfly nymph

Among the many ecosystem components affected by human-induced changes to western Lake Erie (Burns 1985) is the native mussel fauna. Reduced mussel populations that survived degraded conditions of the 1950s have been used in status and trends studies to evaluate traditional forms of pollution in western Lake Erie. Studies in the 1990s have focused on evaluating the effects of exotic species on mussel populations in the Lake Huron–Lake Erie corridor. Exotic species have recently been characterized as "biological pollution," a new concept in evaluating status and trends data.

Plants and Fungi

> I wandered lonely as a cloud
> That floats on high o'er vales and hills,
> When all at once I saw a crowd,
> A host, of golden daffodils;
> Beside the lake, beneath the trees,
> Fluttering and dancing in the breeze.
>
> —William Wordsworth

One way to begin a proper study of plants is to gaze upon, for example, a bouquet of white spider orchids surrounded by sprays of baby's breath with a mixture of bee balm, cornflower, dill, flame lily, ginger, Kansas feather, lilac, mimosa, safflower, and yarrow. Another would be to gaze at Wordsworth's daffodils. Obviously Longfellow felt and understood the bliss and delight flowers bring us when he wrote the following:

> In all places then and in all seasons,
> Flowers expand their light and soul-like wings,
> Teaching us by most persuasive reasons,
> How akin they are to human things.
>
> —Longfellow

This section describes trends in just two of the major kingdoms of life on Earth: the kingdom Plantae (which includes flowers, of course) and the molds, lichens, and mushrooms of the kingdom Fungi (Guntenspergen 1995). In addition to the eye-catching beauty of most flowers (see figure 6.25), members of the plant and fungal kingdoms have both economic and ecological importance. Plants transform solar energy (photosynthesis) into usable economic products essential in our modern society and provide the basis for most life on Earth by generating the air we breathe, the food we eat, and the shelter and clothes that protect us from the elements. Fungi (see figure 6.26), because rarely seen, may be thought of as unimportant. But nothing could be further from the truth: Life on Earth could not exist without fungi. Fungi not only mediate crucial biological and ecological processes, including the breakdown of organic matter and recycling of nutrients, but they also play important roles in mutualistic associations with plants and ani-

Figure 6.25.   Wildflower, George Washington National Forest, Virginia
Photograph by Frank R. Spellman

mals. Members of the kingdom Fungi also produce commercially valuable substances including antibiotics and ethanol, while other fungi are pathogenic and cause damage to crops and trees. Moreover, fungi are essential for proper plant root function. Because fungi and plants play such fundamental roles in our lives, it is important to have a comprehensive knowledge of the taxa comprising these groups. However, at a time when we are increasingly recognizing the importance of these groups, we are impoverishing our biological heritage. Rates of species loss are reaching alarming levels as ecosystems are degraded and habitat is lost. This erosion of biological diversity threatens the maintenance of long-term sustainable development and protection of the Earth's biosphere (Guntenspergen 1995).

FUNGI

Fungi are divided into five classes:

- Myxomycetes, or slime fungi
- Phycomycetes, or aquatic fungi (algae)
- Ascomycetes, or sac fungi

Figure 6.26.   Fungi, George Washington National Forest, Virginia
Photograph by Frank R. Spellman

- Basidiomycetes, or rusts, smuts, and mushrooms
- Fungi imperfecti, or miscellaneous fungi

✔   *Interesting Point:* Although fungi are limited to only five classes, more than 80,000 known species exist.

Fungi differ from bacteria in several ways, including their size, structural development, methods of reproduction, and cellular organization. They differ from bacteria in another significant way as well: Their biochemical reactions (unlike the bacteria) are not important for classification; instead, their structure is used to identify them. Fungi can be examined directly (see figures 6.27 and 6.28), or suspended in liquid, stained, dried, and observed under microscopic examination where they can be identified by the appearance (color, texture, and diffusion of pigment) of their mycelia (filament-like threads; Spellman 1996).

One of the tools available to environmental science students and specialists for use in the fungal identification process is the distinctive terminology used in mycology. Fungi go through several phases in their life cycle; their structural characteristics change with each new phase. Become familiar with the following listed and defined terms. As a further aid in learning how to identify fungi, relate the defined terms to their diagrammatic representations (figure 6.29).

Figure 6.27.   Various mushrooms and moss, George Washington National Forest, Virginia

Photograph by Frank R. Spellman

Figure 6.28.   Mushrooms on deadfall, George Washington National Forest, Virginia
Photograph by Frank R. Spellman

Fungi Key Terms

Blastospore or *bud*—spores formed by budding

Budding—process by which yeasts reproduce

Conidia—asexual spores that form on specialized hyphae called *conidiophores*. Large conidia are called *macroconidia* and small conidia are called *microconidia*.

Hypha (pl. *hyphae*)—a tubular cell that grows from the tip and may form many branches. Probably the best-known example of how extensive fungal hyphae can become is demonstrated in an individual honey fungus, *Armallaria ostoyae*, which was discovered in 1992 in Washington State. This particular fungus has been identified as the world's largest living thing; it covers almost 1,500 acres. Estimations have also been made about its individual network of *hyphae*: it is estimated to be 500 to 1,000 years old.

Mycelium—consists of many branched hyphae and can become large enough to be seen with the naked eye.

Nonseptate or *aseptate*—when crosswalls are not present

Septate hyphae—when a filament has crosswalls

Sexual spores—in the fungi division *Amastigomycota*, four subdivisions are separated on the basis of type of sexual reproductive spores present.

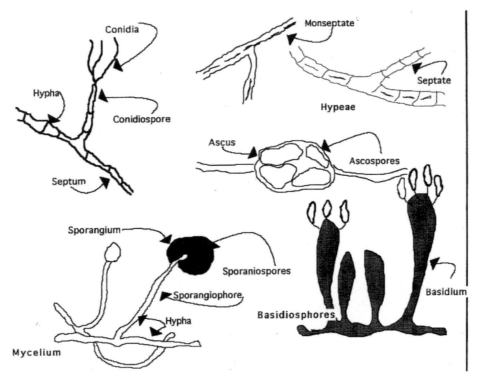

Figure 6.29. Nomenclature of fungi
Adapted from McKinney (1962, 36)

1. Subdivision *Zygomycotina*—consists of nonseptate hyphae and *Zygospores*. Zygospores are formed by the union of nuclear material from the hyphae of two different *strains*.
2. Subdivision *Ascomycotina*—fungi in this group are commonly referred to as the *ascomycetes*. They are also called *sac fungi*. They all have septate hyphae. *Ascospores* are the characteristic sexual reproductive spores and are produced in sacs called *asci* (ascus, singular). The mildews and *Penicillium* with asci in long fruiting bodies belong to this group.
3. Subdivision *Basidiomycotina*—consists of mushrooms, puffballs, smuts, rust, and shelf fungi (found on dead trees). The sexual spores of this class are known as *basidiospores*, which are produced on the club-shaped *basidia*.
4. Subdivision *Deutermycotina*—consist of only one group, the *Deuteromycetes*. Members of this class are referred to as the *fungi imperfecti* and include all the fungi that lack sexual means of reproduction.

Sporangiospores—spores that form within a sac called a *sporangium*. The sporangia are attached to stalks called *sporangiophores*.
Spore—reproductive stage of the fungi

Microfungi

Rossman (1995) and Spellman (1996) describe fungi as a group of organisms that constitute an extremely important and interesting group of eukaryotic, aerobic microbes ranging from the unicellular yeasts to the extensively mycelial molds. Not considered plants, they are a distinctive life form of great practical and ecological importance. Fungi are important because, like bacteria, they metabolize dissolved organic matter; they are the principal organisms responsible for the decomposition of carbon in the biosphere. Fungi, unlike bacteria, can grow in low moisture areas and in low pH solutions, which aids them in the breakdown of organic matter.

✔   *Important Point*: Without fungi, the world would be completely covered with organic debris that would not rot, and nutrients would not be available for plant growth. All plants would die (Rossman 1995).

Microfungi comprise a large group of organisms that include such diverse forms as molds, slime molds, other molds, mushrooms, mildews, puffballs, yeasts as well as rusts, and smuts, which cause plant diseases. They grow in all substrates, including plants, soil, water, insects, cows' rumen, hair, and skin. Because they lack chlorophyll (and thus are not considered plants), they must get nutrition from organic substances. They are either parasites, existing in or on animals or plants, or more commonly are *saporytes*, obtaining their food from dead organic matter. The fungi belong to the kingdom Myceteae. The study of fungi is called *mycology*.

✔   *Interesting Point*: Microfungi are said to be small because only part of the fungus is visible at one time, if at all. The viable parts produce thousands of tiny spores that are carried by the air, spreading the fungus (Rossman 1995).

McKinney (1962), in *Microbiology for Sanitary Engineers*, complains that the study of mycology has been directed solely toward classification of fungi and not toward the actual biochemistry involved with fungi. McKinney goes on to point out that for those involved in the sanitary field it is important to recognize the "sanitary importance of fungi . . . and other steps will follow" (40). For students of ecology understanding the role of fungi is important. Ecologists, for example, need knowledge and understanding of the organism's ability to function and exist under extreme conditions, which make them important elements in biological waste-stream treatment processes and in the degradation that takes place during waste-composting processes.

Among the multitudinous molds are humble servants such as *Penicillium notatum*, the source of penicillin, and *Tolyposporium niveum*, a produce of cyclosporin, the immune-system suppressant used for organ transplant operations. In sustainable agriculture, the fungal performers are agents of biological control and crop nutrition, helping the environment through the reduced use of chemical pesticides and fertilizers. Fungi can stop a hoard of locusts by attacking the chitinous insect exoskeleton or control nematodes that destroy the roots of crop plants (Commonwealth Agricultural Business [CAB] 1993). Although strains of fungi can degrade plastics and break down hazardous wastes such as dioxin (Jong and Edwards 1991), only a fraction of these fungi have been screened as beneficial organisms.

Microfungi can also be harmful, causing the irritating human affliction known as athlete's foot as well as disastrous diseases of crops and trees. The potato famine in Ireland during the mid- to late 1800s was caused by a fungus called *Phytophthora infestans* that rotted the potato crops for several years (Large 1962). Because of this disease, many Irish immigrated to the United States. Once the nature of the disease was determined, a solution based on fungus control was found. Knowing what fungi exist, where they occur, and what they do is essential.

Macrofungi

According to Mueller (1995), macrofungi are a diverse, commonly encountered, and ecologically important group of organisms. Macrofungi may be unicellular or filamentous. They are large, 5–10 microns wide, and can be identified by a microscope. The distinguishing characteristics of the group, as a whole, include the following: (1) they are nonphotosynthetic, (2) lack tissue differentiation, (3) have cell walls of polysaccharides (chitin), and (4) propagate by spores (sexual or asexual).

> ✔ *Important Point*: Macrofungi are vitally significant in forests; many species help break down dead organic material, such as dead tree trunks and leaves, into simple compounds usable by growing plants. Thus, they act as nature's recyclers, without which forests could not function (Mueller 1995).

Fungi can be grown and studied by cultural methods. However, when culturing fungi, use culture media that limits the growth of other microbial types—controlling bacterial growth is of particular importance. This can be accomplished by using special agar (culture media) that depresses pH of the culture medium (usually Sabouraud glucose or maltose agar) to prevent the growth of bacteria. Antibiotics can also be added to the agar that will prevent bacterial growth.

As part of their reproductive cycle, fungi produce very small spores that are easily suspended in air and widely dispersed by the wind. Insects and other animals also spread fungal spores. The color, shape, and size of spores are useful in the identification of fungal species.

Reproduction in fungi can be either sexual or asexual. The union of compatible nuclei accomplishes sexual reproduction. Most fungi form specialized asexual and/or sexual spore-bearing structures (fruiting bodies). Some fungal species are self-fertilizing and other species require outcrossing between different but compatible vegetative thalluses (mycelia).

Most fungi are asexual. Asexual spores are often brightly pigmented and give their colony a characteristic color (green, red, brown, black, blue—the blue spores of *Penicillium roquefort* are found in blue or Roquefort cheese).

Asexual reproduction is accomplished in several ways:

1. Vegetative cells may *bud* to produce new organisms. This is very common in the yeasts.
2. A parent cell can divide into two daughter cells.
3. The most common method of asexual reproduction is the production of spores. Several types of asexual spores are common:
    a. A hypha may separate to form cells (*arthrospores*) that behave as spores.
    b. If a thick wall before separation encloses the cells, they are called *chlamydospores*.
    c. If budding produces the spores, they are called *blastospores*.
    d. If the spores develop within sporangia (sac), they are called *sporangiospores*.
    e. If the spores are produced at the sides or tips of the hyphae, they are called *conidiospores*.

Fungi are found wherever organic material is available. They prefer moist habitats and grow best in the dark. Most fungi are *saprophytes*, acquiring their nutrients from dead organic matter, gained when the fungi secrete hydrolytic enzymes, which digest external substrates. They are able to use dead organic matter as a source of carbon and energy. Most fungi use glucose and maltose (carbohydrates) and nitrogenous compounds to synthesize their own proteins and other needed materials. Knowing from what materials fungi synthesize their own protein and other needed materials in comparison to what bacteria is able to synthesize is important to those who work in the environmental disciplines for understanding the growth requirements of the different microorganisms (Spellman 1996).

Lichens

Bennett points out that lichens are a unique life form because they are actually two separate organisms, a fungus and an alga, living together in a symbiosis. Lichens seem to reproduce sexually, but what appears to be a fruiting structure is actually that of the fungal component. Consequently, lichens are classified by botanists as fungi, but are given their own lichen names.

The best estimate of the number of U.S. lichen species are between 3,500 and 4,000, grouped in about 400 genera. The current checklist for the United States and Canada is probably in excess of 3,600 (Egan 1987).

Lichens are small plantlike organisms that grow just about everywhere: soils, tree trunks and branches, rocks and artificial stones, roofs, fences, walls, and even underwater. Various lichens, mosses, and fungi are shown in figures 6.30A, 6.30B. They are famous for surviving climatic extremes and are even the dominant vegetation in those habitats. Some lichens, however, are only found in very specialized habitats. The diversity of lichens in an area, therefore, is highly dependent on habitat diversity. Many special habitats across the United States are declining or

Figure 6.30A. Lichen and fungi, George Washington National Forest, Virginia
Photograph by Frank R. Spellman

Figure 6.30B.   Mushrooms, lichen, and moss, First Landing State Park, Virginia
Photograph by Frank R. Spellman

disappearing because of human activities, and some lichen species are consequently in decline (Bennett 1995).

Bryophytes

> The beauty there is in mosses must be considered from the holi-
> est, quietist nook.
>
> —Henry David Thoreau (1842)

Merrill (1995) points out that nonvascular bryophytes (mosses, liverworts, and hornworts) are small green plants that reproduce by means of spores (or vegeta-tively) instead of seeds. They are not considered to have given rise to the vascular plants but they probably were the earliest land plants (Qui and Palmer 1999). They evolved, like the rest of the land plants, from green algal ancestors. They are a group of simple land plants, usually only a few centimeters high, that are well-adapted to moist habitats. Although often small and inconspicuous, bryophytes are remarkably resilient and successful. They are sensitive indicators of air and water pollution, and play important roles in the cycling of water and nutrients and in relationships with many other plants and animals. With its varied landscape habitats, it is not surprising that about one third of the world's bryophyte species are found in tropical America, with high levels of endemism (Gradstein, Churchill, and Salazar Allen 2001). Information about bryophytes and their ecology is essential to develop comprehensive conservation and management polices and to restore degraded ecosystems (Merrill 1995).

There are three groups of bryophytes: mosses (approximately 10,000 species), liverworts (hepatics), and hornworts. Bryophytes rank second (after the flowering plants) among major groups of green land plants, with an estimated 15,000–18,000 species worldwide.

Mosses are the best known and most abundant and conspicuous in moist habitats, but are also found in grasslands and deserts, where they endure prolonged dry periods. Hepatics also include some arid-adapted species, but most are plants of humid environments. In mosses and leafy hepatics, the conspicuous plant body is leafy; in some liverworts and all hornworts, the plant is a flattened, ribbonlike "thallus" that lies flat on the ground. Bryophytes have no roots but are anchored by slender threads called rhizoids, which also play a role in the absorption of water and mineral nutrients.

Bryophytes have successfully exploited many environments, perhaps partly because they are rarely in direct competition with higher plants (Anderson 1980). For such small organisms, the climate near the ground (microclimate) is often very different from conditions recorded by standard meteorological methods, and shifts in temperature and humidity are often extreme. A remarkable adaptation of bryophytes is their ability to remain alive for long periods without water, even under high temperatures, and then resume photosynthesis within seconds after being moistened by rain or dew (Merrill 1995).

Bryophytes perform many environmental roles: (1) In forest ecosystems, they act like a sponge retaining and slowly releasing water; (2) many are pioneer plants, growing on bare rock and contributing to soil development; (3) in bogs and mountain forests, they form a thick carpet, reducing erosion; (4) they provide habitat for other plants and small animals as well as microorganisms such as nitrogen-fixing blue-green bacteria; (5) lacking a cuticle and transport tissue, they readily absorb whatever is around them and can serve as bioindicators of pollution and environmental degradation; and (6) they are also closely associated with organisms as diverse as protozoa, rotifers, nematodes, earthworms, mollusks, insects, and spiders (Gerson 1982), as well plants and fungi. Direct interactions of bryophytes include providing food, shelter, and nesting materials for small mammals and invertebrates; indirectly, they serve as a matrix for a variety of interaction between organisms (Spellman 1996).

NATIVE VASCULAR PLANTS

Morse, Kartesz, and Kutner (1995) point out that most of the familiar flora of the American landscape, such as trees, shrubs, herbs, vines, grasses, and ferns, are known as vascular plants. These plants have systems for transporting water and photosynthetic products and are differentiated into stems, leaves, and roots (nonvascular plants include algae, fungi, and mosses and lichens). Except in arctic and alpine areas, vascular plants dominate nearly all of North America's natural plant communities. About 17,000 species of vascular plants are native to one or more of the 50 U.S. states, along with several thousand additional native subspecies, varieties, and named natural hybrids (Kartesz 1994).

Human activities have expanded the geographical distributions of many plant species, particularly farm crops, timber trees, garden plants, and weeds. When a nonnative plant species is found growing outside cultivation, it is considered an exotic species in that area. About 5,000 exotic species are known outside cultivation in the United States. While many exotic plant species are desirable in some contexts (such as horticulture), hundreds of invasive nonnatives have become major management problems when established in places valued as natural areas (McKnight 1991; U.S. Congress 1993). A few particularly troublesome non-natives are regulated under specific federal or state laws as noxious weeds.

Summary of Key Terms

- *Autotroph* (primary producer)—any green plant that fixes energy of sunlight to manufacture food from inorganic substances
- *Appendage*—any extension or outgrowth from the body.
- *Bacteria*—among the most common microorganisms in water. Bacteria are primitive, single-celled microorganisms (largely responsible for decay and decomposition of organic matter) with a variety of shapes and nutritional needs.
- *Biochemical oxygen demand* (BOD)—a widely used parameter of organic pollution applied to both wastewater and surface water, involving the measurement of the dissolved oxygen used by microorganisms in the biochemical oxidation of organic matter.
- *Biotic index*—the diversity of species in an ecosystem is often a good indicator of the presence of pollution. The greater the diversity, the lower the degree of pollution. The biotic index is a systematic survey of invertebrate aquatic organisms, which is used to correlate with river quality.
- *Climax community*—the terminal stage of ecological succession in an area.
- *Competition*—is a critical factor for organisms in any community. Animals and plants must compete successfully in the community to stay alive.
- *Dissolved oxygen* (DO)—the amount of oxygen dissolved in a stream is an indication of the degree of health of the stream and its ability to support a balanced aquatic ecosystem.
- *Eutrophication*—the natural aging of a lake or land-locked body of water that results in organic material being produced in abundance due to a ready supply of nutrients accumulated over the years.
- *Fungi*—a group of organisms that lack chlorophyll and obtain nutrients from dead or living organic matter.
- *Habitat*—ecologists use the term "habitat" to mean the place where an organism lives.

- *Heterotroph* (living organisms)—any living organism that obtains energy by consuming organic substances produced by other organisms.
- *Invertebrates*—organisms without a backbone.
- *Larva* (larvae, plural)—juvenile form of many insects and other organisms that become different in form when changed into adults.
- *Limiting factor*—is a necessary material that is in short supply and because of the lack of it an organism cannot reach its full potential.
- *Macroinvertebrates*—animals that have no backbone and are visible without magnification.
- *Macrophytes*—big plants.
- *Microhabitat*—describes very local habitats.
- *Point source*—source of pollutants that involves discharge of pollutants from an identifiable point, such as a smokestack or sewage treatment plant.
- *Pollution*—an averse alteration to the environment by a pollutant.
- *Population*—a group of organisms of a single species that inhabit a certain region at a particular time.
- *Trophic level*—the feeding position occupied by a given organism in a food chain, measured by the number of steps removed from the producers.

Chapter Review Questions

6.1   Native American name for elk: _____.

6.2   Poor population health across many populations in a species eventually results in the _____ of that species.

6.3   Duck population loss is primarily attached to _____ and wintering habitats.

6.4   Colonial-nesting water birds can be divided into two major groups depending on where they _____.

6.5   Bitterns, herons, and egrets are known as _____.

6.6   Determining and monitoring the population status of raptors are necessary steps in the wise management of our _____.

6.7   The _____ is the most important U.S. game bird in terms of numbers harvested.

6.8   A large passerine bird: _____.

6.9   Mammal order named after mermaid sightings: _____.

6.10   Mammal revered by Native Americans: _____.

6.11   Sometimes called the silvertip bear: _____.

6.12   In order to be self-sustaining, the _____ requires a healthy prairie dog community for its habitat.

6.13   The first vertebrates to live on land were _____.

6.14   Name the five species of turtles.

6.15   Three angles of a triangle, according to Walton: _____.

6.16   A _____ is covered with patches of miniscule dermal denticles.

6.17   Another word for an animal with no backbone is _____

6.18   The _____ can leap 20 times the length of its own body.

6.19   _____ rest with their wings open.

6.20   When polluted, streams go through a cycle called a/an _____.

Cited References and Recommended Reading

Ainley, D. G., and Boekelheide, R. J. eds. 1990. *Seabirds of the Farallon Islands: Ecology, dynamics, and structure of an upwelling-system community*. Palo Alto, CA: Stanford University.

Ainley, D. G., and Hunt, G. W., Jr. 1991. *The status and conservation of seabirds in California: Seabird status and conservation, a supplement*, ed. J. P. Croxall, 103–14. International Council of Bird Preservation Technical Bulletin 11.

Ainley, D. G., and Lewis, T. J. 1974. The history of Farallon Island marine bird population, 1854–1972. *Condor* 76:432–46.

Aldrich, J. W. 1993. Classification and distribution. In *Ecology and management of the mourning dove*, ed. T. S. Baskett, M. W. Sayre, R. E. Tomlinson, and R. E. Mirarchi, 47–54. Harrisburg, PA: Stackpole Books.

Allendorf. F. W., Ryman, N., and Utter, F. M. 1987. Genetics and fishery management past, present, and future. In *Population genetics and fishery management*, ed. N. Ryman and F. M. Utter. Seattle: Washington Sea Grant Program.

Altukhov, Y. P., and Salmenkova, E. A. 1987. Stock transfer relative to natural organization, management and conservation of fish populations. In *Population genetics and fishery management*, ed. N. Ryman and F. M. Utter, 333–44. Seattle: Washington Sea Grant Program.

American-badger.com. 2007. *American badgers*. www.american-badgers.com (accessed March 16, 2007).

Anderson, L. E. 1980. Cytology and reproductive biology of mosses. In *The mosses of North America*, ed. R. J. Taylor, 37–76. San Francisco: Pacific Division of the American Association for the Advancement of Science.

Anderson, M. G. 1989. *Species closures: A case study of wintering waterfowl on San Francisco Bay, 1988–1990*. M.S. thesis, Humbolt State University, Acadia, CA.

Anderson, M. G., Rhymer, J. M., and Rohwer, F. C. 1992. Philopatry, dispersal, and the genetic structure of waterfowl populations. In *Ecology and management of breeding waterfowl*, ed. B. D. Batt, A. D. Kadlee, and G. L. Krapu, 365–95. Minneapolis: University of Minnesota Press.

Artyukhin, Y. K., Sukhoparova, A. D., and Fimukhira, L. G. 1978. The gonads of sturgeon in the littoral zone below the dam of the Volograd water engineering system. *Journal of Ichthyology* 18:912–23.

Austin, J. E., and Miller, M. R. 1995. Northern pintail. In *The Birds of North America*, no. 163, ed. A. Poole and F. Gill. Washington, DC: Academy of Natural Sciences of Philadelphia and the American Ornithologists Union.

Avianweb. 2007. *Dabbling ducks*. http://www.avainweb.com/dabblingducks.html (accessed March 7, 2007).

Baker, P. M. 1993. Resource management: A shell exporter's perspective. In *Conservation and management of freshwater mussels*, ed. K. S. Cummings, A. C. Buchanan, and L. M. Koch, 69–71. Proceedings of a symposium. Champaign: Illinois Natural History Survey.

Barrows, C., Muth, A. Fisher, M., and Lovich, J. 1995. Coachella Valley fringe-toed lizards. In

Our living resources. Washington, DC: U.S. Department of the Interior, National Biological Service.

Bartonek, J. C., and Nettleship, D. N., eds. 1979. *Conservation of marine birds of northern North America.* U.S. Fish & Wildlife Service, Wildlife Research Report 11.

Bartsch, A. F., and Ingram, W. M. 1959. Stream life and the pollution environment. *Public Works* 90:104–10.

Beebee, T. J. C., Flower, F. J., Stevenson, A. C., Patrick, S. T., Appleby, P. G., Fletcher, C., Marsh, C., Natkanski, J., Rippey, B., and Battarbee, R. W. 1990. Decline of the natterjack toad *Bufo calamita* in Britain: Palaeoecological, documentary, and experimental evidence for breeding site acidification. *Biological Conservation* 53:1–20.

Beeton, A. M. 1961. Environmental changes in Lake Erie. *Transactions of the American Fisheries Society* 90:153–59.

Bellrose, F. C. 1976. *Ducks, geese and swans of North America.* Harrisburg, PA: Stackpole Books.

Bellrose, F. C. 1980. *Ducks, geese and swans of North America* (3rd ed.). Harrisburg, PA: Stackpole Books.

Benedict, A. 2007. *Of fog larks and sea quail: Sea birds at risk in a changing world.* http://seawolf-adventures.com/seabirds.html (accessed March 10, 2007).

Bennett, J. P. 1995. Lichens. In *Our living resources.* Washington DC: U.S. Department of the Interior, National Biological Service.

Berry, K. H., and Medica, P. 1995. Desert tortoises in the Mojave and Colorado deserts. In *Our living resources.* Washington, DC: U.S. Department of the Interior, National Biological Service.

Biggins, D., and Godbey, J. 1995. Black-footed ferrets. In *Our living resources.* Washington, DC: U.S. Department of the Interior, National Biological Service.

Biggins, D., Miller, B. J., Hanebury, L., Oakleaf, R., Farmer, A., Crete, R., and Dood, A. 1993. A technique for evaluating black-footed ferret habitat. In *Management of prairie dog complexes for the reintroduction of the black-footed ferret,* ed. J. L. Oldemeyer, D. E. Biggins, B. J. Miller, and R. Crete, 73–88. Washington, DC: U.S. Fish & Wildlife Service Biological Report No. 13.

Biggins, D., Schroeder, M. H., Forrest, S. C., and Richardson, L. 1985. Movements and habitat relationships of radio-tagged black-footed ferrets. In *Proceedings of the Black-footed Ferret Workshop,* ed. S. H. Anderson and D. B. Inkley, 11.1–11.17. Cheyenne: Wyoming Game and Fish Department.

Bishop, C. A., and Petit, K. E. eds. 1992. Declines in Canadian amphibian populations: Designing a national monitoring strategy. *Canadian Wildlife Service Occasion Paper* 76:1–120.

Black-footed Ferret Recovery Program (BFFRP). 2005. *Ferret facts.* http://www.blackfootedferret.org/facts.html (accessed March 6, 2007).

Blaustein, A. R. 1994. Chicken Little or Nero's fiddle? A perspective on declining amphibian populations. *Herpetogolgica* 50:85–97.

Blaustein, A. R., Hoffman, P. D., Hokit, D. G., Kiesecker, J. M., Walls, S. C., and Hays, J. B. 1994. UV repair and resistance to solar UV-B in amphibian eggs: A link to population declines? *Proceedings of the National Academy of Science* 91:1791–95.

Blaustein, A. R., Wake, D. B., and Sousa, W. P. 1994. Amphibian declines: Judging stability, persistence, and susceptibility of populations to local and global extinction. *Conservation Biology* 8:60–71.

Blohm, R. J. 1989. Introduction to harvest understanding surveys and season setting. In *Sixth International Waterfowl Proceedings,* ed. K. H. Beattie, 118–29. Washington, DC: Sixth International Waterfowl Symposium.

Boarman, W. I., and Berry, K. H. 1995. Common ravens in the southwestern United States,

1968–1992. In *Our living resources*. Washington, D.C.: U.S. Department of the Interior, National Biological Service.

Bowen, F., Avise, J. C., Richardson, J. I., Meylan, A., Margaritoulis, D., and Hopkins-Humpy, S. R. 1993. Population structure of loggerhead turtles in the northwestern Atlantic Ocean and Mediterranean Seas. *Conservation Biology* 7:834–44.

Bowen, B. W., Meylan, A. B., Rose, J. P., Limpus, C. J., Balazs, G. H., and Avise, J. C. 1992. Global population structure and natural history of the green turtle in terms of matriarchal phylogeny. *Evolution* 46:865–81.

Bradford, D. F. 1983. Winterkill, oxygen relations, and energy metabolism of a submerged dormant amphibian, *Rana muscosa*. *Ecology* 64:1171–83.

Bradford, D. F. 1989. Allotopic distribution of native frogs and introduced fishes in high Sierra Nevada lakes of California: implication of the negative effect of fish introductions. *Copeia* 1989:775–78.

Bradford, D. F. 1991. Mass mortality and extinction in a high-elevation population of *Rana mucosa*. *Journal of Herpetology* 5:174–77.

Brady, J. T., LaVal, R. K., Kuntz, T. H., Tuttle, M. D., Wilson, D. E., and Clawson, R. L. 1983. *Recovery plan for the Indiana bat*. Washington, DC: U.S Fish and Wildlife Service.

Brooks, R .J., Brown, G. P., and Galbraith, D. A. 1991. Effects of a sudden increase in natural mortality of adults on a population of the common snapping turtle. *Canadian Journal of Zoology* 69:1314–20.

Brooks, R. J., Galbraith, D. A., Nancekivell, E. G., and Bishop, C. A. 1988. Developing guidelines for managing snapping turtles. In *Management of amphibians, reptiles, and small mammals in North America*, ed. R. C. Szaro, K. E. Severson, and D. R. Patton, 174–79. U.S. Forest Service General Technical Report RM-166.

Brown, J. W., and Opler, P. A. 1990. Patterns of butterfly species density in peninsular Florida. *Journal of Biogeography* 17:615–22.

Bugbios. 2007. *Butterfly wing patterns*. http://www.insects.org/class (accessed April 19, 2007).

Burger, G. V. 1978. Agriculture and wildlife. In *Wildlife and America*, ed. H. P. Brokaw, 89–107. Washington, DC: Council on Environmental Quality.

Burns, N. M. 1985. *Erie: The lake that survived*. Totowa, NJ: Rowman & Allenhead.

Bury, R. B., Corn, O. S., Dodd, C. K. Jr., McDiarmid, R. W., and Scott, N. J. 1995. Amphibians. In *Our living resources*. Washington, DC: U.S. Department of the Interior, National Biological Service.

Bury, R. B., Dodd, Jr., C. K., and Fellers, G. M. 1980. Conservation of the amphibia of the United States: A review. *U.S. Fish and Wildlife Service Research Publication* 134:1–34.

Byrd, G. V., Day, R. H., and Knudson, E. P. 1983. Patterns of colony attendance and censusing of auklets at Buldir Island, Alaska. *Condor* 85:274–80.

Caithamer, D. E., and Smith, G. W. 1995. North American ducks. In *Our living resources*. Washington, DC: U.S. Department of the Interior, National Biological Service.

Callahan, E. V. 1993. *Indiana bat summer habitat requirements*. M.S. thesis, University of Missouri, Columbia, MO.

Campton, D. E. 1987. Natural hybridization and introgression in fishes: Methods of detection and interpretation. In *Population genetics and fishery management*, ed. N. Ryman and F. M. Utter, 161–92. Seattle: Washington Sea Grant Program.

Carey, C. 1993. Hypothesis concerning the causes of the disappearance of boreal toads from the mountains of Colorado. *Conservation Biology* 7:355–62.

Carlson, C. A., and Muth, B. T. 1989. The Colorado River: Lifeline of the American Southwest. In *Proceedings of the International Large Rivers Symposium*, ed. D. P. Dodge, 220–39. Canadian Journal of Fisheries and Aquatic Sciences, Special Publication 106.

Carmichael, G. J., Hanson, J. N., Schmidt, M. E., and Morizot, D. C. 1993. Introgression among Apache, cutthroat, and rainbow trout in Arizona. *Transactions of the American Fisheries Society* 122:121–30.

Carpenter, J. W., and Hillman, C. N. 1978. *Husbandry, reproduction, and veterinary care of captive ferrets.* Proceedings of the American Association of Zoo Veterinarians Workshop, Knoxville, TN.

Carter, H. R., Gilmer, D. S., Takekawa, J. E., Lower, R. W., and Wilson, U. W. 1995. Breeding seabirds in California, Oregon, and Washington. In *Our living resources.* Washington, DC: U.S. Department of the Interior, National Biological Service.

Carter, H. R., McChesney, G. J, Jaques, D. L., Strong, C. S., Parker, M. W., Takekawa, J. E., Jory, D. L., and Whitworth, D. L. 1992. *Breeding populations of seabirds in California, 1989–1991*, vol. I. Dixon, CA: U.S. Fish & Wildlife Service.

Carter, H. R., and Morrison, M. I., eds. 1992. Status and conservation of the marbled murrelet in North America. *Proceedings of the 1987 Pacific Seabird Group Symposium.* Camarillo, CA: Proceedings of the Western Foundation of Vertebrate Zoology.

Chmielewski, C. M., and Hall, R. J. 1993. Changes in the emergence of blackflies over 50 years from Algonquin Park streams: Is acidification the cause? *Canadian Journal of Fisheries and Aquatic Sciences* 50:1517–29.

Christensen, N. I. 1981. *Fire regimes in southeastern ecosystems.* U.S. Forest Service General Technical Report WO-26.

Clarke, D. C. 1988. Prairie dog control: A regulatory viewpoint. In *Eight Great Plains Wildlife Damage Control Workshop Proceedings*, ed. D. W. Uresk and G. Schenbeck, 119–20. U.S. Forest Service General Technical Report RM-154.

Cochnauer, T. G. 1981. Survey status of white sturgeon populations in the Snake River, Bliss Dam to C. J. Strike Dam. Boise: Idaho Department of Fish and Game, River and Stream Investigations, Job Performance Report, Project F-73-R-2, Job 1-b.

Cochnauer, T. G. 1983. *Abundance, distribution, growth, and management of white sturgeon in the middle Snake River, Idaho.* Ph.D. dissertation, University of Idaho, Moscow.

Commonwealth Agricultural Business (CAB). 1993. Locust project enters phase two. *Commonwealth Agricultural Bureau International News* (June): 4.

Comstock, A. B. 1986. *Handbook of nature study.* Ithaca, NY: Cornell University Press.

Congdon, J. D., Dunham, A. E., and Van Loben Sels, R. C. 1993. Delayed sexual maturity and demographics of Blanding's turtles: Implications for conservation and management of long-lived organisms. *Conservation Biology* 7:826–33.

Corn, P. S., and Bury, R. B. 1989. Logging in western Oregon: Response of headwater habitats and stream amphibians. *Forest Ecology and Management* 29:39–57.

Corn, P. S., and Fogelman, J. C. 1984. Extinction of montane populations of the northern leopard frog (*Rana pipiens*) in Colorado. *Journal of Herpetology* 18:147–52.

Cornell University. 1999. *Mourning dove.* http://www.birds.cornell.edu.BOW (accessed March 10, 2007).

Craig, J. A., and Hacker, R. L. 1940. The history and development of the fisheries of the Columbia River. *U.S. Bureau of Fisheries Bulletin* 49, no. 32:132–216.

Cunningham, W. P., and Cunningham, M. A. 2002. *Principals of environmental science: Inquiry and applications.* New York: McGraw-Hill.

Deacon, J. E., Kobetich, G., Williams, J. D., Contreras, S., et al. 1979. Fishes of North America endangered, threatened, or of special concern: 1979. *Fisheries* 4, no. 2:30–44.

Delacour, J. T. 1954. *The waterfowl of the world*, vol. 1. London: Country Life.

Dickson, J. G. 1995. Return of wild turkeys. In *Our living resources.* Washington, DC: U.S. Department of the Interior, National Biological Service.

Dodd, C. K., Jr. 1991. The status of the Red Hills salamander *Phaeognathus hubrichti*, Alabama, USA, 1976–1988. *Biological Conservation* 55:57–75.

Dodd, C. K. 1995. Marine turtles in the Southeast. In *Our living resources*. Washington, DC: U.S. Department of the Interior, National Biological Service.

Dolton, D. D. 1995. Mourning doves. In *Our living resources*. Washington, DC: U.S. Department of the Interior, National Biological Service.

Doughty, R. W. 1984. Sea turtles in Texas: A forgotten commerce. *Southwestern Historical Quarterly* 88:43–70.

Drobney, R. D., and Clawson, R. L. 1995. Indiana bats. In *Our living resources*. Washington, DC: U.S. Department of the Interior, National Biological Service.

Droege, S., and Sauer, Jr., C. 1990. *North American breeding bird survey annual summary 1989*. U.S. Fish and Wildlife Service Biological Report 90, no. 8.

Ducks Unlimited. 2007. *Prairie Pothole Region*. http://www.ducks.org/conservation/initiative45. aspx (accessed March 7, 2007).

Dunson, W. A., Wyman, R. L., and Corbett, E. S. 1992. A symposium on amphibian declines and habitat acidification. *Journal of Herpetology* 16:349–52.

England, R. E., and DeVos, A. 1969. Influence of animals on pristine conditions on the Canadian grasslands. *Journal of Range Management* 22:87–94.

Egan, R. S. 1987. A fifth checklist of the lichen-forming, lichenocolous and allied fungi of the continental United States and Canada. *Bryologist* 90:77–173.

Erwin, R. M. 1995. Colonial waterbirds. In *Our living resources*. Washington, DC: U.S. Department of the Interior, National Biological Service.

Erwin, R. M., Frederick, P. C., and Trapp, J. L. 1993. Monitoring of colonial waterbirds in the United States: Needs and priorities. In *Waterfowl and wetland conservation in the 1990s: A global perspective. Proceeding of the International Waterfowl and Wetlands Research Bureau Symposium*, ed. M. Moser, B. C Prentice, and J. van Vessem, 18–22, St. Petersburg Beach, FL. Slimbridge, UK: International Waterfowl and Wetlands Research Bureau Special Publication.

Escambia. 2007. *Turtle types*. http://escambia.ifas.ufl.edu/marine/types_of_sea_turtles.htm (accessed March 18, 2007).

Estes, J. A., Jameson, R. J., Bodkin, J. L., and Carlson, D. R. 1995. California sea otters. In *Our living resources*. Washington, DC: U.S. Department of the Interior, National Biological Service.

FaunaWest Wildlife Consultants.1989. *Relative abundance and distribution of the common raven in the deserts of southern California and Nevada during spring and summer of 1989*. Riverside, CA: Bureau of Land Management.

Federal Register. 1970. 35:16047. Federal endangered species. U.S. Fish & Wildlife Services, p. 25–26.

Federal Register. 1993. Proposal to list the southwestern willow flycatcher as an endangered species, and to designate critical habitat. U.S. Fish and Wildlife Service. *Federal Register* 58:39,495–39,522.

Fellers, G. M., and Drost, C. A. 1993. Disappearance of the Cascades frog *Rana Cascadae* at the southern end of its range. *Biological Conservation* 65, no. 2:177–81.

Finch, D. M., and Stangel, P. W. 1993. *Status and management of Neotropical migratory birds; 1992 September 21–25*. Estes Park, CO: U.S. Forest Service General Technical Report RM-229, Rocky Mountains Forest and Range Experiment Station, Fort Collins, CO.

Flather, C. H., and Hoekstra, T. W. 1989. *An analysis of the wildlife and fish situation in the United States: 1989–2040*. U.S. Department of Agriculture Forest Service General Technical Report RM-178.

Forrest, S. C., Biggins, D. E., Richardson, L., Clark, T. W., Campbell II, T. M., Fagerstone, K. A., and Thorne, E. T. 1988. Population attributes for the black-footed ferret at Meeteetse, Wyoming, 1981–1985. *Journal of Mammalogy* 69, no. 2:261–73.

Franson, J. C., Sileo, L., and Thomas, N. J. 1995. Causes of eagle deaths. In *Our living resources*. Washington, DC: U.S. Department of the Interior, National Biological Service.

Freda, J. 1986. The influence of acidic pond water on amphibians: A review. *Water, Air, and Soil Pollution* 30:439–50.

Fuller, M. R., Henny, C. J., and Wood, P. B. 1995. Raptors. In *Our living resources*. Washington, DC: U.S. Department of the Interior, National Biological Service.

Gardner, J. E., Garner, J. D., and Hofmann, J. E. 1991. *Summer roost selection and roosting behavior of Myotis sodalis in Illinois*. Final report, Illinois Natural History Survey. Champaign: Illinois Department of Conservation.

Gee, G. G., and Hereford, S. G. 1995. Mississippi sandhill cranes. In *Our living resources*. Washington, DC: U.S. Department of the Interior, National Biological Service.

Gerson, U. 1982. Bryophytes and invertebrates. In *Bryophyte ecology*, ed. A. J. E. Smith, 291–322. New York: Chapman & Hall.

Gill, R. E., Jr., Handel, C. M., and Page, G. W. 1995. Western North American shorebirds. In *Our living resources*. Washington, DC: U.S. Department of the Interior, National Biological Service.

Gould, P. J., Forsell, D. J., and Lensink, C. J. 1982. *Pelagic distribution and abundance of seabirds in the Gulf of Alaska and eastern Bering Sea*. U.S. Fish and Wildlife Service FWS/OBS-8248.

Gradstein, S. R., Churchill, S. P., and Salazar Allen, N. 2001. Guides to the bryophytes of tropical America. *Memoirs of the N.Y. Botanical Gardens* 86:1–577.

Grasshopper Facts. 2007. http://www.thaibugs.com/Articles/grasshoppers (accessed April 9, 2007).

Guntenspergen, G. R. 1995. Plants. In *Our living resources*. Washington, DC: U.S. Department of the Interior, National Biological Service.

Haig, S. 1992. The piping plover. In *Birds of North America*, ed. A. Poole, P. Stettenheim, and F. Gill, 1–18. Washington, DC: American Ornithologists Union.

Haig, S., and Plissner, J. H. 1995. Piping plovers. In *Our living resources*. Washington, DC: U.S. Department of the Interior, National Biological Service.

Hale, S. F., Schwalbe, C. R., Jarchow, J. L., May, C. J., Lowe, C. H., and Johnson, T. B. 1995. Disappearance of Tarahumara frog. In *Our living resources*. Washington, DC: U.S. Department of the Interior, National Biological Service.

Halls, L. K., ed. 1984. *White-tailed deer: Ecology and management*. Harrisburg, PA: Stackpole Books.

Harrington, B. A. 1995. Shorebirds: East of the 105th meridian. In *Our living resources*. Washington, DC: U.S. Department of the Interior, National Biological Service.

Hatch, S. A. 1993a. Ecology and population status of northern fulmars *Fulmarus glacialis* of the North Pacific. In *The status, ecology and conservation of marine birds of the North Pacific*, ed. K. Vermeer, K. T. Briggs, and K. H. Morgan, 82–92. Ottawa: Canadian Wildlife Service.

Hatch, S. A. 1993b. Population trends of Alaskan seabirds. *Pacific Seabird Group Bulletin* 20:3–12.

Hatch, S. A., Byrd, G. V., Irons, D. B., and Hunt, G. L, Jr. 1993. Status and ecology of kittiwakes in the North Pacific. In *The status, ecology and conservation of marine birds of the North Pacific*, ed. K. Vermeer, K. T. Briggs, K. H. Morgan, and D. Siegel-Causey, 140–53. Ottawa: Canadian Wildlife Service.

Hatch, S. A., and Hatch, M. A. 1978. Colony attendance and population monitoring of black-legged kittiwakes on the Semidi Islands, Alaska. *Condor* 90:613–20.

Hatch, S. A., and Hatch, M. A. 1989. Attendance patterns of murres at breeding sites: Implications for monitoring. *Journal of Wildlife Management* 53:483–93.

Hatch, S. A., and Platt, J. F. 1995. Seabirds in Alaska. In *Our living resources*. Washington, DC: U.S. Department of the Interior, National Biological Service.

Hatch, S. A., Roberts, B. D., and Fadely, B. S. 1993b. Adult survival of black-legged kittiwakes in a Pacific colony. *Ibis* 135:247–54.

Hayes, M. P., and M. R. Jennings. 1986. Decline of ranid frog species in western North America: Are bullfrogs (*Rana catesbeiana*) responsible? *Journal of Herpetology* 20:490–509.

Heinrich, B. 1989. *Ravens in winter*. New York: Summit Books.

Hestback, J. B. 1995a. Canada geese in the Atlantic flyway. In *Our living resources*. Washington, DC: U.S. Department of the Interior, National Biological Service.

Hestback, J. B. 1995b. Decline of northern pintails. In *Our living resources*. Washington, DC: U.S. Department of the Interior, National Biological Service.

Heyer, W. R., Donnelly, M. A., McDiarmid, R. W., Hayek, L., and Foster, M. S., eds. 1994. *Measuring and monitoring biological diversity: standard methods for amphibians*. Washington, DC: Smithsonian Institution Press.

Hillman, C. N, and Linder, R. L. 1973. The black-footed ferret. In *Proceedings of the Black-footed Ferret and Prairie Dog Workshop*, ed. R. L. Linder and C. N. Hillman, 10–20. Brookings: South Dakota State University Publications.

Hine, R. L., Les, B. L., and Hellmich, B. F. 1981. Leopard frog populations and mortality in *Wisconsin, 1974–1976. Wisconsin Department of Natural Resources Technical Bulletin* 122:1–39.

Hochbaum, G. S., and Bossenmaier, E. F. 1972. Response of pintail to improved breeding habitat in southern Manitoba. *Canadian Field-Naturalist* 86:79–81.

Hodges, R. W. 1995. Diversity and abundance of insects. In *Our living resources*. Washington, DC: U.S. Department of the Interior, National Biological Service.

Hohman, W. L., Haramis, G. M., Jorde, D. G., Korschgen, C. E., and Takekawa, J. 1995. In *Our living resources*. Washington, DC: U.S. Department of the Interior, National Biological Service.

Holden, C. 1989. Entomologists wane as insects wax. *Science* 246:754–56.

Hudson, P. L., Lenat, D. R., Caldwell, B. A., and Smith, D. 1990. *Chironomidae of the southeastern United States: A checklist of species and notes on biology, distribution, and habitat*. U.S. Fish and Wildlife Service Fish and Wildlife Research 7.

Huff, W. R. 1993. Biological indices define water quality standards. *Water Environment and Technology* 5:21–22.

Hulbert, L. C. 1973. Management of Konza Prairie prior to approximate pre-white man influences. In *Third Midwest prairie conference proceedings*, ed. L. C. Hulbert, 14–19. Manhattan: Kansas State University.

Humphrey, S. R., Richter, A. R., and Cope, J. B. 1977. Summer habitat and ecology of the endangered Indiana bat *Myotis sodalis*. *Journal of Mammalogy* 58:334–46.

Hunt, G. L., Pitman, R. L., Naughton, M., Winnett, K, Newman, A., Kelly, P. R., and Briggs, K. T. 1979. Reproductive ecology and forging habits of breeding seabirds. In *Summary of marine mammal and seabird surveys of the southern California Bight area 1975–1978*, vol. 3—Investigators' reports. Part 3. Seabirds—Book 2, 1–399. Santa Cruz: University of California-Santa Cruz, for U.S. Bureau of Land Management, Los Angeles, CA. Contract AA550-CT7-36.

Hupp, J. W., Schmutz, J. A., and Ely, C. D. 2007. Moult migration of emperor geese. *Journal of Avian Biology*.

Hutchinson, G. E. 1981. Thoughts on aquatic insects. *Bioscience* 31:495–500.

Interactive Broadcasting Corporation (IBC). 2007. *White sturgeon*. http://www.bcadventure.com/adventure/angling/game_fish/sturgeon.phtml (accessed March 31, 2007).

Jefferies, M., and Mills, D. 1990. *Freshwater ecology: Principles and applications.* London: Belhaven Press.

Jennings, M. R. 1995. Native ranid frogs in California. In *Our living resources.* Washington, DC: U.S. Department of the Interior, National Biological Service.

Jennings, M. R., and Hayes, M. P. 1993. *Amphibian and reptile species of special concern in California.* Final report, California Department of Fish and Game, Inland Fisheries Division, Rancho Cordova, under Contract (8023).

Johnson, D. H., and Grier, J. W. 1988. Determinants of breeding distributions of ducks. *Wildlife Monograph* 100:1–37.

Johnson, J. E. 1987. *Protected fishes of the United States and Canada.* Bethesda, MD: American Fisheries Society.

Johnson, J. E. 1995. Imperiled freshwater fishes. In *Our living resources.* Washington, DC: U.S. Department of the Interior, National Biological Service.

Jong, S. C., and Edwards, J. J. 1991. *American type culture collection catalogue of filamentous fungi* (18th ed.). Rockville, MD: American Type Culture Collection.

Kagarise Sherman, C., and Morton, M. L. 1993. Population declines of Yosemite toads in the eastern Sierra Nevada of California. *Journal of Herpetology* 27, no. 2:186–98.

Kaminski, R. M., and Weller, M. W. 1992. Breeding habits of Nearctic waterfowl. In *Ecology and management of breeding waterfowl,* ed. B. J. Batt, A. D. Afton, M. G. Anderson, C. D. Ankney, D. H. Johnson, J. A. Kadlec, and G. L. Krapu, 568–89. Minneapolis: University of Minnesota Press.

Kamrin, M. A. 1989. *Toxicology.* Chelsea, MI: Lewis.

Kappes, J. J. 1993. *Interspecific interactions associated with red-cockaded woodpecker cavities at a north Florida site.* M.S. thesis, Gainesville: University of Florida.

Kartesz, J. T. 1994. *A synonymized checklist of the vascular flora of the United States, Canada, and Greenland* (2nd ed.). Portland, OR: Timber Press.

Kenyon, K. W. 1969. The sea otter in the eastern Pacific Ocean. *North American Fauna* 68:1–3.

Kiel, W. H., Jr. 1959. *Mourning dove management units: A progress report.* U.S. Fish and Wildlife Service Special Scientific Report 42.

Kinsinger, A. 1995. Marine mammals. In *Our living resources.* Washington, DC: U.S. Department of the Interior, National Biological Service.

Knight, R., and Call, M. 1980. *The common raven.* Bureau of Land Management Technical Note 344.

Knight, R., and Kawashima, J. 1993. Responses of raven and red-tailed hawk populations to linear right-of-ways. *Journal of Wildlife Management* 57:266–71.

Kurta, A., King, D., Teramino, J. A., Stribley, J. M., and Williams, K. J. 1993. Summer roosts of the endangered Indiana bat (*Myotis sodalis*) on the northern edge of its range. *American Midland Naturalist* 129:132–38.

Kutkuhn, J. H. 1981. Stock definition as a necessary basis for cooperative management of Great Lakes fish resources. *Canadian Journal of Fisheries and Aquatic Sciences* 3:1476–78.

Lamb, T., Avise, J., and Gibbons, J. W. 1989. Phylogeographic patterns in mitochondrial DNA of the desert tortoise and evolutionary relationships among the North American gopher tortoises. *Evolution* 43, no. 1:76–87.

Large, E. C. 1962. *Advance of the fungi.* New York: Dover.

Larsen, K. H., and Dietrich, J. H. 1970. Reduction of a raven population on lambing grounds with DRC-1339. *Journal of Wildlife Management* 34:200–204.

Lee, D. S., Gilbert, C. B., Hocutt, C. H., Jenkins, R. E., McAlister, D. E., and Stauffer, Jr., J. R. 1980. *Atlas of North American freshwater fishes.* North Carolina State Museum of Natural History.

Lennartz, M. R., Hooper, R. G., and Harlow, R. F. 1987. Sociality and cooperative breeding of red-cockaded woodpeckers. *Behavioral Ecology and Sociobiology* 20:77–88.

Levine, M. B., Hall, A. T., Barret, G. W., and Taylor, D. H. 1989. Heavy-metal concentration during ten years of sludge treatment to an old-field community. *Journal of Environmental Quality* 18, no. 4:411–18.

Linder, R. L., Dahlgren, R. B., and Hillman, C. N. 1972. Black-footed ferret-prairie dog inter-relationships. In *Proceedings of the Symposium on Rare and endangered Wildlife of the Southeastern U.S*, 22–37. Santa Fe, NM: New Mexico Department of Game and Fish.

Lindzey, F. G. 1982. The North American badger. In *Wild mammals of North America*, ed. J. A. Chapman and G. A. Feldhammer, 653–63. Baltimore, MD: Johns Hopkins University Press.

Lovich, J. E. 1984. Biodiversity and zoogeography of non-marine turtles in Southeast Asia. In *Biological diversity: Problems and challenges*, ed. S. K. Majumdar, F. J. Brenner, J. E. Lovich, J. F. Schalles, and E. W. Miller, 381–91. Easton: Pennsylvania Academy of Sciences.

Lovich, J. E. 1995. Turtles. In *Our living resources*. Washington, DC: U.S. Department of the Interior, National Biological Service.

Marnell, L. F. 1986. Impacts of hatchery stocks on wild fish populations. In *Fish culture in fisheries management*, ed. R. H. Stroud, 339–47. Bethesda, MA: American Fisheries Society.

Marnell, L. F. 1988. Status of the westslope cutthroat trout in Glacier National Park, Montana. *American Fisheries Society Symposium* 4:61–70.

Marnell, L.F. 1995. Cutthroat trout in Glacier National Park, Montana. In *Our living resources*. Washington, DC: U.S. Department of the Interior, National Biological Service.

Marnell, L. F., Behnke, R. J., and Allendorf, F. W. 1987. Genetic identification of cutthroat trout in Glacier National Park, Montana. *Canadian Journal of Fisheries and Aquatic Sciences* 44:1830–39.

Marsh, R. E. 1984. Ground squirrels, prairie dogs and marmots as pests on rangeland. In *Proceedings of the Conference for Organization and Practice of Vertebrate Pest Control*, 195–208. Fernherst, UK: ICI Plant Protection Division.

Mason, C. F. 1990. Biological aspects of freshwater pollution. In *Pollution: Causes, effects, and control*, ed. R. M. Harrision. Cambridge: Royal Society of Chemistry.

Mason, W. T., Jr. 1975. *Chironomidae* (Diptera) as biological indicators of water quality. In *Organisms and biological communities as indicators of environmental quality*, ed. C. C. King and L. E. Elfner, 40–51. Columbus: Circular 8, Ohio Biological Survey.

Mason, W. T., Jr. 1995. Invertebrates. In *Our living resources*. Washington, DC: U.S. Department of the Interior, National Biological Service.

Mason, W. T., Jr., Fremling, C. R., and Nebeker, A. V. 1995. Aquatic insects as indicators of environmental quality. In *Our living resources*. Washington, DC: U.S. Department of the Interior, National Biological Service.

Mattson, D. J., Wright, R. G., Kendall, K. C., and Martinka, C. J. 1995. Grizzly bears. In *Our living resources*. Washington, DC: U.S. Department of the Interior, National Biological Service.

Maughan, O. E. 1995. Fishes. In *Our living resources*. Washington DC: U.S. Department of the Interior, National Biological Service.

McCabe, T. L. 1995. The changing insect fauna of Albany's pine barrens. In *Our living resources*. Washington, DC: U.S. Department of the Interior, National Biological Service.

McDiarmid, R. W. 1995. Reptiles and amphibians. In *Our living resources*. Washington, DC: U.S. Department of the Interior, National Biological Service.

McKenzie, S. 2007. *Poems on/about water*. www.poemsabout.com (accessed April 19, 2007).

McKinney, R. E. 1962. *Microbiology for sanitary engineers*. New York: McGraw-Hill.

McKnight, B. N., ed. 1991. *Biological pollution: The control and impact of invasive exotic species*. Indianapolis: Indiana Academy of Science.

Mech, L. D., Pletscher, D. H., and Martinka, C. J. 1995. Gray wolves. In *Our living resources*. Washington, DC: U.S. Department of the Interior, National Biological Service.

Merrill, G. L. S. 1995. Bryophytes. In *Our living resources*. Washington, DC: U.S. Department of the Interior, National Biological Service.

Merritt, R. W., and Cummins, K. W. 1984. Introduction. In *An introduction to the aquatic insects of North America* (2nd ed.), ed. R. W. Merritt and R. W. Cummins, 1–3. Dubuque, IA: Kendall/Hunt.

Messer, J. J., Linthurst, R. A., and Overton, W. S. 1991. An EPA program for monitoring ecological status and trends. *Environmental Monitoring and Assessment* 17:67–78.

Meyer, E. 1989. *Chemistry of hazardous materials* (2nd ed.). Englewood Cliffs, NJ: Prentice-Hall.

Miller, R. R. 1961. Man and the changing fish fauna of the American Southwest. *Papers of the Michigan Academy of Science, Arts, and Letters* 46:365–404.

Miller, R. R. 1972. Threatened freshwater fishes of the Untied States. *Transactions of the American Fisheries Society* 101, no. 2:239–52.

Miller, A. I., Counihan, T. D., Parsley, M. J., and Beckman, L. G. 1995. Columbia River Basin white sturgeon. In *Our living resources*. Washington, DC: U.S. Department of the Interior, National Biological Service.

Minckley, W. I., and Deacon, J. E. 1968. Southwestern fishes and the enigma of endangered species. *Science* 159:1424–32.

Minckley, W. I., and Deacon, J. E., eds. 1991. *Battle against extinction: Native fish management in the American West*. Tucson: University of Arizona Press.

Missouri Department of Conservation. 1991. *Endangered bats and their management in Missouri*. Jefferson City: Missouri Department of Conservation.

Morse, L. E., Kartesz, J. T., and Kutner, L. S. 1995. Native vascular plants. In *Our living resources*. Washington, DC: U.S. Department of the Interior, National Biological Service.

Moyle, P. B., and Cech, J. J., Jr. 1988. *Fishes: An introduction to ichthyology*. Englewood Cliffs, NJ: Prentice Hall.

Moyle, P. B., Li, H., and Baron, B. A. 1986. The Frankenstein effect: Impact of introduced fishes in North America. In *Fish culture in fisheries management*, ed. R. H. Stroud, 415–26. Bethesda, MD: American Fisheries Society.

Mueller, G. M. 1995. Macrofungi. In *Our living resources*. Washington, DC: U.S. Department of the Interior, National Biological Service.

Nagel, H. 1992. The link between Platte River flows and the regal fritillary butterfly. *Braided River* 4:10–11.

Nagel, H. G., Nightengale, T., and Dankert, N. 1991. Regal fritillary butterfly population estimation and natural history on Rowe Sanctuary, Nebraska. *Prairie Naturalist* 23:145–52.

National Academy of Sciences. 1978. *Eutrophication: Causes, consequences, correctives*. Washington, DC: Author.

National Oceanic and Atmospheric Administration (NOAA). 2007. *Pinnipeds, whales, dolphins, porpoises and sirenia order*. http://mnmml.afsc.noaa.gov/education/sierenia.htm (accessed March 13, 2007).

National Park Service. 1995. *The tall grass prairie*. Washington, DC: U.S. Department of Interior.

National Research Council. 1990. *Decline of the sea turtles: Causes and prevention*. Washington, DC: National Academy Press.

National Wildlife Refuge. 2007. *Mississippi sandhill cranes*. http://www.fws.gov/Mississippi sandhillcrane/mscranes (accessed March 11, 2007).

Nature. 2007. Ravens. http://www.pbs.org/wnet/nature/Ravens (accessed April 15, 2007).

Neal, W. 1985. Endangered and threatened wildlife and plants: Reclassification of the American alligator in Florida to threatened due to similarity of appearance. *Federal Register* 50(119):25,672–25,678.

Nettleship, D. N., Sanger, G. A., and Springer, P. F., eds. 1984. *Marine birds: Their feeding ecology and commercial fisheries relationships*. Proceedings of the Pacific Seabird Group symposium. Ottawa, Ontario: Canadian Wildlife Service Special Publication.

Novak, R. M. 1994. Another look at wolf taxonomy. In *Ecology and conservation of wolves in a changing world*, ed. L. D. Carbyn, S. H. Fritts, and D. R. Seip. Edmondson, AB: Canadian Circumpolar Institute.

Obbard, M. E., Jones, J. G., Newman, R., Booth, A., Satterthwaite, A. J., and Linscombe, G. 1987. Furbearer harvests in North America. In *Wild furbearer management and conservation in North America*, ed. M. Novak, J. A. Baker, M. E. Obbard, and B. Malloch, 1007–38. Toronto: Ontario Trappers Association and the Ministry of Natural Resources.

O'Brien, S., Martenson, J. S., Eichelberger, M. A., Thorne, E. T., and Wright, F. 1989. Biochemical genetic variation and molecular systematic of the back-footed ferret. In *Conservation biology and the black-footed ferret*, 21–33. New Haven, CT: Yale University Press.

Opler, P. A. 1995. Species richness and trends of western butterflies and moths. In *Our living resources*. Washington, DC: U.S. Department of the Interior, National Biological Service.

O'Toole, C., ed. 1986. *The encyclopedia of insects*. New York: Facts on File.

Otte, D. 1995. Grasshoppers. In *Our living resources*. Washington, DC: U.S. Department of the Interior, National Biological Service.

Paine, G. W., and Gaufin, A. R. 1956. Aquatic Diptera as indicators of pollution in a Midwestern stream. *Ohio Journal of Science* 56:291–304.

Patrick, R., and Palavage, D. M. 1994. The value of species as indicators of water quality. *Proceedings of the Academy of Natural Sciences Philadelphia* 145:55–92.

Pattee, O. H., and Mesta, R. 1995. California condors. In *Our living resources*. Washington, DC. U.S. Department of the Interior, National Biological Service.

Pattee, O. H., and Wilbur, S. R. 1989. Turkey vulture and California condor. In *Proceedings of the Western Raptor Management Symposium and Workshop*, 61–65. Washington, DC: National Wildlife Federation.

Pechmann, J. H. K., Scott, D. E., Semlitsch, R. D., Caldwell, J. P., Vitt, L. J., and Gibbons, J. W. 1991. Declining amphibian populations: The problem of separating human impacts from natural fluctuations. *Science* 253:892–95.

Pechmann, J. H. K., and Wilbur, H. M. 1994. Putting declining amphibian populations in perspective: Natural fluctuations and human impacts. *Herpetologica* 50:65–84.

Peek, J. M. 1995. North American elk. In *Our living resources*. Washington, DC: U.S. Department of the Interior, National Biological Service.

Pennak, R. W. 1978. *Freshwater invertebrates of the United States* (2nd ed.). New York: John Wiley & Sons.

Peterjohn, B. G., and Sauer, J. R. 1993. North American Breeding Bird Survey annual summary 1990–1991. *Bird Populations* 1:52–67.

Petranka, J. W., Eldridge, M. E., and Haley, K. E. 1993. Effects of timber harvesting on southern Appalachian salamanders. *Conservation Biology* 7:363–70.

Philipp, D. P., and Clausen, J. E. 1995. Loss of genetic diversity among managed populations. In *Our living resources*. Washington, DC: U.S. Department of the Interior, National Biological Service.

Philipp, D. P., Epifanio, J. M., and Jennings, J. J. 1993. Conservation genetics and current stocking practices: Are they compatible? *Fisheries* 18:14–16.

Pianka, E., and Vitt, L. 2006. *Lizards: Windows to the evolution of diversity*. Los Angeles: University of California Press.

Pohly, J. n.d. Grasshopper facts. *Colorado State University*. http://www.coopext.colostate.edu/4DMG/Garden/Amazing/grasfact.htm (accessed April 9, 2007).

Porter, W. F. 1991. *White-tailed deer in eastern ecosystems: implications for management and research in national parks*. National Resources Report NPS/NRSUNY/NRR-91/05. Denver, CO: National Park Service.

Pounds, J. A., and Crump, M. L. 1994. Amphibian declines and climate disturbance: The case of the golden toad and harlequin frog. *Conservation Biology* 8:72–85.

Powell, J. A. 1995. Lepidoptera inventories in the continental United States. In *Our living resources*. Washington, DC: U.S. Department of the Interior, National Biological Service.

Pulliam, H. R. 1988. Sources, sinks, and population regulations. *American Naturalist* 132:652–61.

Qui, Y.-L., and Palmer, J. D. 1999. Phylogeny of early land plants: Insights from genes and genomes. *Trends in Plant Science* 4:26–30.

Richards, S. J., McDonald, K. R., and Alford, R. A. 1993. Declines in populations of Australia's endemic tropical rainforest frogs. *Pacific Conservation Biology* 1:66–77.

Ricker, W. E. 1963. Big effects from small causes: Two examples from fish population dynamics. *Journal of the Fisheries Research Board of Canada* 20:257–64.

Rieman, B. E., and Beamesderfer, R. C.. 1990. White sturgeon in the lower Columbia River: Is the stock overexploited? *North American Journal of Fisheries Management* 10:388–96.

Risser, P. G., Birney, E. C., Blocker, H. D., May, S. W., Parton, W. J., and Wiens, J. A. 1981. *The true prairie ecosystem*. Stroudsburg, PA: Hutchinson Ross.

Roback, S. S. 1957. *The immature tendipedids of the Philadelphia area*. Academy of Natural Sciences of Philadelphia Monograph 9.

Robbins, C. S., Bruun, B., and Zim, H. S. 1966. *Birds of North America*. New York: Golden Press.

Robison, H. W. 1986. Zoogeographic implications of the Mississippi River Basin. In *The zoogeography of North American freshwater fishes*, ed. C. H. Hocutt and E. O. Wiley, 267–85. New York: John Wiley & Sons.

Rocky Mountain Elk Foundation. 1989. Wapiti across the west. *Bugle* 6:138–40.

Rogers, L. L. 2002. *Black bear facts*. http://www.bear.org/Black/black_bear_Facts.html. (accessed March 15, 2007).

Root, T. L, and McDaniel, L. 1995. Winter population trends of selected songbirds. In *Our living resources*. Washington, DC: U.S. Department of the Interior, National Biological Service.

Rossman, A. Y. 1995. Microfungi: Molds, mildews, rusts, and smuts. In *Our living resources*. Washington, DC: U.S. Department of the Interior, National Biological Service.

Rusch, D. H., Malecki, R. E., and Trost, R. 1995. Canada geese of North America. In *Our living resources*. Washington, DC: U.S. Department of the Interior, National Biological Service.

Sargeant, A. B., and Raveling, D. G. 1992. Mortality during the breeding season. In *Ecology and management of breeding waterfowl*, ed. B. J. Batt, A. D., Afton, M. G. Anderson, C. D. Ankney, D. H. Johnson, J. A. Kadlec, and G. I. Krapu, 296–422. Minneapolis: University of Minnesota Press.

Sauer, C. 1950. Grassland climax, fire and management. *Journal of Range Management* 3:16–20.

Sault Ste. Marie Horticultural Society (SSMHS). 2007. *Woolly bear caterpillars.* www.backyard wildlifehabitat.info/index.htm (accessed April 17, 2007).

Schloesser, D. W., and Nalepa, R. F. 1995. Freshwater mussels in the Lake Huron-Lake Erie corridor. In *Our living resources.* Washington, DC: U.S. Department of the Interior, National Biological Service.

Scott, B. 2001. *Macroinvertebrates.* http://www.wavcc.org/wvc/cadre/waterQuality/Index.html (accessed April 19, 2007).

Scott, W. B., and Crossman, E. J. 1973. *Freshwater fishes of Canada.* Fisheries Research Board of Canada Bulletin 184.

Sealy, S. G., ed. 1990. *Auks at sea: Proceedings of an International symposium of the Pacific Seabird Group.* Studies in Avian Biology 14.

Serie, J. 1993. *Waterfowl harvest and population survey data.* Laurel, MD: Office of Migratory Bird Management, U.S. Fish & Wildlife Service.

Sharp, J. 1988. Politics, prairie dogs, and the sportsman. In *Eight Great Plains wildlife damage control workshop proceedings,* ed. D. W. Uresk and G. Schenbeck, 117–18. U.S. Forest Service General Technical Report RM-154.

Smith, R. I. 1968. The social aspects of reproductive behavior in the pintail. *Auk* 85:381–96.

Smith, R. I. 1970. Response of pintail breeding populations to drought. *Journal of Wildlife Management* 34:943–46.

Sogge, M. K. 1995. Southwestern willow flycatchers in the Grand Canyon. In *Our living resources.* Washington, DC: U.S. Department of the Interior, National Biological Service.

Soule, M. E. ed. 1987. *Conservation biology: The science of scarcity and diversity.* Sunderland, MA: Sinauer.

Sowls, A. L., DeGange, A. R., Nelson, J. W., and Lester, G. S. 1980. *Catalog of California seabird colonies.* U.S. Fish and Wildlife Service FWS/OBS 37/80.

Sowls, A. L., Hatch, S. A., and Lensink, C. J. 1978. *Catalog of Alaskan seabird colonies.* U.S. Fish and Wildlife Service FWS/OBS-7878.

Speich, S. M, and Wahl, T. R. 1989. *Catalog of Washington seabird colonies.* U.S. Fish and Wildlife Service Biological Rep 88(6).

Spellman, F. R. 1996. *Stream ecology and self-purification: An introduction for wastewater and water specialists.* Lancaster, PA: Technomic.

Spellman, F. R. 2003. *Handbook of water/wastewater treatment plant operations.* Boca Raton, FL: CRC Press.

Stalmaker, C. B., and Holden, P. B. 1973. Changes in native fish distribution in the Green River system, Utah-Colorado. *Utah Academy of Science, Arts, Letters Proceedings* 50:25–32.

Starnes, W. C. 1995. Colorado River Basin fishes. In *Our living resources.* Washington, DC: U.S. Department of the Interior, National Biological Service.

Starnes, W. C., and Etnier, D. A. 1986. Drainage evolution and fish biogeography of the Tennessee and Cumberland rivers drainage realm. In *The zoogeography of North American freshwater fishes,* ed. C. H. Hocutt and E. O. Wiley, 325–61. New York: Wiley & Sons.

Steeg, B. V., and Warner, R. E. 1995. American badgers in Illinois. In *Our living resources.* Washington, DC: U.S. Department of the Interior, National Biological Service.

Stevens, W. 2004. Frogs eat butterflies. Snakes eat frogs. Hogs eat snakes. Men eat hogs. *Poetry X,* ed. J. Dempsey. http://poetry.poertyx.com/Poems/5301/ (accessed March 24, 2007).

Stockley, C. 1981. *Columbia River sturgeon.* Olympia: Washington Department of Fisheries Progress Report 150.

Storm, G. L., and Palmer, G. L. 1995. White-tailed deer in the Northeast. In *Our living resources.* Washington, DC: U.S. Department of the Interior, National Biological Service.

Swengel, A. B. 1990. Monitoring butterfly populations using the 4th of July butterfly count. *American Midland Naturalist* 124:395–406.

Swengel, A. B. 1993. Permutations of painted ladies. *American Butterflies* 1, no. 2:34.

Swengel, A. B. 1995. Fourth of July butterfly count. In *Our living resources.* Washington, DC: U.S. Department of the Interior, National Biological Service.

Swengel, A. B., and Swengel, S. R. 1995. The tall-grass prairie butterfly community. In *Our living resources.* Washington, DC: U.S. Department of the Interior, National Biological Service.

Taylor, J. N., Courtenay, Jr., W. R., and McCann, J. A. 1984. Known impacts of exotic fishes in the continental United States. In *Distribution, biology and management of exotic fishes,* ed. W. R. Courtenay, Jr. and J. R. Stauffer, 322–73. Baltimore, MD: Johns Hopkins University Press.

Templeton, A. R. 1987. Coadaptation and outbreeding depression. In *Conservation biology: The science of scarcity and diversity,* ed. M. E. Soule, 105–16. Sunderland, MA: Sinauer.

Texas Parks and Wildlife Department (TPWD). 2007a. *Red-cockaded woodpecker.* www.tpwd .state.tx.us/huntwild/wild/species/rcw (accessed March 11, 2007).

Texas Parks and Wildlife Department (TPWD). 2007b. *Stocking public waters.* http://www.tp-wd.state.tx.us/fishboat/fish/management/stocking/ (accessed March 29, 2007).

Tomlinson, R. E., Dolton, D. D., Reeves, H. M., Nichols, J. D., and McKibben, L. A. 1988. *Migration, harvest, and population characteristics of mourning doves banded in the Western Management Unit,* 164–77. U.S. Fish and Wildlife Service Technical Report 13.

Tomlinson, R. E., and Dunks, J. H. 1993. Population characteristics and trends in the Central Management Unit. In *Ecology and management of the mourning dove,* ed. T. S. Baskett, M. W. Sayre, R. E. Tomlinson, and R. E. Mirarchi, 305–40. Harrisburg, PA: Stackpole Books.

Tracy, C. A. 1993. *Status of white sturgeon resources in the main stem: Columbia River.* Final Report, Dingell/Johnson-Wallop/Breaux Project F-77-R, Washington Department of Fisheries.

Tuggle, B. N. 1995. Mammals. In *Our living resources.* Washington, DC: U.S. Department of the Interior, National Biological Services.

Turgeon, D. D., Bogan, A. E., Coan, F. V., Emerson, W. K., Lyons, W. G., Pratt, W. L. Roper, C. F. E., Scheltema, A., Thompson, F. G., and Williams, J. D. 1988. *Common and scientific names of aquatic invertebrates from the United States and Canada: Mollusks.* American Fisheries Society Special Publication 16.

U.S. Congress, Office of Technology Assessment. 1993. *Harmful non-indigenous species in the United States.* Washington, DC: U.S. Government Printing Office OTA-F-565.

U.S. Department of the Interior, National Biological Service. 1995a. *Our living environment.* Washington, DC: Author.

U.S. Department of the Interior, National Biological Service. 1995b. *Our living resources: Birds.* http://biology.U.S. Department of the Interior, National Biological Service.gov/t + s / index (accessed March 6, 2007).

U.S. Environmental Protection Agency (USEPA). 1986. *Superfund public health evaluation manual.* Washington, DC: Office of Emergency and Remedial Response.

U.S. Fish and Wildlife Service (USFWS). 1982. *Mexican wolf recovery plan.* Albuquerque, NM: Author.

U.S. Fish and Wildlife Service (USFWS). 1987. *Northern Rocky Mountain wolf recovery plan.* Denver, CO: Author.

U.S. Fish and Wildlife Service (USFWS). 1992a. *Alaska seabird management plan.* Anchorage, AK: Division of Migratory Birds.

U.S. Fish and Wildlife Service (USFWS). 1992b. *Recovery plan for the eastern timber wolf.* Twin Cities, MN: Author.

U.S. Fish and Wildlife Service (USFWS). 1994. *Desert tortoise (Mojave population) Recovery Plan.* Portland, OR: U.S. Fish and Wildlife Service.

U.S. Fish and Wildlife Service (USFWS). 1998. Gray wolf. http://ww.fws.gov/species/ species_accounts/ bio_gwol.html (accessed March 14, 2007).

U.S. Fish and Wildlife Service (USFWS). 2002. Colonial-nesting waterbirds. http://www.fws.-gov/birds/waterbirds (accessed April 15, 2007).

U.S. Fish and Wildlife Service (USFWS). 2004. USFWS. Southwestern willow flycatcher. http://www.fws.gov (accessed March 11, 2007).

U.S. Fish and Wildlife Service (USFWS). 2007a. *All about piping plovers*. http://www.fws.gov/ plover /facts.html (accessed March 11, 2007).

U.S. Fish and Wildlife Service (USFWS). 2007b. *Discover freshwater mussels: America's Hidden Treasure*. http://www.fws.gov.news/mussels.html (accessed April 11, 2007).

U.S. Fish and Wildlife Service (USFWS). 2007c. *Pintails! What are they?* http://www.fws.gov (accessed April 15, 2007).

U.S. Fish and Wildlife Service (USFWS). 2007d. *Seabirds*. http://alaska.fws.gov/mbsp/mbm/ seabirds/seabirds.htm (accessed March 9, 2007).

U.S. Fish and Wildlife Service (USFWS). 2007e. *Tarahumara frog*. www.fws.gov/southwest/ es.arizona/T_Frog_SpeciesAccount.htm (accessed March 26, 2007)

U.S. Fish and Wildlife Service (USFWS) and Canadian Wildlife Service. 1986. *North American Waterfowl Management Plan*. Washington, DC: U.S. Fish and Wildlife Service.

U.S. Fish and Wildlife Service (USFWS) and Canadian Wildlife Service. 1994. *North American Waterfowl Management Plan 1994 update, expanding the commitment*. Washington, DC: U.S. Fish and Wildlife Service.

Vaughan, M. R., and Pelton, M. R. 1995. Black bears in North America. In *Our living resources*. Washington, DC: U.S. Department of the Interior, National Biological Service.

Vermeer, K., Briggs, K. T., Morgan, K. H., and Siegel-Causey, E., eds. 1993. *The status, ecology, and conservation of marine birds of the North Pacific*. Proceedings of a Pacific Seabird group Symposium. Canadian Wildlife Service Special Publication, Ottawa, Ontario.

Vogl, R. J. 1974. Effect of fire on grasslands. In *Fire and ecosystems*, ed. T. T. Kozlowski and C. E. Ahlgren, 139–94. New York: Academic Press.

Wake, D. B. 1991. Declining amphibian populations. *Science* 253, no. 5022:860.

Wake, D. B., and Morowitz, H. H. 1991. Declining amphibian populations—a global phenomenon? Findings and recommendations. *Alytes* 9, no. 1:33–42.

Walton, I. 1652. In *Anna Comstock*. 1986. Handbook of nature study. Ithaca, New York: Cornell University Press.

Warren, M., and Burr, H. M. 1994. Status of freshwater fishes of the United States: Overview of an imperiled fauna. *Fisheries* 19, no. 1:6–18.

Webb, D. W. 1995. Biodiversity degradation in Illinois stoneflies. In *Our living resources*. Washington, DC: U.S. Department of the Interior, National Biological Service.

Welch, D. W., and Beamesderfer, R. C. 1993. Maturation of female white sturgeon in lower Columbia River impoundments. In *Status and habitat requirements of the white sturgeon populations in the Columbia River downstream form McNary Dam*, vol. 2, ed. R. C. Beamesderfer and A. A. Nigro, 89–108. Final report. Portland, OR: Bonneville Power Administration.

Wheeler, Q. D. 1990. Insect diversity and cladistic constraints. *Annals of the Entomological Society of America* 83, no. 6:1031–47.

Wilbur, S. R. 1978. *The California condor, 1966–76: A look at its past and future*. U.S. Fish and Wildlife Service North American Fauna 72.

Williams, E. S., Thorne, E. T., Appel, M. J. G., and Belitsky, D. W. 1988. Canine distemper in black-footed ferrets from Wyoming. *Journal of Wildlife Diseases* 24:385–98.

Williams, J. E., Johnson, J. E., Hendrickson, D. A., Contreras-Balderas, S., Williams, J. D., Navarro-Mendoza, M., McAllister, D. E., and Deacon, J. E. 1989. Fishes of North America endangered, threatened, or of special concern: 1989. *Fisheries* 14, no. 6:2–20.

Williams, J. D., and Neves, R. J. 1995. Freshwater mussels: A neglected and declining aquatic resource. In *Our living resources*. Washington, DC: U.S. Department of the Interior, National Biological Service.

Williams, J. D., Warren, M. L., Jr., Cummings, K. S., Harris, J. L., and Neves, R. J. 1993. Conversation status of freshwater mussels of the United States ad Canada. *Fisheries* 18, no. 9:6–22.

Wilson, D. E., Bogan, M. A., Brownell, R. L., Jr., Burdin, A. M., and Maminov, M. K. 1991. Geographic variation in sea otters: *Enhydra luris*. *Journal of Mammalogy* 2, no. 1:22–26.

Woodward, A.R., and Moore, C.T. 1995. American alligators in Florida. In *Our living resources*. Washington, DC: U.S. Department of the Interior, National Biological Service.

Wooten, A. 1984. *Insects of the world*. New York: Facts on File.

Wright, S. 1931. Evolution in Mendelian populations. *Genetics* 16:97–159.

Wright, S. 1978. *Evolution and the genetics of populations*. Vol. 4, Variability within and among natural populations. Chicago: University of Chicago Press.

Sawyer Glacier, Alaska.
Photograph by Frank R. Spellman

CHAPTER 7

Terrestrial Ecosystems

The more fundamental conception is . . . the whole system (in the sense of physics) including not only the organism-complex, but also the whole complex of physical factors forming what we call the environment. . . . We cannot separate (the organisms) from their special environment with which they form one physical system. . . . It is the system so forced which (provides) the basic units of nature on the face of the earth. . . . These ecosystems, as we may call them, are of the most various kinds and sizes.

—A. G. Tansley (1935)

Topics

Note: Portions of this chapter are adapted from or based on information from U.S. Geological Survey's (USGS) *Our Living Environment* (1995).

Boyd (1995) explains that terrestrial ecosystems include a rich variety of types and cover a range extending from nearly aquatic wetlands along our coasts and myriad rivers, lakes, and streams to mountain tops and arid, desert locations. Smith and Smith (2006) point out that terrestrial ecosystems "can be classified into broad categories called *biomes*." Biomes are defined as "the world's major communities, classified according to the predominant vegetation and characterized by adaptations of organisms to that particular environment" (Campbell 1996). Biomes generally include tropical forests, tropical savannas, deserts, shrubland, temperate forests, temperate grasslands, conifer forests, and tundra. The diversity of these ecosystems offers both challenge and opportunity. The challenge stems from the sheer number of potential ecosystems to be studied. Grossman and Goodin (1995) point out that there are a total of somewhere between 2,500 and 3,500 individual terrestrial ecosystem (community) types. Obviously, a single chapter cannot hope to address more than a few of these many terrestrial ecosystems and communities. With this in mind, the common thread in this chapter, as in all chapters in this book, is that if unchecked, human activities will continue to result in an upset balance of species interactions, alteration of ecosystems, and extensive habitat loss (Boyd 1995).

Discussions of biological diversity have traditionally revolved around the protection of individual species. More recently, because large numbers of natural ecosystems are now in danger, we have begun to realize that the protection of community or ecosystem diversity is equally important. Patchwork conversions of natural landscapes for agriculture, silviculture, and development result in a fragmentation that leaves small remnant areas of natural ecosystems (Burgess and Sharpe 1981). As these natural patches become smaller and more isolated, their ability to maintain healthy populations of many plant and animal species is reduced (Harris 1984). As individual species are lost from each fragment, the community changes and both species and ecosystem diversity are reduced.

In this chapter, one of the imperiled ecosystems selected and discussed deals with the whitebark pine ecosystem of the western mountains. This ecosystem is endangered because of the combined effects of introduced disease and fire suppression (Kendall 1995). The effects of introduced diseases on natural species and ecosystems have been well documented. Several species, such as the American chestnut, have been virtually eliminated and other species have been greatly reduced by introduced diseases. The effects on ecosystems where these species were previously found have been dramatic (Shugart and West 1977).

Another factor that has played a major role in the reshaping of natural ecosystems is the alteration of natural fire regimes. In many systems, a reduction in fire frequency can lead to invasion by fire-intolerant species and eventual loss of the original ecosystem, demonstrated in this chapter by Henderson and Epstein (1995) in their discussion of how fire suppression and other factors caused tremendous losses of oak savannas throughout the Midwest.

In other systems, an increase in fire frequency can also lead to changes in ecosystem structure and function. Although we now realize that fire is a natural and necessary part of many ecosystems, it was not until after the devastating fires of Yellowstone National Park that the general public was alerted to the benefits of such fires (Elfring 1989). An effective fire-suppression program can allow accumulation of vast amounts of detritus (dead organic material such as leaves, branches, and stems). If this material is not consumed periodically by small fires burning along the forest floor, it will accumulate to the point of providing raw materials for an exceptionally intense fire that can burn tree crowns and destroy the existing forest (Boyd 1995). Ferry et al. (1995) discuss four fire-adapted ecosystems that have been affected by modified fire regimes and conclude, "Managers must balance the suppression program with a program of prescribed fire applied on a landscape scale if we are to meet our stewardship responsibilities."

In addition to the effects fire and disease have on our natural resources, numerous variables also have an effect. These variables include pollution (Peterson 1995; Nash, Tonnessen, and Flores 1995), conversions to other uses, harvesting activities such as logging, and global climate change. In this chapter, Cole (1995b) demonstrates that over the past 5,000 years change has been a natural part of our terrestrial ecosystems. Within a given ecosystem, some species decline in importance while others increase over time, resulting in a change in the overall character of the ecosystem. A key feature to stand out in the 5,000-year chronology developed by Cole is that the current rates of change are about ten times higher than presettlement rates. Human intervention in one form or another is now the principal agent of change. Another feature of this chapter, provided by Keeland, Allen, and Burkett (1995), is a review of U.S. Forest Service data; changes brought about by forestry management practices are also discussed. Moreover, at a reduced spatial scale, this review also discusses changes within the forested wetlands of the southeastern United States. Forested wetlands have been especially reduced and fragmented as a result of land-use conversions, predominately to agricultural activities (Boyd 1995).

U.S. Forest Resources

According to the U.S. Department of Agriculture (USDA), it is estimated that in 1630 the area of forest land in the United States was just over 1 billion acres or about 46 percent of the total land area (2004). By 1907, the area of forest land had declined to an estimated 759 million acres or 34 percent of the total land area. Forest area has been relatively stable since 1907. In 2002, forest land comprised 749 million acres, or 33 percent of the total land area of the United States. Since 1630, there has been a net loss of 297 million acres of forest land, predominantly due to agricultural conversions. Nearly two-thirds of the net conversion to other uses occurred in the last half of the 19th century when an average of 13 square miles of forest was cleared every day for 50 years.

Darr (1995) points out that the secretary of agriculture is directed by law to make and keep current a comprehensive inventory and analysis of the present and prospective conditions of and requirements for the renewable resources of U.S. forests and rangelands. This inventory includes all forests and rangelands, regardless of ownership. The work is carried out by people in the Forest Inventory and Analysis program of the U.S. Department of Agriculture Forest Service (USFS).

> ✔ *Important Point*: Inventories provide key forest resource information for planners and policymakers. Increasingly, people turn to these inventories for information on biological diversity, forest health, and developmental decisions.

According to the USFS (1992), information is collected from over 130,000 permanent sample plots selected to ensure statistical reliability. Vegetation on the plots is measured on average about every ten years. Characteristics of the vegetation and land are measured, including ownership, productivity for timber production, the kinds and sizes of trees, how fast trees are growing, whether any trees have died from natural causes, and whether they have been cut.

It is important to note that stable forest area does not mean that there has been no change in the character of the forest. There have been shifts from agriculture to forests and vice versa. Some forest land has been converted to more intensive uses, such as urban. Even where land has remained in forest use, there have been changes as forests respond to human manipulation, aging, and other natural processes (USDA 2004).

As mentioned, over the years, the U.S. forest cover has changed because of the way people use and manage forest land. Today, about 33 percent of the U.S. land area, or 298 million hectares (737 million acres), is forest land, about two-thirds of the forested area in 1600. Since 1600, some 124 million hectares (307 million acres) of forest land have been converted to other uses, mainly agricultural. More than 75 percent of this conversion occurred in the 19th century, but by 1920, clearing forests for agriculture had largely halted (Darr 1995).

Some 34 percent of all forest land is federally owned and managed by the U.S. Forest Service, the Bureau of Land Management, and other federal agencies. The rest is owned by nonfederal public agencies, forest industry, farmers, and other private individuals. About 19 million hectares (47 million acres; 6 percent of all U.S. forest land) are reserved from commercial timber harvest in wilderness, parks, and other land classifications.

Forest land is widely but unevenly distributed. North Dakota has the smallest percentage of forest cover (1 percent) and Maine has the greatest (8 percent). Forest areas vary greatly from sparse scrub forests of the arid interior west to the highly productive forests of the Pacific Coast and the South, and from pure hardwood forests to multispecies mixtures and coniferous forests. In total, 57 percent of the forest land is east of the Great Plains states. In the East, the oak-hickory forest type group is the most common. Figure 7.1 shows a portion of the Blue Ridge Parkway in the Shenandoah National Forest, Virginia. Shenandoah National Forest is home to a mix of plant life, from algae to oak trees. The northern Blue Ridge Mountains have about 1,600 different species of higher plants. Fewer than one hundred of these are trees and shrubs that make up the dominant vegetation that is visible year round. The Blue Ridge region provides the requirements of deciduous trees such as oaks, hickories, and maples. Much of Shenandoah today is an oak-hickory forest. The forest would be incomplete, however, without rose azalea, jack-in-the-pulpit, interrupted fern, lady slipper orchid, and British soldiers lichen (*Shenandoah National Park* 2002). In the West, the category referred to as "other softwoods" is most common (Darr 1995).

Timberland forests are logged for timber, plywood, and paper products. This timberland is generally the most productive and capable of producing at least 1.4 cubic meters of industrial wood per hectare a year (20 cubic feet per acre) and is not reserved for timber harvest (Powell, Faulkner, Darr, Zhu, and MacCleery 1993). Two-thirds of the nation's forested ecosystems (198 million hectares or 490

Figure 7.1.   Blue Ridge Mountains, Shenandoah National Park, Virginia
Photograph by Frank R. Spellman

million acres) are classed as timberland. Because of historical interest in timber production, more information is available for the characteristics of timber inventories on timberland than for other forest land.

USDA (2004) points out that U.S. forests provide wildlife habitat and thereby support biodiversity, take carbon out of the air and thus serve as carbon sinks, and provide the outdoor environments desired by many people for recreation. Moreover, gathering nontimber forest products is a significant use of the nation's forests that affects forest ecosystem. These products include medicinals, food and forage species, floral and horticultural species, resins and oils, art and craft species (leaves and branches), game animals, and fur bearers. Harvest of these products from forest ecosystems is a significant and very important activity for many Americans, for recreational, commercial, subsistence, and cultural uses.

> Maybe a vision of the original longleaf pine flatwoods has been endowed to me through genes, because I seem to remember their endlessness. I seem to recollect when these coastal plains were one big, brown-and-tan, daybreak-to-dark longleaf forest. It was a monotony one learned to love . . . with the passing years. A forest never tells its secrets but reveals them slowly over time—and a longleaf forest is full of secrets.
>
> —Janisse Ray, *Ecology of a Cracker Childhood*, 1999

Forested Wetlands

If the two words "forested wetlands" are separated, they are somewhat distinct; once separate, we tend to define each in stand-alone fashion. On the surface, "forested" is rather easy to comprehend, and to define and to accept without explanation: a place with a lot of trees. "Wetland," however, is a matter (a place or region) of a much different sort. The term *wetland* conjures up many internal manifestations of descriptive thoughts (mostly not pleasant): Our human microchip warehouse displays for us, in regard to wetlands, a scene of mosquito- and yellow fly–infested dreariness. A soggy piece of ground too deep for a human to wade in, too shallow for a boat to draw, too tangled for passage (see figure 7.2). As Ray (2005) puts it, "a wetland is simply a natural feature full of natural features." The problem is most of us have difficulty recognizing, understanding, and/or defining natural features.

European settlers encountered these "natural features" in many parts of the southern United States where the landscape is largely comprised of forested wetlands. These wetlands were a major feature of river floodplains and isolated depressions or basins or pocosins from Virginia to Florida, west to eastern Texas and Oklahoma, and along the Mississippi River to southern Illinois. Based on the accounts of pre-20th-centrury naturalists such as Audubon, Banister, John and Wil-

Figure 7.2.   First Landing State Park, Virginia Beach, Virginia
Photograph by Frank R. Spellman

liam Bartram, Brickell, and Darby, the flora and fauna of many wetlands were unusually rich even by precolonial standards (Wright and Wright 1932). These early travelers described vast unbroken forests of oaks, ashes, maples, and other tree species, many with an almost impassable understory of saplings, shrubs, vines, switch cane, and palmetto. Low swampy areas with deep, long-term flooding were dominated by bald cypress and tupelo and typically had sparse understories (Keeland et al. 1995).

Most southern forested wetlands fall in the broad category of bottomland hardwoods, characterized and maintained by a natural hydrological regime of alternating annual wet and dry periods and soils that are saturated or inundated during a portion of the growing season. Variations in elevation, hydroperiod, and soils result in a mosaic of plant communities across a floodplain. Wharton, Kitchens, and Sipe (1982) classified bottomland hardwoods into 75 community types, including forested wetland types such as Atlantic white cedar bogs, red maple and cypress-tupelo swamps, pocosins, hydric hammocks, and Carolina bays.

✔ *Interesting Point*: Albemarle/Pamlico coastal region in North Carolina is home to an Atlantic white cedar bog region. This is a unique habitat that has naturally acidic waters and is cooler than surrounding hardwood swamps or pinelands. Cedar bogs support large breeding bird populations (U.S. Fish and Wildlife Service [USFWS] 1998).

✔ *Interesting Point*: A pocosin, or swamp on a hill, is a seemingly flat area that rises slightly in the center, forming a raised bog (U.S. Army Corps of Engineers 1998).

✔ *Interesting Point*: Hydric hammocks are forested wetlands (swamps) that are dominated by a mixture of primarily hardwood tree species with sabal palms. Hydric hammocks are typically found in areas where limestone is close to the soil surface. Soils are variable; often a clay layer or limestone layer helps keep the soil saturated for long periods (Spellman 1998).

Realistic estimates of the original extent of forested wetlands are not available because accurate records of wetlands were not maintained until the early 20th century, and many accounts of wetland size were little more than speculation (Dahl 1990). Klopatek, Olson, Emerson, and Jones (1979) estimated the precolonial forested wetland area of the United States to be about 27.2 million hectares (67.2 million acres), but Abernathy and Turner (1987) suggested that this figure was low because it ignored small isolated wetlands.

Fire Regimes within Fire-Adapted Ecosystems

Fires ignited by people or through natural causes have interacted over evolutionary time with ecosystems, exerting a significant influence on numerous ecosystem functions (Pyne 1982). Fire recycles nutrients, reduces biomass, influences insect and disease populations, and is the principal change agent affecting vegetative structure, composition, and biological diversity. As humans alter fire frequency and intensity, many plant and animal communities experience a loss of species diversity, site degradation, and increases in size and severity of wildfires (Ferry et al. 1995). This section examines the role fire plays in the ecological process around which most North America ecosystems evolved.

In order to understand the influence fire can have on an ecosystem, it is important to know the basics of fire regime. Fire regimes are considered as the total pattern of fires over time that is characteristic of a region or ecosystem (Kilgore and Heinselman 1990). A fire regime describes the pattern that fire follows in a particular ecosystem. It consists of the following components (Bond and Keeley 2005):

1. *Fuel consumption and spread patterns.* Fire can burn at three levels. Ground fires burn through soil that is rich inorganic matter. Surface fires burn through dead plant materials that are on the ground. Crown fires burn in the tops of shrubs and trees. An ecosystem may experience mostly one level of fire or a mix of the three.
2. *Intensity.* Intensity is defined as the energy release per unit length of fireline. It can be estimated as the product of linear spread rate, low heat of combustion, and combusted fuel mass per unit area.
3. *Severity.* Severity is a term used by ecologists to refer to the impact that a fire has on an ecosystem (estimate of plant mortality).
4. *Frequency.* This is a measure of how common fires are in a given ecosystem (the interval between fires at a given site, or the amount of time it takes to burn an equivalent of a specified area).
5. *Seasonality.* This refers to the time of year during which fires are most common.

In the following, five plant communities are discussed: the sagebrush steppe, juniper woodlands, ponderosa pine forest, lodgepole pine forest, and southern pineland. Status and trends of altered fire regimes in fire-adapted ecosystems highlight the role that fire plays in wildland stewardship. Fire regimes are considered as the total pattern of fires over time that is characteristic of a region or ecosystem (Kilgore and Heinselman 1990).

1. *Sagebrush-grass plant communities.* Ferry et al. (1995) point out that a greater frequency of fire has seriously affected the sagebrush steppe during the past 50 years. One such community, the semiarid intermountain sagebrush steppe, encompasses about 45 million hectares (112 million acres). After repeated fires, nonnative European annual grasses such as cheat-grass and medusa-head now dominate the sagebrush steppe (West and Hassan 1985). It is unclear whether cheat-grass invasion, heavy grazing pressure, or shorter fire return intervals initiated the replacement of perennial grasses and shrubs by the non-native annual grasses. It is clear, however, that wildfires aid in replacing native grasses with

cheat-grass, as well as causing the loss of the native shrub component (Whisenant 1990). Inventories show that cheat-grass is dominant on about 6.8 million hectares (17 million acres) of the sagebrush steppe and that it could expand into an additional 25 million hectares (62 million acres) in the sagebrush steppe and the Great Basin sagebrush type (Pellant and Hall 1994).

2. *Western juniper woodlands.* Juniper woodlands occupy 17 million hectares (42 million acres) in the intermountain region (West 1988). Juniper species common to this region are western juniper, Utah juniper, single-seeded juniper, and Rocky Mountain juniper. Presettlement juniper woodlands were usually savanna-like or confined to rocky outcrops not typically susceptible to fire (Nichol 1937). Juniper woodlands began increasing in both density and distribution in the late 1800s (R. F. Miller; unpublished data) because of climate, grazing, and lack of fire (Miller and Waigand 1994). Warm and wet climate conditions then were ideal for juniper and grass seed production. Fire frequency had decreased because the grazing of domestic livestock had greatly reduced the grasses and shrubs that provided fuel, and the relocation of Native Americans eliminated an important source of ignition. Continued grazing and 50 years of attempted fire exclusion have allowed juniper expansion to go unchecked.

3. *Ponderosa pine forest.* Decreases in fire frequency are also seriously affecting the ponderosa pine forest, a common component on about 16 million hectares (40 million acres) in the western United States. Historically, the ponderosa pine ecosystem had frequent, low-intensity, surface fires that perpetuated parklike stands with grassy undergrowth (Barrett 1980). Ponderosa pine is a tree that is well adapted to fire. As ponderosa pines get older, their bark gets thicker. Thick bark protects the cambium layer from being burned. They generally have fewer branches low on their trunks so fire can't burn into the tops of the trees. The trees are spaced widely so one tree can't catch another on fire, and have a deep taproot (Boise National Forest [BNF] 2004). For six decades, humans attempted to exclude fire on these sites (Office of Technology Assessment [OTA] 1993). Fifty years ago, Weaver (1943) stated that complete prevention of forest fires in the ponderosa pine region had undesirable ecological effects and that already-deplorable conditions were becoming increasingly serious. Today, many ponderosa pine forests are overstocked, plagued by epidemics of insects and disease, and subject to severe stand-destroying fires (Mutch et al. 1993).

4. *Lodgepole pine forest.* Like ponderosa pine forests, lodgepole pine forests are experiencing a change in the structure, distribution, and functioning of natural processes because of fire exclusion and increases in disease. Wildfire may be the most important factor responsible for the establishment of existing stands (Wellner 1970). Historical stand-age distributions in lodgepole pine forests indicated an abundance of younger age classes resulting from periodic fires. Fire exclusion, by precluding the initiation of new stands, is responsible for a marked change in distribution of age classes in these forests. Dwarf mistletoe, the primary disease of lodgepole pine, also has a profound effect on forest struc-

ture and function, although it occurs slowly. Data show that chronic increases of dwarf mistletoe are partly due to the exclusion of fire (Zimmerman and Laven 1984) because fire is the natural control of dwarf mistletoe and has played a major role in the distribution and abundance of current populations and infection intensities (Alexander and Hawksworth 1975). As the frequency and extent of fires have decreased in lodgepole pine stands over the past 200 years, dwarf mistletoe infection intensity and distribution are clearly increasing (Zimmerman and Laven 1984).

5. *Southern pinelands.* In contrast to the juniper, ponderosa pine, and lodgepole pine communities, fire frequencies have not drastically decreased in the 78 million hectares (193 million acres) of southern pinelands. These pinelands are composed of diverse plant communities associated with longleaf, slash, loblolly, and shortleaf pins. Fire has continued on an altered basis as an ecological process in much of the southern pinelands; historically, fire burned 10 to 30 percent of the forest annually (Wright and Bailey 1982); the southern culture never effectively excluded fire for its pinelands (Pyne 1982), although human-ignited fires have partially replace natural fires. Consequently, the amount of fire has been reduced and the season of burns has changed from predominately growing-season to dormant-season (fall or winter) fires (Robbins and Myers 1992). Altering the burning season and frequency has significantly affected southern pineland community structure, composition, and biological diversity.

The role of fire becomes more complex as it interacts with land management. Maintaining interactions between disturbance processes and ecosystem functions is emphasized in ecosystem management. It is vital for mangers to recognize how society influences fire as an ecological process. In addition, managers must uniformly use information on fire history and fire effects to sustain the health of ecosystems that are both fire-adapted and fire-dependent. Managers must balance the suppression program with a program of prescribed fire applied on a landscape scale if we are to meet our stewardship responsibilities (Ferry et al. 1995).

Vegetation Change in National Parks

Natural ecosystems are always changing, but recent changes in the United States have been startlingly rapid, driven by 200 years of disturbances accompanying settlement by an industrialized society. Logging, grazing, land clearing, increased or decreased frequency of fire, hunting of predators, and other changes have affected even the most remote corners of the continent. Recent trends can be better understood by comparison with more natural past trends of change, which can be reconstructed from fossil records. Conditions before widespread impacts in a region are termed "presettlement"; conditions after the impacts are "postsettlement" (Cole 1995b).

Fossil plant materials from the past few thousand years are used to study past changes in many natural areas. Pollen buried in wetlands, for example, can reveal past changes in vegetation (Faegri and Iverson 1989), and larger fossil plant parts can be studied in deserts where the fossilized plant collections of packrats, called packrat middens, have been preserved (Betancourt, Van Devender, and Martin 1990).

This section summarizes the rates of vegetation change in four national park areas over the past 5,000 years as reconstructed from fossil pollen and packrat middens. These four national park areas from different ecological regions demonstrate the flexibility of these paleoecological techniques and display similar results (Cole 1995b).

1. *Northern Indiana prairie.* A 4,500-year history of vegetation change was collected from Howes Prairie Marsh, a small marsh surrounded by prairie and oak savanna in the Indiana Dunes National Lakeshore near the southern tip of Lake Michigan. Only 40 kilometers (25 miles) from Chicago, this area has been affected by numerous impacts from settlements but still supports comparably pristine tall-grass prairie vegetation as well as the endangered Karner blue butterfly (an endangered species native to the Great Lakes region of the United States). Although this site has experienced more disturbances than any of the others described here, it is a most valuable site because of its many species (Wilhelm 1990) and its tall-grass prairie vegetation that has been nearly eliminated elsewhere.

2. *Northern Michigan forest.* A similar analysis was carried out on pollen from a small bog (12-Mile Bog) surrounded by pine forest along the southern shore of Lake Superior. This site, within Pictured Rocks National Lakeshore, was more severely affected by logging and slash burning in the 1890s than by the periodic wildfires that characterized this forest earlier, but it has been protected for the past 80 years. The magnitude of change caused by the crude logging and slash burning of the logging era was far greater than any recorded during the 2,500 years since Lake Superior receded to create the forest of white and red pine (Cole 1995b).

3. *California coastal sage scrub.* Fossil pollen was analyzed from an estuary on Santa Rosa Island off the coast of southern California (Cole and Liu 1994). The semiarid landscape around the estuary is covered with coastal sage scrub, chaparral, and grassland. This site, within Channel Islands National Park, is one of the least-affected areas in this region of rapidly expanding urbanization, although the island's native plants and animals were not well adapted to withstand the grazing of the large animals introduced with the ranching era of the 1800s. This island, which had no native large herbivores, became populated with thousands of sheep, cattle, horses, goats, pigs, deer, and elk. The National Park Service is removing many of the large herbivores, although most of the island remains an active cattle ranch (Cole 1995b).

4. *Southern Utah desert.* Because fossil pollen is usually preserved in accumulating sediments of wetlands, different paleoecological techniques are necessary in arid areas. In western North America, fossil deposits left by packrats have proven a

useful source of paleoecological data (Betancourt et al. 1990). Past desert vegetation can be reconstructed by analyzing bits of leaves, twigs, and seeds collected by these small rodents and incorporated into debris piles in rocks shelters or caves. These debris piles can be collected, analyzed, and radiocarbon-dated (Cole 1995b).

The vegetation history of a remote portion of Capitol Reef National Park (Harnett Draw) was reconstructed through the analysis of eight packrat middens ranging in age from 0 to 5,450 years (Cole 1995a). The vegetation remained fairly stable throughout this period until the past few hundred years. The most recent deposits contain many plants associated with overgrazed areas such as whitebark rabbitbrush, snakeweed, and greasewood, which were not recorded at the site before settlement.

Conversely, other plants that are extremely palatable to grazing animals were present throughout the past 5,450 years, only to disappear since settlement. Plant species preferred by sheep and cattle, such as winterfat and rice grass, disappeared entirely, while many other palatable plant species declined in abundance after 5,000 years of comparative stability.

The past rates of vegetation change for this site were calculated in a manner similar to the fossil pollen records. Although the rate of change calculation is less precise than the fossil pollen records because there were fewer samples, the results show a similar pattern. The rate of vegetation change is highest between the two most recent records (Cole 1995b).

Although this area is still grazed by cattle today through grazing leases to private ranchers from the National Park Service, much of the severe damage was probably done by intensive sheep grazing during the late 1800s when the entire region was negatively affected by open-land sheep ranching. We cannot yet demonstrate whether the grazing effects are continuing or if the site is improving, although reinvasion of palatable species is unlikely in the face of even light grazing. Severe overgrazing is required to eliminate abundant palatable species, but once they are eliminated, even light grazing can prevent their restoration (Cole 1995b).

To successfully meet human needs, competing demands for the use of the land's resources must be resolved. Experience has shown that wise land management decisions are more likely to be made if land managers understand a site and are able to place the status quo into a historical perspective. Because the ultimate goal for the management of many areas is to mitigate settlement impacts and return the land to its presettlement status, detailed knowledge of the effects of settlement is imperative.

Whitebark Ecosystem

The USGS (2003) reports that whitebark pine trees produce large seeds high in fats that are an important source of food for many animal species. Red squirrels and Clark's nutcrackers usually harvest the lion's share of whitebark pine seeds. Grizzly bears and black bears raid squirrel caches that contain whitebark cones to get pine seeds, one of their favorite foods (Kendall and Arno 1990). Other mammals, large and small, and many species of birds also feed on whitebark pine seeds, or pine nuts as they commonly are called. Because whitebark pine are long-lived and can grow large trunks, they provide valuable cavities for nesting squirrels, northern flickers (a type of woodpecker; it is the only ground-feeding woodpecker), and mountain bluebirds.

In addition to its importance as a wildlife resource, whitebark pine trees play other pivotal roles in the ecosystems where they occur. Because whitebark pine can grow in dry, windy, and cold sites where no other trees can establish, it pioneers many harsh subalpine and alpine sites and commonly dominates treeline communities in the northern Rocky Mountains. Its presence can modify the microclimate in alpine ecotone communities enough to allow other trees to establish, such as subalpine fir. Whitebark pine communities also influence the hydrology of the drainages where they occur. Because they grow on windy ridges and in other mountainous areas and have broad crowns, whitebark pine trees tend to act as snow fences and are responsible for significant amounts of high-elevation snow accumulation. This results in delayed snow melt and extends ephemeral stream flow periods (USGS 2003).

Whitebark pine is well-suited to harsh conditions and populates high-elevation forests in the northern Rocky Mountain, North Cascade, and Sierra Nevada ranges. These pine trees are adapted to cold, dry sites and pioneer burns and other disturbed areas. At timberline, they grow under conditions tolerated by no other tree species, thus playing an important role in snow accumulation and persistence. Because few roads occur in whitebark pine ecosystems and because the tree's wood is of little commercial interest, information on the drastic decline of this picturesque tree has only recently emerged (Kendall 1995).

Whitebark pine is threatened by introduced disease and by fire suppression. In its northern range, many whitebark pine stands have declined by more than 90 percent. The most serious threat to the tree is from white pine blister rust, a nonnative fungus that has defied control. Fewer than one whitebark pine tree in 10,000 is rust-resistant. Mortality has been rapid in areas such as western Montana, where 42 percent of whitebark pine trees have died from the disease in the past 20 years; 89 percent of the remaining trees are infected with rust (Keane and Arno 1993). Although drier conditions have slowed the spread of blister rust in whitebark pine's southern range, infection rates there are increasing and large dieoffs are eventually expected to occur.

Before fire suppression, whitebark pine stands burned every 50–300 years. Under current management, they will burn at 3,000-year intervals. Without fire, serial whitebark pine trees are replaced by shade-tolerant conifers and become more vulnerable to insects and disease (Kendall 1995).

The alarming loss of whitebark pine has broad repercussions: mast for wildlife is diminished and the number of animals the habitat can support is reduced. Such results hinder grizzly bear recovery and may be catastrophic to Yellowstone grizzlies for whom pine seeds are a critical food. Predicted changes in whitebark pine communities include the absence of reforestation of harsh sites after disturbance and the lowering of treelines. In addition, stream flow and timing will be altered as snowpack changes with vegetation.

Whitebark pine will be absent as a functional community component until rust-resistant strains evolve. Natural selection could be speeded with a breeding program like that developed for western white pine, which also suffers from rust. In some areas where whitebark pine is regenerating, its competitors should be eliminated. To perpetuate whitebark pine at a landscape scale, fires must be allowed to burn in whitebark pine ecosystems.

Wisconsin Oak Savanna

The term *savanna* has never been well defined, maybe loosely at best. Cole (1960) summed up the situation this way: "Perhaps of all types of vegetation the savanna is the most difficult to define, the least understood, and the one whose distribution and origin is the most subject to controversy." Henderson and Epstein (1995) point out that even though not well defined, an oak savanna is well recognized. It is a class of North American plant communities that were part of a large transitional complex of communities between the vast treeless prairies of the West and the deciduous forests of the East. This system or mosaic was driven by frequent fires and possibly influenced by large herbivores (ungulates) such as bison and elk. Oaks were the dominant trees, hence the term *oak savanna*. A wide range of community types was found within this transitional complex; collectively, they represented a continuum from prairie to forest, with no clear dividing lines between savanna and other community types (Curtis 1959).

Savannas all have a partial canopy of open grown trees and a varied ground layer of prairie and forest herbs, grasses, and shrubs, as well as plants restricted to the light shading and mix of shade and sun so characteristic of a savanna. Definitions of savanna tree cover range from 5–80 percent canopy; however, the lower canopy covers of 5–50 percent or 5–30 percent are more widely used criteria. Savanna types range from those associated with dry, gravelly, or sandy soils; those on rich, deep soils; and those on poorly drained, moist soils (Henderson and Epstein 1995).

Oak savannas have probably been in North America for some 20–25 million years (Barry and Spicer 1987), expanding and contracting with climatic changes and gaining and losing species (on a geologic time scale) through evolution and extinction. For the past several thousand years, such savannas have existed as a relatively stable band of varying width and continuity, form northern Minnesota to central Texas.

At the time of European settlement (ca. 1830), oak savanna covered many millions of hectares. It varied somewhat in species composition from north to south and east to west, but structure and functions were probably similar throughout. In the upper Midwest (Minnesota, Wisconsin, Michigan, Iowa, Illinois, Indian, and Missouri), there were an estimated 12 million hectares (29.6 million acres; Curtis 1959). As the Midwest's rich soils were used for agriculture and fire was suppressed, this ecosystem all but disappeared from the landscape throughout its range. Today, oak savannas are a globally endangered ecosystem (Henderson and Epstein 1995).

Summary of Key Terms

Biome—regional land-based ecosystem type such as a tropical rainforest, taiga, temperate deciduous forest, tundra grassland, or desert. Biomes are characterized by consistent plant forms and are found over a large climatic area.

Bole—trunk of a tree above the root collar and extending along the main axis.

Decomposers—heterotrophic organisms that break down dead protoplasm and use some of the products and release others for use by consumer organisms.

Ecosystem—a functioning unit of nature that combines biotic communities and the abiotic environments with which they interact. Ecosystems vary greatly in size and characteristics.

Ecotones—a transitional area between two (or more) distinct habitats or ecosystems, which may have characteristics of both or its own distinct characteristics. The edge of woodland, next to a field or lawn, is an ecotone, as are some savanna areas between forests and grasslands.

Extinction—the dying out of a species, or the condition of having no remaining living members; also the process of bringing about such a condition.

Forest land—land at least 10 percent stocked by forest trees of any size, including land that formerly had such tree cover and that will be naturally or artificially regenerated. The minimum area for classification of forest land is one acre.

Hardwood—a dicotyledonous tree, usually broad-leaved and deciduous.

Herbivore—an animal that feeds on plants.

Mortality—the volume of sound wood in growing stock trees that died from natural causes during a specified year.

National forest—an ownership class of federal lands, designated by executive order or statute as national forests of purchase units, and other lands under the administration of the USDA Forest Service.

Silviculture—management of forest land for timber.

Subalpine—describing the region, the climate, the vegetation, or all three found just below alpine regions, usually on mountainsides at 1,300 to 1,800 meters in elevation. Subalpine vegetation is that just below the treeline, often dominated by pine or spruce trees.

Softwood—a coniferous tree, usually evergreen, having needles or scalelike leaves.

Timberland—forest land that is capable of producing crops of industrial wood and not withdrawn form timber utilization by statute or administrative regulation. (Note: Areas qualifying as timberland are capable of producing in excess of 20 cubic feet per acre per year of industrial wood in natural stands.)

Chapter Review Questions

7.1   The world's major communities are known as _____.

7.2   The U.S. state with the smallest percentage of forest cover is _____.

7.3   Much of this famous U.S. valley region is an oak-hickory forest: _____.

7.4   A wetland is simply a natural feature full of _____ features.

7.5   Swamp on a hill: _____.

7.6   _____ recycles nutrients, reduces biomass, influences insect and disease populations, and is the principal change agent affecting vegetative structure and composition.

7.7   A _____ describes the pattern that fire follows in a particular ecosystem.

7.8   The components of a fire regime are _____.

7.9   The primary disease of lodgepole pine is _____.

7.10 Favorite/critical food of grizzly bears: _____.

Cited References and Recommended Reading

Abernathy, Y., and Turner, R. E. 1987. U.S. forested wetlands: 1940–1980, *Bioscience* 37:721–27.

Alexander, M. E., and Hawksworth, F. G. 1975. *Wildland fires and dwarf mistletoe: A literature review of ecology and prescribed burning*. Fort Collins, CO: U.S. Forest Service General Technical Report RM-14, Rocky Mountain Forest Range Experiment Station.

Arno, S. F., and Hoff, R. J. 1989. *Silvics of whitebark pine*. U.S. Forest Service General Technical Report INT-235.

Barrett, S. W. 1980. Indians and fire. *Western Wildlands* 6, no. 3:17–21.

Barry, A. T., and Spicer, R. A. 1987. *The evolution and palaeobiology of land plants*. London: Croom Helm.

Betancourt, J. L., Van Devender, T. R., and Martin, P. S., eds. 1990. *Packrat middens: The last 40,000 years of biotic change*. Tucson: University of Arizona Press.

Boise National Forest (BNF). 2004. *Trees: Ponderosa pine*. http://www.Fs.fed.us/r4/boise/local-resources/trees/p-pine.shtml (accessed May 4, 2007).

Bond, W. J., and Keeley, J. E. 2005. Fire as a global "herbivore": The ecology and evolution of flammable ecosystems. *Trends in Ecology and Evolution* 20:387–94.

Boyd, R. J. 1995. Terrestrial ecosystems. In *Our living resources*. Washington, DC: U.S. Department of the Interior, National Biological Service.

Burgess, R. I., and Sharpe, D. M., eds. 1981. *Forest Island dynamics in man-dominated landscapes*. New York: Springer-Verlag.

Campbell, N.A. 1996. *Biology* (4th ed.). Menlo Park, CA: Benjamin/Cummings.

Cole, K. 1995a. *A survey of the fossil packrat middens and reconstruction of the pregrazing vegetation of Capitol Reef National Park*. National Park Service Res. Rep.

Cole, K. 1995b. Vegetation change in national parks. In *Our living resources*. Washington, DC: U.S. Department of the Interior, National Biological Service.

Cole, K. L., Engstrom, D. R., Futyma, R. P., and Stottlemyer, R. 1990. Past atmospheric deposition of metals in northern Indiana measured in a peat core from Cowles Bog. *Environmental Science and Technology* 24:543–49.

Cole, K. L., and Liu, G. 1994. Holocene paleoecology of an estuary on Santa Rosa Island, California. *Quaternary Research* 41:326–35.

Cole, M. M. 1960. Cerrado, caating and pentanal: The distribution and origin of the savanna vegetation of Brazil. *Geography Journal* 126:168–79.

Curtis, J. T. 1959. *The vegetation of Wisconsin*. Madison: University of Wisconsin Press.

Dahl, T. E. 1990. *Wetlands losses in the United States 1780s to 1980s*. Washington, DC: U.S. Fish and Wildlife Service.

Darr, D. R. 1995. U.S. forest resources. In *Our living resources*. Washington, DC: U.S. Department of the Interior, National Biological Service.

Elfring, C. 1989. Yellowstone: Fire storm over fire management. *Bioscience* 39, no. 10:667–72.

Faegri, R., and Iversen, I. 1989. *Textbook of pollen analysis*. New York: Wiley & Sons.

Ferry, G. W., Clark, R. G., Montgomery, R. E., Mutch, R. W., Leenhouts, W. P., and Zimmerman, G. T. 1995. Altered fire regimes within fire-adapted ecosystems. In *Our living resources*. Washington, DC: U.S. Department of the Interior, National Biological Service.

Grossman, D. H., and Goodin, K. L. 1995. Rare terrestrial ecological communities of the United States. In *Our living resources*. Washington, DC: U.S. Department of the Interior, National Biological Service.

Harris, L. D. 1984. *The fragmented forest*. Chicago: University of Chicago Press.

Henderson, R. A., and Epstein, E. J. 1995. Oak savannas in Wisconsin. In *Our living resources*. Washington, DC: U.S. Department of the Interior, National Biological Service.

Jacobson, G. L., Jr., and Grimm, E. C. 1986. A numerical analysis of Holocene forest and prairie vegetation in central Minnesota. *Ecology* 67:958–66.

Keane, R. R., and Arno, S. F. 1993. Rapid decline of whitebark pine in western Montana: Evidence from 20-year remeasurements. *Western Journal of Applied Forestry* 8, no. 2:44–47.

Keeland, B. D., Allen, J. A., and Burkett, V. V. 1995. Southern forested wetlands. In *Our living resources*. Washington, DC: U.S. Department of the Interior, National Biological Service.

Kendall, K. C. 1995. Whitebark pine: Ecosystem in peril. In *Our living resources*. Washington, DC: U.S. Department of the Interior, National Biological Service.

Kendall, K. C., and Arno, S. F. 1990. *Whitebark pine: An important but endangered wildlife resource*. U.S. Forest Service General Technical Report INT-270, 264–73.

Kerry, G. W., Clark, R. G., Montgomery, R. E., Mutch, R. W., Leenhouts, W. P., and Zimmerman, G. T. 1995. Altered fire regimes: Within fire-adapted ecosystems. In *Our living resources*. Washington, DC: U.S. Department of the Interior, National Biological Service.

Kilgore, B. M., and Heinselman, M. L. 1990. Fire in wilderness ecosystems. In *Wilderness management*, ed. J. C. Hendee, G. M. Stankey, and R. C. Lucas, 297–335. 2nd ed. Golden, CO: North American Press.

Klopatek, J. M., Olson, R. J., Emerson, C. J., and Jones, J. L. 1979. Land use conflicts with natural vegetation in the United States. *Environmental Conservation* 6:192–200.

Miller, R. F., and Waigand, P. E. 1994. Holocene changes in semi-arid pinion-juniper woodlands: Response to climate, fire and human activities in the great Basin. *Biological Science* 44, no. 7.

Mutch, R. W., Arno, S. F., Brown, J. K., Carlson, C. E., Ottmar, R. D., and Peterson, J. L. 1993. *Forest health in the Blue Ridge Mountains: A management strategy for the fire-adapted ecosystems*. U.S. Forest Service General Technical Report PNW-310.

Nash, B. L., Tonnessen, K., and Flores, D. J. M. 1995. Air quality in the national park system. In *Our living resources*. Washington, DC: U.S. Department of the Interior, National Biological Service.

Nichol, A. A. 1937. *The natural vegetation of Arizona.* University of Arizona Technical Bulletin 68.

Office of Technology Assessment (OTA). 1993. *Preparing for an uncertain climate.* 2 vols. Washington, DC: U.S. Congress, OTA.

Olgilvie, R. T. 1990. *Distribution and ecology of whitebark pine in western Canada.* U.S. Forest Service General Technical Report INT-270, 54–60.

Pellant, M., and Hall, C. 1994. Distribution of two exotic grasses on public lands in the Great Basin: status in 1992. In *Proceedings of a symposium on ecology, management and restoration of intermountain annual rangelands.* Ogden, UT: Intermountain Forest and Range Experiment Station.

Peterson, D. L. 1995. Air pollution effects on forest ecosystems in North America. In *Our living resources.* Washington, DC: U.S. Department of the Interior, National Biological Service.

Powell, D. S., Faulkner, J. L., Darr, D.R., Zhu, Z., and MacCleery, D. W. 1993. *Forest resources of the United States, 1992.* General Technical Report RM-234, U.S. Forest Service, Rocky Mountain Forest and Range Experiment Station, Fort Collins, CO.

Pyne, S. J. 1982. *Fire in America: A cultural history of wildland and rural fire.* Princeton, NJ: Princeton University Press.

Ray, J. 1999. *Ecology of a cracker childhood.* Minneapolis, MN: Milkweed Productions.

Ray, J. 2005. *Pinhook: Finding wholeness in a fragmented land.* White River Junction, VT: Chelsea Green.

Robbins, L. E., and Myers, R. L. 1992. *Seasonal effects of prescribed burning in Florida: A review.* Tall Timbers Research, Inc. Miscellaneous Publication 8.

Shenandoah National Park. 2002. http://www.shenandoah.national-park.com/nat.hm (accessed May 2, 1007).

Shugart, H. H., and West, D. C. 1977. Development of an Appalachian deciduous forest succession model and its application to assessment of the impact of the chestnut blight. *Journal of Environmental Management* 5:161–79.

Smith, T. M., and Smith, R. L. 2006. *Elements of ecology* (6th ed.). Menlo Park, CA: Benjamin/ Cummings.

Spellman, F. R. 1998. *The science of environmental pollution.* Lancaster, PA: Technomic.

Tansley, A. G. 1935. *Changing the face of the Earth* (2nd ed.). Oxford: Blackwell.

U.S. Army Corp of Engineers. 1998. *Management of peatland shrub and forest-dominated communities for threatened and endangered species.* Washington, DC: Author.

U.S. Department of Agriculture (USDA). 2004. *U.S. forest resource facts and historical trends.* http://fia.fs.fed.us (accessed April 27, 2007).

U.S. Fish and Wildlife Service (USFWS). 1998. *Restoring an Atlantic white cedar bog.* http:// www.fws.Gov/nc-es/coastal/awhitecedar.htm (accessed May 3, 2007).

U.S. Forest Service (USFS). 1992. *Forest Services resources inventories: An overview. Forest inventory, economics, and recreation research.* Washington, DC: U.S. Forest Service.

U.S. Geological Survey (USGS). 1995. *Our living environment.* Washington, DC: Author.

U.S. Geological Survey (USGS). 2003. *Whitebark pine communities.* http://www.nmsc.usgs.gov (accessed May 6, 2007).

Waldrop, T. A., Van Lear, D. H., Lloyd, R. T., and Harms, W. R. 1987. *Long-term studies of prescribed burning of loblolly pine forests of the southeastern Coastal Plain.* U.S. Forest Service Southeastern Forest Experiment Station General Technical Report SE-45, Asheville, NC.

Weaver, H. 1943. Fire as an ecological and silvicultural factor in the ponderosa-pine region of the Pacific Slope. *Journal of Forestry* 41:7–14.

Wellner, C. A. 1970. Fire history in the northern Rocky Mountains. In *The role of fire in the Intermountain West*, 41–64. Proceedings of the Intermountain Fire research Council symposium, Missoula, MT.

West, N. E. 1988. Intermountain deserts, shrub steppes, and woodlands. In *North American terrestrial vegetation*, ed. M. B. Barbour and W. D. Billings, 209–30. New York: Cambridge University Press.

West, N. E., and Hassan, M. A. 1985. Recovery of sagebrush-grass vegetation following wildlife. *Journal of Range Management* 38:131–34.

Wharton, C. H., Kitchens, W. M., and Sipe, T. W. 1982. *The ecology of bottomland hardwood swamps or the Southeast: A community profile*. U.S. Fish and Wildlife Service General Technical Report FWS/OBS-81/37.

Whisenant, S. G. 1990. Changing fire frequencies on Idaho's Snake River Plains: Ecological and management implications. In *Proceedings of a symposium on cheat-grass invasion, shrub die-off, and other aspects of shrub biology and management*, ed. E. D. McArthur, E. M. Romney, S. D. Smith, and P. T. Tueller, 4–10. Ogden, UT: U.S. Forest Service General Technical Report INT-276, Intermountain Forest and Range Experiment Station.

Wilhelm, G. S. 1990. *Special vegetation of the Indiana Dunes National Lakeshore*. Porter, IN: Indian Dunes National Lakeshore Research Program, Report 90–02.

Wright, H. A., and Bailey, S. W. 1982. *Fire ecology: United States and southern Canada*. New York: John Wiley & Sons.

Wright, A. H., and Wright, A. A. 1932. The habitats and composition of the vegetation of Okefenokee Swamp, Georgia. *Ecological Monographs* 2:109–232.

Zimmerman, G. T., and Laven, R. D. 1984. Ecological interrelationship of dwarf mistletoe and fire in lodgepole pine forests. In *Biology of dwarf mistletoes: Proceedings of the symposium*, ed. F. G. Hawksworth and R. F. Scharpf, 123–31. Fort Collins, CO: U.S. Forest Service General Technical Report RM-111, Rocky Mountain Forest Range Experiment Station.

Abbot Lake and Sharp Top Mountain, Peaks of Otter, Virginia.
Photograph by Frank R. Spellman

CHAPTER 8

Aquatic Ecosystems

Little brook, sing a song of a leaf that sailed along,
Down the golden braided center of your current swift and strong.

—J. W. Riley

Topics

The Balanced Aquarium
Freshwater Ecology
Habitat Changes in the Upper Mississippi River Floodplain
Fish Populations in the Illinois River
The Great Lakes Ecosystem
Gastropod Fauna in the Mobile Bay Basin
Protozoa
Aquatic Algae
Diatoms
Riparian Corridors
Coastal and Marine Ecosystems
Summary of Key Terms
Chapter Review Questions

Chapter 7 noted that ecologists classify terrestrial ecosystems according to their dominant plant life forms; however, classification of aquatic ecosystems is largely based on features of the physical environment. Smith and Smith (2006) point out "one of the major features that influence the adaptations of organisms to the aquatic environment is water salinity. For this reason, aquatic ecosystems fall into two major categories: freshwater or marine."

Freshwater ecosystems have been especially subjected to the environmental degradation that has occurred over the past century in this country. Nearly every activity that occurs on land ultimately affects the receiving waters in that drainage. Whether it is pesticides and herbicides applied to crops, silt washed away because of vegetation removal, or even atmospheric deposition, freshwater ecosystems are a product of all local disturbances regardless of where they occur. In addition, waterways have been used for numerous activities other than providing habitat to aquatic organisms. They have been altered for transportation, diverted for agricultural and municipal needs, dammed for energy, borrowed as an industrial coolant, and straightened for convenience. These abuses have taken their toll, as evidenced by worldwide declines in fisheries, monumental floods, an ever-growing list of endangered aquatic species, and communities trying to deal with finite water supplies (Mac 1995).

The traits that make freshwater (and other aquatic) ecosystems particularly vulnerable also make them useful for monitoring environmental quality. Water serves to integrate these impacts by distributing them among the elements within aquatic ecosystems. Although dilution is occurring, subtle changes can be detected in habitats or organisms over a much larger area that may be the result of a single point source. A clean freshwater ecosystem with a healthy biological community will be indicative of the condition of the terrestrial habitat in the watershed, whereas the reverse may not necessarily be true (Mac 1995).

After a brief discussion of the basics of freshwater ecology, this chapter features accounts of the alterations of freshwater, riparian, and marine habitats and their impacts on the biota. For example, evidence is presented documenting habitats destroyed by dams or channelization, contaminants affecting organism health, wetlands affected by water-level control, reduced water quality, and introductions of exotic species. These kinds of changes have caused declining biodiversity in many groups of aquatic species ranging from freshwater mussels to waterfowl.

This chapter discusses some encouraging trends that are emerging. Persistent organic contaminants in the Great Lakes have declined, and marginal water quality improvement has been accompanied by increased diversity of the fish community. Despite these achievements, much needs to be done to effectively manage and conserve aquatic resources. As is evident from the response on diatoms, algae, and protozoa, little is known of the national trends in their populations, diversity, or biomass. Our knowledge of these groups is poor even though they provide basic functions of photosynthesis, production, and decomposition critical to the normal functioning of aquatic ecosystems.

Without increased monitoring, some very basic attributes of aquatic systems may be unknowingly lost or severely degraded. Groups of species that seem insignificant may actually be critical parts of a food web that supports valuable commercial and sport species. Subtle changes such as losses of island habitat and constant water depth or level may lead to drastic declines in productivity or diversity. The loss of some of these integral pieces of ecosystems may be impossible to restore. The unsuccessful attempt to restore self-sustaining lake trout populations in the Great Lakes, despite massive efforts, exemplifies this (Mac 1995).

Note: Portions of this chapter are adapted from or based on information from U.S. Geological Survey's (USGS) *Our Living Environment* (1995).

The Balanced Aquarium

Ours is a water planet. Water covers three quarters of its surface, makes up two-thirds of our bodies. It is so vital to life we can't live more than four days without it. If all the earth's water—an estimated 325 trillion gallons—were squeezed into a gallon jug and you poured off what was not drinkable (too salty, frozen or polluted) you'd be left with one drop. And even that might not pass U.S. water quality standards. (Narr 1990)

Normal stream life can be compared to that of a "balanced aquarium" (American Society for Testing and Materials [ASTM] 1969). That is, nature continuously strives to provide clean, healthy, normal streams. This is accomplished by maintaining the stream's flora and fauna in a balanced state. Nature balances stream life by maintaining both the numbers and the types of species present in any one part of the stream. Such balance ensures that there is never an overabundance of one species compared to another. Nature structures the stream environment so that plant and animal life is dependent upon the existence of others within the stream. Thus, nature has structured an environment that provides for interdependence, which leads to a balanced aquarium in a normal stream (Spellman 1996).

To this point in the text, the fundamental concepts of ecology, which are generally related to both terrestrial and aquatic habitats, have been discussed. Before narrowing the focus to the topic of aquatic ecology, this might be the place in which to contrast the two different ecosystems: land and aquatic (freshwater) habitats.

The major difference between land and aquatic habitats is in the medium in which both exist. The land or terrestrial habitat is enveloped in a medium of air, the atmosphere. The aquatic habitat, on the other hand, exists in a water medium. Although the two ecosystems are different, they both use oxygen. How the oxygen is formulated in each system and how organisms utilize it is where a major contrast exists.

The following data clearly illustrate this contrast. Atmospheric air contains at least 20 times more oxygen than water. Air has approximately 210 milliliters (ml) of oxygen per liter; water contains 3 to 9 ml per liter, depending on temperature, presence of other solutes, and degree of saturation. Moreover, aquatic organisms must work harder for their oxygen. That is, they must expend far more effort extracting oxygen from water than land animals expend removing oxygen from air (Spellman 1996).

Other contrasts between the land and water ecosystems can be seen. Water, for example, is approximately 1,000 times denser than air and approximately 50 times more viscous. Additionally, natural bodies of water have tremendous thermal capacity, with little temperature fluctuation as compared to atmospheric air.

In this discussion of aquatic ecology, freshwater ecology is addressed first. Freshwater ecology is the branch of ecology that deals with the biological aspect of limnology. Limnology, as defined by Welch, "deals with biological productivity of inland waters and with all the causal influences which determine it" (Welch 1963). Limnology divides freshwater ecosystems into two groups or classes: lentic and lotic habitats. The lentic (*lenis* = calm) or standing water habitats are represented by lakes, ponds, and swamps. The lotic (*lotus* = washed) or running water habitats are represented by rivers, streams, and springs. On occasion, these two different habitats are not well differentiated. This can be seen in the case of an old, wide, and deep river where water velocity is quite low and the habitat, therefore, becomes similar to that of a pond.

Freshwater Ecology

Lentic Habitat

Lakes and ponds range in size from just a few square feet to thousands of square miles. Scattered throughout the Earth, many of the first lakes evolved during the Pleistocene ice age. Many ponds are seasonal, just lasting a couple of months, such as sessile pools, while lakes last many years. There is not that much diversity in species since lakes and ponds are often isolated from one another and from other water sources such as streams and oceans.

Lakes and ponds are divided into four different "zones" that are usually determined by depth and distance from the shoreline. The four distinct zones— littoral, limnetic, profundal, and benthic—are shown in figure 8.1. Each zone "provides a variety of ecological niches for different species of plant and animal life" (Miller 1988).

The littoral zone is the topmost zone near the shores of the lake or pond with light penetration to the bottom. It provides an interface zone between the land and the open water of lakes. This zone contains rooted vegetation such as grasses, sedges, rushes, water lilies, and waterweeds, and a large variety of organisms. The littoral zone is further divided into concentric zones, with one group

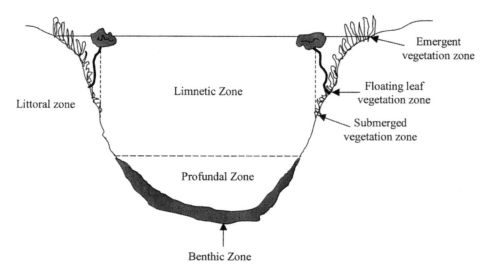

Figure 8.1.   Vertical section of a pond showing major zones of life
Modified from Enger, Kormelink, Smith, and Smith. 1989. *Environmental Science: An Introduction*, p. 77.

replacing another as the depth of water changes. Figure 8.1 also shows these concentric zones: emergent vegetation, floating leaf vegetation, and submerged vegetation zones, which proceed from shallow to deeper water.

The littoral zone is the warmest zone since it is the area that light hits, contains flora such as rooted and floating aquatic plants, and contains a very diverse community, which can include several species of algae (such as diatoms), grazing snails, clams, insects, crustaceans, fishes, and amphibians. The aquatic plants aid in providing support by establishing "excellent habitats for photosynthetic and heterotrophic (requires organic food from the environment) microflora as well as many zooplankton and larger invertebrates" (Wetzel 1983). In the case of insects, such as dragonflies and midges, only the egg and larvae stages are found in this zone. The fauna includes such species as turtles, snakes, and ducks that feed on the vegetation and other animals.

From figure 8.2, it can be seen that the limnetic zone is the open-water zone up to the depth of effective light penetration; that is, the open water away from the shore. The community in this zone is dominated by minute suspended organisms, the plankton, such as phytoplankton (plants) and zooplankton (animals), and some consumers such as insects and fish. Plankton are small organisms that can feed and reproduce on their own and serve as food for small chains.

✔   *Important Point*: Without plankton in the water, there would not be any living organisms in the world, including humans.

In the limnetic zone, the population density of each species is quite low. The rate of photosynthesis is equal to the rate of respiration; thus, the limnetic zone is at compensation level. Small shallow ponds do not have this zone; they have a littoral zone only. When all lighted regions of the littoral and limnetic zones are discussed as one, the term *euphotic* is used for both, designating that these zones

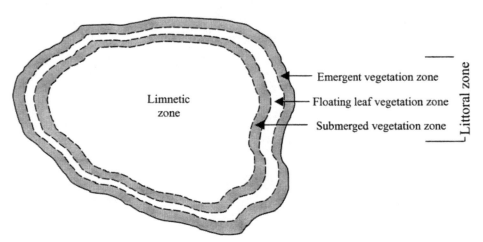

Figure 8.2.   View looking down on concentric zones that make up the littoral zone

as having sufficient light for photosynthesis and the growth of green plants to occur.

The small plankton do not live for a long time. When they die, they fall into the deep-water part of the lake or pond, the profundal zone. The profundal zone, because it is the bottom or deep-water region, is not penetrated by light. This zone is primarily inhabited by heterotrophs adapted to its cooler, darker water and lower levels of oxygen.

The final zone, the benthic zone, is the bottom region of the lake. It supports scavengers and decomposers that live on sludge. The decomposers are mostly large numbers of bacteria, fungi, and worms that live on dead animal and plant debris and wastes that find their way to the bottom.

LOTIC HABITAT

Lotic (washed) habitats are characterized by continuously running water or current flow (see figure 8.3). These running water bodies, rivers, and streams, have typically three zones: riffle, run, and pool. The riffle zone contains faster flowing, well-oxygenated water, with coarse sediments. In the riffle zone, the velocity of current is great enough to keep the bottom clear of silt and sludge, thus providing a firm bottom for organisms. This zone contains specialized organisms that are adapted to life in running water. For example, organisms adapted to live in fast streams or rapids (trout, for example) have streamlined bodies, which aids in their respiration and in obtaining food (Smith 1974). Stream organisms that live under rocks to avoid the strong current have flat or streamlined bodies. Others have hooks or suckers to cling or attach to a firm substrate to avoid the washing-away effect of the strong current (Allen 1996).

The run zone (or intermediate zone) is the slow-moving, relatively shallow part of the stream with moderately low velocities and little or no surface turbulence.

The pool zone of the stream is usually a deeper water region where velocity of water is reduced and silt and other settling solids provide a soft bottom (more homogeneous sediments), which is unfavorable for sensitive bottom-dwellers. Decomposition of some of these solids causes a lower amount of dissolved oxygen (DO). It is interesting to note that some stream organisms spend some of their time in the rapids part of the stream and other times can be found in the pool zone (trout, for example). Trout typically spend about the same amount of time in the rapid zone pursuing food as they do in the pool zone pursuing shelter.

Organisms are sometimes classified based on their mode of life. The following section provides a listing of the various classifications based on mode of life.

CLASSIFICATION OF AQUATIC ORGANISMS BASED ON MODE OF LIFE

1. *Benthos* (mud dwellers): The term *benthos* originates from the Greek word for bottom and broadly includes aquatic organisms living on the bottom or on submerged vegetation. They live under and on rocks and in the sedi-

Figure 8.3.   Falling spring, Virginia
Photograph by Frank R. Spellman

ments. A shallow sandy bottom has sponges, snails, earthworms, and some insects. A deep, muddy bottom will support clams, crayfish, and nymphs of damselflies, dragonflies, and mayflies. A firm, shallow, rocky bottom has nymphs of mayflies and stone flies, and larvae of water beetles.

2.   *Periphytons* or *Aufwuchs*: The first term usually refers to microfloral growth upon substrata (i.e., benthic attached algae). The second term, *aufwuchs*

(pronounce: OWF-vooks; German: "growth upon") refers to the fuzzy, sort of furry-looking, slimy green coatings that attach or cling to stems and leaves of rooted plants or other objects projecting above the bottom without penetrating the surface. It consists not only of algae such as Chlorophyta but also diatoms, protozoa, bacteria, and fungi.

3. *Planktons* (drifters): Planktons are small, mostly microscopic plants and animals that are suspended in the water column; movement depends on water currents. They mostly float in the direction of current. There are two types of planktons: (1) phytoplanktons are assemblages of small plants (algae) and have limited locomotion abilities; they are subject to movement and distribution by water movements. (2) Zooplankton are animals that are suspended in water and have limited means of locomotion. Examples of zooplanktons include crustaceans, protozoa, and rotifers.

4. *Nektons* or *Pelagic Organisms* (capable of living in open waters): Nektons are distinct from other planktons in that they are capable of swimming independently of turbulence. They are swimmers that can navigate against the current. Examples of nektons include fish, snakes, diving beetles, newts, turtles, birds, and large crayfish.

5. *Neustons*: Neustons are organisms that float or rest on the surface of the water. Some varieties can spread out their legs so that the surface tension of the water is not broken, for example, water striders.

6. *Madricoles*: Madricoles are organisms that live on rock faces in waterfalls or seepages.

LIMITING FACTORS

An aquatic community has several unique characteristics. The aquatic community operates under the same ecologic principles as terrestrial ecosystems, but the physical structure of the community is more isolated and exhibits limiting factors that are very different from the limiting factors of a terrestrial ecosystem.

Certain materials and conditions are necessary for the growth and reproduction of organisms. If, for instance, a farmer plants wheat in a field containing too little nitrogen, it will stop growing when it has used up the available nitrogen, even if the wheat's requirements for oxygen, water, potassium, and other nutrients are met. In this particular case, nitrogen is said to be the limiting factor.

A limiting factor is a condition or a substance (the resource in shortest supply) that limits the presence and success of an organism or a group of organisms in an area (Spellman 1996). There are two well-known laws about limiting factors:

1. *Liebig's Law of the Minimum*: Odum (1971) has modernized Liebig's law in the following: "Under steady state conditions the essential material available in amounts most closely approaching the critical minimum needed, will tend to be the limiting one." Liebig's law is normally restricted to chemicals that

limit plant growth in the soil, for instance, nitrogen, phosphorus, and potassium. It does not deal with the excess of a factor as limiting.

2. *Shelford's Law of Tolerance*: Although Liebig's law does not deal with the excess of a factor as limiting, excess is or can be a limiting factor. The presence and success of an organism depends on the completeness of a complex set of conditions. Odum (1971) describes Shelford's law of tolerance as follows: "Absences or failure of an organism can be controlled by the qualitative and quantitative deficiency or excess with respect to any one of the several factors which may approach the limits of tolerance for that organism." For instance, too much and too little heat, light, and moisture can be limiting factors for some plants.

Price (1984) states that "these two laws actually relate to individual organisms, and the survival of an individual in a given set of conditions, independent of others in the same niche." Expressed differently, both these laws state that the presence and success of an organism or a group of organisms depend upon a complex set of conditions, and any condition that approaches or exceeds the limits of tolerance is said to be a limiting condition or factor.

Several factors affect biological communities in streams. These include:

- water quality
- temperature
- turbidity
- dissolved oxygen
- acidity
- alkalinity
- organic and inorganic chemicals
- heavy metals
- toxic substances
- habitat structure
- substrate type
- water depth and current velocity
- spatial and temporal complexity of
- physical habitat
- flow regime
- water volume
- temporal distribution of flows
- energy sources
- type, amount, and particle size of
- organic material entering stream
- seasonal pattern of energy
- availability
- biotic interactions
- competition
- predation
- disease
- parasitism
- mutualism

The common physical limiting factors in freshwater ecology important to this discussion include:

1. *Temperature*
2. *Light*
3. *Water movement—stream currents, especially rapids*
4. *Turbidity*
5. *Dissolved atmospheric gases, especially oxygen*
6. *Biogenic salts in macro and micronutrient forms*
     —macronutrients, such as nitrogen, phosphorus, potassium, calcium, and sulfur
     —micronutrients such as iron, copper, zinc, chlorine, and sodium

Temperature

Aquatic organisms are very sensitive to temperature change, as water temperature generally does not change rapidly. It should be noted, however, that surface waters can be subject to great temperature variations. Tchobanoglous and Schroeder (1985) point out that across the United States, for example, surface water temperatures can vary from 0.5°C to 27°C (33°F to 81°F). Water has some unique properties such as very high molar heat of fusion (1.44 kcal) and molar heat of vaporization (9.70 kcal), which allow a very slow change in water temperature.

Aquatic organisms often have narrow temperature tolerance and are known as stenothermal (narrow temperature range). The limits for abrupt changes in water temperature are − 20°C (− 5°F). Water has its greatest density (1 g/cm3) at 4°C (39°F). Above and below this temperature, it is lighter. Temperature changes, therefore, produce a characteristic pattern of stratification of lakes and ponds in tropical and temperate regions, which helps the aquatic life to survive under severe winter and summer conditions. Temperature has an effect on thermal stratification, causing turnover of lakes and ponds.

During the summer, turning over occurs because the top layer of water becomes warmer than the bottom; and as a result, there are two layers of water, the top one lighter and the bottom one heavier. With the further rise in temperature, the top layer becomes even lighter than the bottom layer, and a middle layer with medium density is created. These layers, from top to bottom, are known as epilimnion, thermocline and hypolimnion. They are lightest and warmest, medium weight and warmer, and heaviest and cool, respectively. There is a strong drop in temperature at the thermocline. There is no circulation of water in these three layers. If the thermocline is below the range of effective light penetration, which is quite common, the oxygen supply becomes depleted in the hypolimnion, since both photosynthesis and the surface source of oxygen are cut off. This state is known as summer stagnation.

During the fall, as the air temperature drops, so does the temperature of the epilimnion until it is the same as that of the thermocline. At this point, the two layers mix. The temperature of the whole lake is now the same, and there is a complete mixing. As the temperature of the surface water reaches 4°C (39°F), it becomes more dense than the water below, which is not in direct contact with the air and does not cool as rapidly at the lower levels. The denser oxygen-rich surface layer stirs up organic matter as the water sinks to the bottom; this is known as fall turnover (Northington and Goodin 1984).

During the winter, the epilimnion, which is icebound, is at the lowest temperature and thus lightest; the thermocline is at medium temperature and medium weight; and the hypolimnion is at about 4°C and heaviest. This is winter stratification. In winter, the oxygen supply is usually not greatly reduced, as the low temperature solubility of oxygen is higher and bacterial decomposition along with other life activities are operating at a low rate. When there is too much ice with heavy snow accumulation, light penetration is reduced. This reduces the rate of photosynthesis, which, in turn, causes oxygen depletion in the hypolimnion, resulting in winter kill of fish.

In spring, as ice of the epilimnion melts (often aided by warm rains), there is a mixing of the top two layers, and as it reaches 4°C it sinks down to cause spring overturn. Odum (1971) describes this spring overturn phenomenon as being analogous to the lake taking a "deep breath."

Light

Viewed physically, light is part of the radiant energy of the electromagnetic spectrum. It is capable of doing work and of being transformed from one form into another. Light as radiant energy is transformed into potential energy by biochemical reactions, such as photosynthesis, or as heat. As the source of energy for photosynthesis, light is a very important factor for aquatic life. The rate of photosynthesis depends on the intensity of light and photoperiod (light hours per day). The amount of biomass and oxygen production corresponds to the rate of photosynthesis. There is a daily cycle of the amount of dissolved oxygen in water bodies based on the photosynthetic activity of plants. The amount of dissolved oxygen (DO) is maximum at 2 p.m. and minimum at 2 a.m.

Water Movements

Water movement or current is a very important limiting factor in lakes and streams. Water movements, such as wave action in lakes and the current in streams, mix the dissolved oxygen (DO) from the interphase of air and water into deeper layers. This increases the rate of absorption of oxygen from the atmosphere. The current also helps to keep the bottom clean by washing away settleable solids, thus creating a proper habitat for a large number of benthic species. Where stream current is strong, such as in riffles, specialized organisms become firmly attached to the bottom. For example, the caddisfly larvae attach themselves to the bottom substrate. In the case of fish, trout and other varieties that can swim against the current may also occupy the rapids zone of streams. Due to current flow, streams and rivers seldom have a complete depletion of dissolved oxygen (DO) in spite of organic pollution, whereas lakes and ponds can go anaerobic.

Turbidity

When one looks upon the waters of a lake or stream, perhaps the first thing noticed is the transparency of the water. The rate of penetration of light is affected inversely by the amount of turbidity in the water. Turbidity, or degree of clarity, is caused by the suspended particles that block the passage of light; it often fluctuates (in surface waters) with the amount of precipitation. Turbidity, therefore, affects photosynthesis and thus lowers the number of organisms by reducing productivity. Turbidity also causes the growth of slime on the body surface of aquatic organisms, which damages the respiratory organs, such as gills in the case of fish.

Turbidity is usually measured as NTU. NTUs are nephelometric turbidity units, as measured with a nephelometric turbidimeter. In fieldwork, it is more common to measure turbidity by using a Secchi disk. The Secchi disk is a white disk (sometimes checkered black and white) lowered into the water column until it just disappears from view (see figure 8.4). The depth of visual disappearance becomes the Secchi disk transparency light extinction coefficient, which will range from a few centimeters in very turbid water to 35 meters in a very clear lake, such as Lake Tahoe.

Dissolved Respiratory Gases

Oxygen and carbon dioxide are often limiting factors in freshwater ecosystems. Carbon dioxide is produced during respiration and is essential for photosynthesis, whereas oxygen, which is produced during photosynthesis, is essential for respiration.

Several factors account for the amount of oxygen and carbon dioxide gases in water. These factors include water movements, photosynthesis, temperature, and biodegradable organics. The higher the water temperature, the lower the solubility of a gas in water, and vice versa. Biodegradable organics reduce the amount of dissolved oxygen (DO) in water due to biological oxygen demand (BOD) for their composition. The amount of DO is affected inversely by the amount of carbon dioxide in the water. Because oxygen is an essential ingredient for life, therefore, it becomes a very important limiting factor for aquatic life in water bodies receiving human wastes. The minimum amount of DO for normal aquatic communities is 5 mg/liter. Organic pollution can dramatically reduce the amount

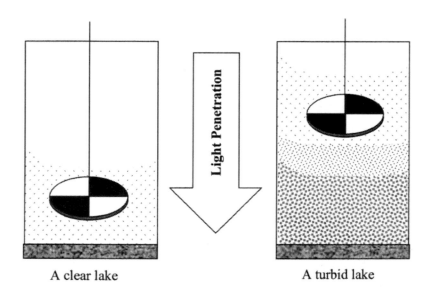

A clear lake   A turbid lake

Water clarity governs the depth of light penetration in lakes.
Periodic testing with a Secchi disk may show seasonal variations in clarity.

Figure 8.4.   Measuring water clarity with a Secchi disk

of DO in a stream. A discussion of pollution-induced oxygen sag will be presented later.

Biogenic Salts

Chlorides, sulfates, nitrates, phosphates of calcium, magnesium, and potassium are biogenic salts that are common to some extent in all freshwater ecosystems. They are essential for protoplasm synthesis. Nitrogen and phosphorous are common limiting factors in freshwater ecology. In streams and rivers, the three primary sources of basic nutrients are runoff, dissolution of rocks, and sewage discharge.

In attempting to determine limiting factors for a stream ecosystem, both field observation and laboratory experimentation are necessary. Due to small variations in the natural environment, a laboratory situation is an undependable guide for determining limiting factors. Only in the field can seasonal changes in population sizes be observed.

The best way in which to determine limiting factors affecting a lake or stream is to use a combination of field and laboratory observation and experiment. An aquatic bioassay is one technique that can be used. In a bioassay, the scale or degree of response is determined by the rate of growth or decrease of population. Bioassays are important in evaluation of water quality, since they determine the effects of liquid waste on the aquatic environments in which experimental organisms, such as fish, may be subjected to concentrations of known or suspected toxicants.

As previously stated, fish are the organisms most often studied. The species chosen should be representative of the water being studied, however. Capture of specimens taken specifically from the water source under study is recommended. Additionally, the species chosen to be studied should be the one that is most susceptible to environmental change. For example, if one is to conduct a bioassay study on a particular stream organism, it would not be appropriate to choose a species that lives in a saltwater environment.

CLASSIFICATION OF LAKES

Lakes can be classified in several ways. For example, Kevern, King, and Ring (1999) classify lakes in three ways. One classification is based on productivity of the lake (or its relative richness). This is the trophic basis of classification. A second classification is based on the times during the year that the water of a lake becomes mixed and the extent to which the water is mixed. And a third classification is based on the fish communities in the lakes.

For the purpose of this discussion, we use a somewhat different classification scheme than the ones just described. That is, we classify lakes based on eutrophication, special types of lakes, and impoundments.

Eutrophication is a natural aging process that results in organic material being produced in abundance due to a ready supply of nutrients accumulated over time. Through natural succession, eutrophication causes a lake ecosystem to turn into a bog and eventually to a terrestrial ecosystem. Eutrophication has received a great amount of publicity lately. In recent years, humans have accelerated the eutrophication of many surface waters by the addition of wastes containing nutrients. This accelerated process is called cultural eutrophication. Sources of human wastes and pollution are sewage, agricultural runoff, mining, industrial wastes, urban runoff, leaching from cleared land, and landfills.

Classification of Lakes Based on Eutrophication

Lakes can be classified into three types based on their eutrophication stage.

1. *Oligotrophic lakes* (few foods): They are young, deep, crystal-clear water, nutrient-poor lakes with little biomass productivity. Only a small quantity of organic matter grows in an oligotrophic lake; the phytoplankton, the zooplankton, the attached algae, the macrophytes (aquatic weeds), the bacteria, and the fish are all present as small populations. It's like planting corn in sandy soil: not much growth. Lake Superior is an example from the Great Lakes.

2. *Mesotrophic lakes*: It is hard to draw distinct lines between oligotrophic and eutrophic lakes, and often the term *mesotrophic* is used to describe a lake that falls somewhere between the two extremes. Mesotrophic lakes develop with the passage of time. Nutrients and sediments are added through runoffs, and the lake becomes more productive biologically. There is a great diversity of species with very low populations at first, but a shift toward higher and higher populations with fewer and fewer species occurs. Sediments and solids contributed by runoffs and organisms make the lake shallower. At an advanced mesotrophic stage, a lake has undesirable odors and colors in certain parts. Turbidity increases and the bottom has organic deposits. Lake Ontario has reached this stage.

3. *Eutrophic lakes* (good foods): This is a lake with a large or excessive supply of nutrients. As the nutrients continue to be added, large algal blooms occur, fish types change from sensitive to more pollution-tolerant ones, and biomass productivity becomes very high. Populations of a small number of species become very high. The lake takes on undesirable characteristics such as offensive odors, very high turbidity, and a blackish color. This high level of turbidity can be seen in studies of Lake Washington in Seattle, Washington. Laws reports that "Secchi depth measurements made in Lake Washington from 1950 to 1979 show an almost fourfold reduction in water clarity" (Laws 1993). Along with the reduction in turbidity, the lake becomes very shallow. Lake Erie is at this stage. Over a period of time, a lake eventually becomes filled with sediments as it evolves into a swamp and finally into a land area.

Special Types of Lakes

Odum (1971) refers to several special lake types.

1. *Dystrophic* (such as bog lakes): They develop from the accumulation of organic matter from outside of the lake. In this case, the watershed is often forested and there is an input of organic acids (e.g., humic acids) from the breakdown of leaves and evergreen needles. There follows a rather complex series of events and processes resulting finally in a lake that is usually low in pH (acid) and often is moderately clear, but color ranges from yellow to brown. Dissolved solids, nitrogen, phosphorus, and calcium are low, and humic matter is high. These lakes are sometimes void of fish fauna; other organisms are limited. When fish are present, production is usually poor. They are typified by the bog lakes of northern Michigan.
2. *Deep ancient lakes*: These lakes contain animals found nowhere else (endemic fauna), for example, Lake Baikal in Russia.
3. *Desert salt lakes*: These are specialized environments such as the Great Salt Lake, Utah, where evaporation rates exceed precipitation rates, resulting in salt accumulation.
4. *Volcanic lakes*: These are lakes on volcanic mountain peaks, such as in Japan and the Philippines.
5. *Chemically stratified lakes*: Examples include Big Soda Lake in Nevada. These lakes are stratified due to different densities of water caused by dissolved chemicals. They are meromictic, which means partly mixed.
6. *Polar lakes*: These are lakes in the polar regions; their surface water temperature is mostly below 4°C (39°F).
7. *Marl lakes* (Kevern et al. 1999): These lakes are different in that they generally are very unproductive, yet they may have summertime depletion of dissolved oxygen in the bottom waters and very shallow Secchi disk depths, particularly in the late spring and early summer. These lakes gain significant amounts of water from springs that enter at the bottom of the lake. When rainwater percolates through the surface soils of the drainage basin, the leaves, grass, and other organic materials incorporated in these soils are attacked by bacteria. These bacteria extract the oxygen dissolved in the percolating rainwater and add carbon dioxide. The resulting concentrations of carbon dioxide can get quite high, and when they interact with the water, carbonic acid is formed.

    As this acid-rich water percolates through the soils, it dissolves limestone. When such groundwater enters a lake through a spring, it contains very low concentrations of dissolved oxygen and is supersaturated with carbon dioxide. The limestone that was dissolved in the water reforms very small particles of solid limestone in the lake as the excess carbon dioxide is given off from the lake to the atmosphere. These small particles of limestone are marl and, when formed in abundance, cause the water to appear turbid, yielding a shallow Secchi disk depth. The low dissolved oxygen in the water entering from the springs produces low dissolved oxygen concentrations at the lake bottom.

Impoundments (Shut-Ins)

These are artificial lakes made by trapping water from rivers and watersheds. They vary in their characteristics according to the region and nature of drainage. They have high turbidity and a fluctuating water level. The biomass productivity, particularly of benthos, is generally lower than that of natural lakes (Odum 1971).

LENTIC COMMUNITIES

It was pointed out earlier that lakes are inland depressions containing standing water. Most lakes have outlet streams, but the lake community is quite different from the typical stream community due to the lack of current in its environment. Lentic (still water) communities are inhabited by three different classes of organisms: producers, consumers, and decomposers.

Producers are represented by rooted plants of the littoral zone and phytoplanktons of the limnetic zone. The emergent vegetation of the littoral zone consists of plants such as reeds, cattails, arrowheads, and bulrushes. The floating leaf vegetation is represented by plants such as the water buttercup and water lily. The submerged vegetation is formed of pondweeds and hornworts. These waterweeds have leaves that are thin and finely divided and provide food and resting places for clamberers.

The nonrooted plants of the limnetic zone consist of the phytoplanktons. Although these producers are microscopic and are not readily visible, they often add the distinctive green color to the water. These plants are represented by various types of algae such as blue-green algae, diatoms, filamentous green algae, and flagellate algae such as Euglena.

Consumers in the lentic habitat are represented mainly by crustaceans, insects, mollusks, fish, annelids, helminths, rotifers, and protozoa. All five classes of the aquatic organisms (described earlier) are presented in lentic communities. Benthos (macrobenthos) are represented by sprawlers such as crayfish, mayflies, dragonflies, clams, sludgeworms, and bloodworms. Periphytons (microbenthos) are formed of green algae, protists, diatoms, and clamberers such as mayflies, damselflies, caddisflies, and some beetles. Making up the nektons are organisms such as fish, amphibians, reptiles, large crustaceans, water beetles, and water scorpions. Neustons are represented by water striders and whirligig beetles. Planktons make up the final class. Actually, these planktons are zooplanktons such as water fleas, copepods, rotifers, and protozoa.

Decomposers consist of fungi and bacteria and are generally in the bottom sediments. This is not always the case, however, especially with fungi. For example, when a dead fly falls upon the surface of the water, it soon is enveloped by a halo of white fungi filaments. Fungi and bacteria work together to reduce and transform dead animals and plants into humic substances.

As stated previously, all five classes of organisms are found in lentic communities. Their distribution by zone varies, however. For example, most of the organisms are found in the littoral zone, whereas the limnetic zone has phytoplanktons, zooplanktons, and fish. The profundal zone has some benthos varieties, such as sludgeworms and annelids.

MAJOR DIFFERENCES BETWEEN LOTIC AND LENTIC SYSTEMS

As pointed out in the above sections, there are major differences between lotic (running water) and lentic (standing water) systems. In the following, we highlight these major differences.

- Current
- Open system (lotic) vs. closed system (lentic)
- Temperature and oxygen stratification (lentic)
- Bottom (substrate) types
  - Lotic substrate generally more coarse due to current
  - Lentic substrate generally finer due to deposition
  - Plankton community is an important biological component of lentic systems; usually of minor importance in lotic systems
  - Filter feeders important component of lotic systems; usually minor in lentic

Habitat Changes in the Upper Mississippi River Floodplain

Wlosinski et al. (1995) point out the U.S. Congress recognized the Upper Mississippi River (UMR) as a nationally significant ecosystem in 1986. The UMR extends northward from the confluence of the Mississippi and Ohio rivers to the Twin Cites, Minnesota, a distance of more than 850 miles. The floodplain (area between the bluffs) of the UMR includes 2,110,000 acres of land and water. The Mississippi River is a major migration corridor for waterfowl and provides habitat for more than 150 fish and 40 freshwater mussel species.

Since 1824, the federal government has implemented numerous changes on the UMR. The river was first modified by removing snags and then sandbars, with changes progressing to rock excavation, elimination of rapids, closing of side channels, and the construction of hundreds of wing dams, 27 navigation dams, and hundreds of kilometers of levees. Reservoirs forced by the navigation dams are known locally as pools, which are numbered from north to south. Construction of the dams (mostly during the 1930s) significantly altered the northern 650 miles of the UMR (north of St. Louis, Missouri) by increasing the amount of open water and marsh areas. Wing dams and levees have altered aquatic habitats south of St. Louis (the open river) by reducing open-water habitats and isolating the river from much of the floodplain. Most of the changes to the river ecosystem were either designed for navigational improvements or to control the movement of river water.

Comparison of the land-cover and land-use data between 1891 and 1989 in the dammed portion of the UMR showed that open water and marsh habitats generally increased, mostly at the expense of grass/forb, woody terrestrial, and agricultural classes.

In many pools, inundation created an impounded area with a mosaic of islands, open water, and marsh, which, in general, increased aquatic habitats for fish and wildlife. Although dam construction has benefited aquatic habitats in many pools, the reservoir aging process has reduced these benefits, especially in areas just upriver of the dams.

Sedimentation is also a major concern on the UMR; rates of 0.4–1.2 inches per year have been measured (McHenry, Ritchie, Cooper, and Verdon 1984). Erosion and sedimentation were both detected in comparisons between present elevation data and surveys before dam construction. Erosion was more prevalent in shallow areas and sedimentation more prevalent at greater depths. Erosion and sedimentation cover at depths of three to five feet. This has resulted in a more homogeneous distribution of depth, which is dominated by areas three to five feet in depth. Similar frequency distributions of water depth were observed for lower portions of some pools. Comparison of historical and present bottom geometry revealed the loss of elevational diversity (Wlosinski et al. 1995).

In the areas of the UMR unaffected by navigation dams (the 40-mile stretch of river near Cape Girardeau), there was a 28 percent reduction in open water and a 38 percent reduction in woody and terrestrial habitat between 1891 and 1989. Agricultural areas increased by 15,700 acres. The 4,710-acre reduction of open water can be explained by the construction of levees and wing dams (also known as pile dikes). One large side channel that existed in 1891 was cut off by construction of a levee, reducing the area of water by 1,350 acres. In all, nearly 1,240 miles of levees now isolate more than 988,000 acres from the river during all but the highest discharge rates.

Wing dams and levees, along with other changes to the watershed, have also had a major effect on habitats by changing the relationship between discharge and water-surface elevations. Wing dams have narrowed and deepened the main channel so that water elevations at low discharges are now lower than they were historically. Levees restrict flows and result in higher water elevations during high discharges (Wlosinski et al. 1995).

BIOTA OF THE UPPER MISSISSIPPI RIVER ECOSYSTEM

The Mississippi River is one of the world's major river systems in size, habitat and biotic diversity, and biotic productivity. The navigable Upper Mississippi River, extending 850 miles from St. Anthony Falls (Minnesota) to the confluence with the Ohio River, has been impounded by 27 locks and dams to enhance commercial navigation. The reach between two consecutive locks and dams is termed a "pool." [Note: In the discussion that follows, the analysis is spatially constrained by available data to the reach of river extending from Pool 2 (near Minneapolis-St. Paul, Minnesota) to Pool 19 (near Keokuk, Iowa)]. The upstream portions of many pools are similar to the unimpounded river, whereas the downstream reaches are similar to reservoirs.

The Upper Mississippi River contains a diverse array of wetland, open-water, and floodplain habitats, including extensive national wildlife and fish refuges. Human activities, though, have greatly altered this river ecosystem; much of the watershed is intensively cultivated, and many tributary streams deliver substantial loads of nutrients, pesticides, and sediment from farmland. Pollutants also enter the river from point sources.

In 1992, benthic macroinvertebrates were sampled in soft sediments of five reaches of the Upper Mississippi River and one reach of the Illinois River to estimate densities of fingernail clams and the burrowing mayfly. In 1975 and 1990, benthic macroinvertebrates were extensively sampled in five habitats (marsh, bay, open water, side channel, and dredged side channel) in Pool 8 of the Upper Mississippi (near La Crosse, Wisconsin) to examine changes in abundance, biomass, and community structure (Brewer 1992).

Benthic Macroinvertebrates

Densities of fingernail clams declined significantly in five of eight pools examined along 435 miles of river from Hastings, Minnesota, to Keokuk, Iowa. Densities in Pool 19, which had the longest historical record on fingernail clams, averaged 12,800 per square foot in 1985 and decreased to zero in 1990. In 1992, densities of fingernail clams were still low in sampled areas on the Upper Mississippi and Illinois rivers, averaging 5–94 individuals per square mile. Only 8 percent of 721 samples taken in 1992 had densities exceeding 100 fingernail clams per square mile (9.3 per square foot). Corresponding mean densities of burrowing mayflies in these areas ranged from 0.9–9.2 per square foot.

Wilson et al. (1994) hypothesized that the declines in fingernail clams in Pools 2 to 9 were linked to point-source pollution, and that the declines in Pool 19 were linked to low-flow conditions during drought. The causal mechanisms by which low flow influences fingernail clam abundance may involve unfavorable changes in the chemistry of sediment pore water.

The biodiversity of the unionid mussel fauna in the Upper Mississippi River drainage has declined from about 50 to 60 species in the early 1920s to about 30 species in the mid-1980s. Many of these species are commercially important; others are threatened or endangered. Unionid mussels are further imperiled by the zebra mussel, which recently invaded the Illinois and Upper Mississippi rivers (Wiener et al. 1995).

Rooted Aquatic Plants

The abundance of submerged aquatic plants—including wild celery, which produces a vegetative tuber important as food for certain migratory waterfowl—declined along extensive reaches of the Upper Mississippi River in the late 1980s. This decline has been attributed to changing environmental conditions caused by Pool 7; the abundance of wild celery was fairly stable during 1980–1984, but declined greatly after the dry summer of 1988. In Pools 5 through 9, more than 4,000 hectares (10,000 acres) of wild celery beds were lost (C. E. Korschgen, Upper Mississippi Science Center, unpublished data). Overall, the abundance of wild celery and many other submerged plants declined along 600 kilometers (375 miles) of river from Pool 5 to Pool 19. Coincidentally, the abundance of the exotic plant Eurasian watermilfoil has seemingly increased, particularly in locations formerly occupied by wild celery or other native submerged plants.

Migratory Birds

Millions of migratory birds use the Mississippi River corridor during fall and spring migration. The river is critical in the life cycle of many migratory birds because of its north-to-south orientation and its nearly contiguous habitat. Diving ducks, swans, pelicans, and cormorants use the river's open waters. Dabbling ducks, geese, herons, egrets, terns, bitterns, rails, and many resident and Neotropical songbirds use the shallow riverine wetlands. Bottomland forests support migrating and nesting songbirds, and nesting raptors, herons, egrets, and waterfowl.

The primary factor affecting the use of the river ecosystem by birds is the production of food by various plants and animals. The number of birds in riverine habitats decreases rapidly if preferred food resources are unavailable. The use of Lake Onalaska (Pool 7) by canvasback ducks, for example, decreased greatly when the abundances of their preferred foods, wild celery and benthic invertebrates (Korschgen 1989), declined in the late 1980s. A gradual increase in foods in 1992 resulted in increased use by canvasbacks (Wiener et al. 1995).

Numbers of other migratory waterfowl have also decreased along the river corridor, reflecting deterioration of habitat on the breeding grounds and the river. The decrease in the abundance of fingernail clams has adversely affected waterfowl that feed heavily on the small mollusk, particularly lesser scaup (a small diving duck).

Mink

The abundance of mink on the Upper Mississippi River Refuge declined precipitously during 1959–1965, remained low until about 1970, and then began to slowly increase to numbers that are now less than half those of the 1950s (Dahlgren 1990). In contrast, mink populations in the adjoining states of Iowa, Minnesota, and Wisconsin were relatively stable during this period and did not exhibit the pattern of decline and partial recovery seen in populations on the refuge. These patterns indicate that some factor unique to the river corridor, not present in the mostly agricultural watersheds of the adjoining states, caused the decline of mink populations on the refuge.

The survival and reproduction of mink are adversely affected by dietary exposure to small doses of polychlorinated biphenyls, (PCBs; Aulerich and Ringer 1977; Wren 1991). The decline of mink on the refuge coincided with the probable period of most severe PCB contamination in the river. Conversely, the partial recovery of mink populations that began in the later 1970s coincided with a period of declining PCB levels in riverine fishes (Hora 1984). In 1989–1991, PCB concentrations in mink from the Upper Mississippi River in Minnesota exceeded those in mink from all other areas of the state except Lake Superior (Ensor, Pitt, Helwig 1993). Recent studies show that PCBs continue to enter or cycle within the riverine ecosystem and that they are transferred from the sediment to higher trophic levels via the benthic food chain (Steingraeber, Schwartz, Wiener, and Lebo 1994).

ECOSYSTEM HEALTH

The declines in these riverine flora and fauna signal deterioration in the health of this ecosystem. In recent decades, populations of fingernail clams, unionid mussels, certain other invertebrates, submerged vegetation, migratory waterfowl, and mink have decreases along extensive reaches of the river. The Upper Mississippi River is often heralded as a multiple-use resource, and human use of the river for navigation, hydropower, discharge of wastes, and other purposes may increase while input of sediment, nutrients, and chemicals from the watershed continues. Yet the cumulative impacts of humans may already exceed the assimilative capacity of this ecosystem (Wiener et al. 1995).

Many complex questions concerning environmental degradation, declining flora and fauna, and human impact on this ecosystem need objective analysis and resolution. It is suspected that mink populations declined in response to PCB contamination and that fingernail clams declined in response to sediment toxicity, perhaps linked to low-flow conditions during droughts (Wilson et al. 1994). The factors causing most of the observed biotic declines are largely unknown, however, and are thereby hampering the application of corrective measures. Several factors, for example, are suspected of contributing to declines in the unionid mussel fauna, including habitat modification and degradation, contaminants, overharvest, commercial and recreational navigation, and poor water quality (Williams, Warren, Cummings, Harris, and Neves 1993). The need for scientifically based, integrated resource management of the Upper Mississippi River is illustrated by the economic and ecological effects of the flood of 1993 on the river floodplain and its inhabitants. Federal and state agencies involved with resource management need integrated, proactive polices based on an understanding of the ecological structure and functioning of this complex ecosystem

Fish Populations in the Illinois River

The Illinois River is formed by the confluence of the Des Plaines and Kankakee rivers, about 50 miles southwest of Chicago, Illinois. It then flows 273 miles to join with the Mississippi River about 31 miles northwest of St. Louis, Missouri. The Illinois River has been extensively modified and degraded by industrial and municipal pollution for most of the 20th century (Mills, Starrett, and Bellrose 1966). The upper river reaches above the Stared Rock Dam became the most degraded because most of this pollution originated in the densely populated and heavily industrialized Chicago metropolitan area. In fact, by the late 1920s, the upper river was thought devoid of fish (Thompson 1928). Soon after this period, as pollution-control efforts began to have an effect, fish gradually returned.

Changes in the composition of a fish community in a polluted environment can be a useful index for assessing environmental health and the effectiveness of pollution control because different fish species vary in the ability to tolerate effects of pollution. In 1957, the Illinois Natural History Survey (INHS) initiated an annual electrofishing survey of the Illinois River to monitor fish populations. A central purpose of the survey was to relate changes in fish populations to environmental conditions.

Because recovery of fish populations in the upper Illinois Waterway appears to be a response to pollution-control efforts, definite restoration goals should be identified to help guide further recovery and to determine expectations. In addition, the specific causes for the high incidence of abnormalities in benthic fishes need to be explicitly identified (Lerczak and Sparks 1995).

The Great Lakes Ecosystem
CONTAMINANT TRENDS IN GREAT LAKES FISH

The Great Lakes region is home to many large, industrialized cities and extensive agricultural areas that produce and use an array of potentially toxic chemicals. Hesselberg and Gannon (1995) point out some of these chemicals entering the lakes' food chain have been related to environmental health problems, including poor egg-hatching success, reproductive abnormalities, and birth defects in fish as well as fish-eating birds and mammals. Tumors and other deformities in some fish and wildlife species are also attributed to exposure to toxic contaminants. In addition, fish consumption advisories are issued annually by the Great Lakes states and the province of Ontario for certain fish species and larger sizes of Great Lakes fish that accumulate toxic contaminants.

To measure progress in reducing chemicals in the Great Lakes ecosystem, the National Biological Service's (NBS) Great Lakes Science Center began a contaminant trend-monitoring program in Lake Michigan in 1969. The program was expanded in 1977 to include all of the Great Lakes and additional species of fish through a cooperative agreement between the NBS Great Lakes Science Center and the U.S. Environmental Protection Agency (USEPA), Great Lakes National Program Office. Fish are sampled for this program from 12 sites. All sites were sampled annually through 1982 and thereafter were divided into odd- and even-year sampling regimes. Results from these long-term monitoring programs are extremely valuable in understanding the dynamics of contaminants, developing predictive models for contaminant trends, and determining the effectiveness of regulatory programs.

✔ *Important Point*: In regard to contaminant trends, concentrations of contaminants in fish consistently declined until the mid-1980s, but since then the downward trend has leveled off. Similar trends have been observed in fish in Canadian waters of the Great Lakes (Baumann and Whittle 1988).

Lake Michigan

Contaminants were higher in Lake Michigan lake trout than in fish of any of the other Great Lakes. Both total DDT and PCBs declined; total PCBs did not decline after the voluntary control in 1972 but did after the mandatory ban in 1976 (Hesselberg and Gannon 1995).

In lake trout, dieldrin reached a high in 1978 and a low in 1987. Dieldrin is higher in Lake Michigan fish than in fish from the other Great Lakes, and changes in fish tissue concentrations do not follow use patterns for reasons that are not well understood.

Lake Superior

Total DDT and PCB concentrations in lake trout from Lake Superior were the lowest of all the Great Lakes and generally declined form 1977 to 1990. Dieldrin was always low and varied little from 1977 to 1990. Contaminant concentrations are lowest in Lake Superior because of the low density of agriculture and industry in the lake basin.

Lake Huron

Concentrations of total DDT and PCBs in lake trout from Lake Huron were intermediate between Lake Michigan and Lake Superior. Similar trends of declining concentrations of these chemicals were observed in Lake Huron. Dieldrin concentrations were similar to Lake Superior but declined from a high in 1979 to a low by 1988. With the exception of the Saginaw Valley, both agriculture and industry are much less developed surrounding Lake Huron than Lake Michigan, thereby resulting in lower contaminants in Lake Huron fish.

Lake Ontario

The contaminants in Lake Ontario fish are relatively high, second only to Lake Michigan. Trends in total DDT concentrations in lake trout from Lake Ontario were fairly constant from 1977 to 1990. Total PCBs in lake trout declined significantly from a high in 1977 to a low in 1990, a slower decline than in Lake Michigan. The relatively high contaminant concentrations in Lake Ontario fish are a result of the highly urbanized, industrial, and agricultural basin. In addition, it is the lowermost of the Great Lakes, receiving pollutants from upstream through the Niagara River. Dieldrin concentrations in lake trout from Lake Ontario reached a high in 1979 and then declined to a low by 1988.

Lake Erie

Total DDT, PCB, and dieldrin concentrations in Lake Erie walleye were lower and more similar to concentrations in lake trout in Lake Superior than those of other Great Lakes. Total DDT and PCBs peaked in 1977 and declined to a low in 1982; no consistent trend was noted for dieldrin. Low concentrations of contaminants in Lake Erie were similar to those in Lake Superior even though Lake Erie is surrounded by the largest urbanized, industrial, and agricultural basin of all the Great Lakes. Lake Erie, however, is the shallowest of all the Great Lakes and contains the highest amount of particulate matter. Contaminants flush more quickly through the shallow lake and are removed from the water column as they adhere to particulate matter and settle to the bottom. These factors work together in reducing the amount of contaminants available to fish in Lake Erie (Hesselberg and Gannon 1995).

CONTAMINANT EFFECTS

Reduced reproductive success in fish-eating birds has been linked with DDT and PCBs (Giesy, Ludwig, and Tillitt 1994). As the concentrations of these contaminants have declined, populations of fish-eating birds such as the bald eagle are beginning to recover in the Great Lakes basin. In lake trout, PCBs are also linked to reduced egg hatchability and may also be responsible for fry deformities and mortality (Mac, Schwartz, Edsall, and Frank 1993). In spite of reductions in PCBs in lake trout in all of the Great Lakes, substantial natural reproduction occurs only in Lake Superior (Mac and Edsall 1991). The role of contaminants and other factors in lake trout reproductive problems in the other four Great Lakes is still under investigation.

Another fish health problem associated with toxic chemicals is found in Great Lakes harbors and tributaries where heavy industry was located (Baumann, Mac, Smith, and Harshbarger 1991). Bottom sediments in these areas are heavily contaminated with polycyclic aromatic hydrocarbons (PAHs). Presence of liver tumors and other deformities such as lip papillomas (tumors), stubbed barbels (whisker-like organs), or skin discolorations in bottom-feeding fishes such as the brown bullhead have been linked to the presence of PAHs in the sediment (Baumann et al. 1991; Smith et al. 1994). Tumors and other deformities have been detected in 15 locations (Hartig and Mikol 1992).

Reproductive problems, tumors, and other deformities are still being detected in certain fish and wildlife populations in most of the Great Lakes. Similarly, consumption advisories recommending restrictions on eating certain species and sizes of Great Lakes fish still remain. The United States and Canada have agreed upon a virtual elimination policy for toxic contaminants under the auspices of the Great Lakes Water Quality Agreement. Remedial action plans are being developed by federal and state agencies in cooperation with local municipalities and local citizens to eliminate beneficial use impairments in the most contaminated rivers, harbors, and bays in the Great Lakes. Continued long-term monitoring of contamination in fish is required to determine the success of these programs and to guide where further corrective actions may be necessary.

LAKE TROUT IN THE GREAT LAKES

Hansen and Peck (1995) point out that lake trout populations in the Great Lakes collapsed catastrophically during the 1940s and 1950s because of excessive predation by the sea lamprey and exploitation by fisheries. The lake trout was the top-level predator in most of the Great Lakes as well as an important species harvested by commercial fisheries. Interagency efforts to restore lake trout into the Great Lakes included comprehensive control of sea lamprey populations (Smith 1971), regulation of commercial and recreational fishers, and stocking (Eschmeyer 1968). It was hoped that without sea lamprey predation and fishery exploitation, stocked lake trout could reproduce and eventually restore wild lake trout populations in each of the Great Lakes. Lake trout restoration began during the 1950s in Lake Superior (Hansen et al. 1995), the 1960s in Lake Michigan (Holey et al. 1995), the 1970s in Lake Huron (Eshenroder, Payne, Johnson, Bowen, and Ebener 1995) and Lake Ontario (Elrod et al. 1995), and the 1980s in Lake Erie (Cornelius, Muth, and Kenyon 1995).

Long-term monitoring of lake trout populations relied on catch records of commercial fisheries before the populations collapsed. Later monitoring of lake trout populations relied on assessment fisheries to measure the increase in abundance of stocked fish and, subsequently, on naturally produced fish. At present, natural reproduction by lake trout has been widespread only in Lake Superior. In contrast, lake trout reproduced in only limited areas of Lakes Huron, Michigan, and Ontario, and only in Lake Huron have progeny survived to adulthood.

Lake Superior

Abundance of wild lake trout in the state of Michigan declined from stable levels in the 1930s to nearly zero in the late 1960s (Hansen et al. 1995). In the 1970s and 1980s, abundance of wild lake trout increased steadily, but in the late 1980s and early 1990s decreased slowly because of increased commercial fishing and sea lamprey predation. The abundance of stocked fish increased in the later 1960s well beyond the 1929–1943 average and remained there during most of the 1970s.

Lake trout reestablished self-sustaining populations in much of Lake Superior, though few have reached former levels of abundance. Still, most of these populations are sufficiently large to support limited commercial and sport fishing. Current or proposed strategies for restoring wild lake trout in Lake Superior include controlling fishery exploitation, reducing sea lamprey populations, and reducing or eliminating stocking where self-sustaining populations exist.

Lake Michigan

Wild lake trout populations collapsed in Lake Michigan during the 1940s and the species became extirpated in the 1950s (Holey et al. 1995). Stocking began in the 1960s. The abundance of stocked lake trout increased in the late 1970s, then decreased in the northern part of the lake because of excessive fishery exploitation. Scattered evidence of lake trout reproduction, including eggs deposited on spawning grounds and newly hatched juvenile lake trout, has been found since the 1970s, although the only production of wild lake trout more than one year old was in Grand Traverse Bay during the late 1970s and early 1980s. Unfortunately, excessive fishery exploitation destroyed the wild lake trout produced in Grand Transverse Bay, preventing the establishment of a self-sustaining population (Holey et al. 1995). Current efforts to restore lake trout to Lake Michigan focus on stocking a variety of lake trout strains in offshore refuges that may afford protection from fishery exploitation, allowing restoration of wild populations to occur.

Lake Huron

Wild lake trout populations collapsed in Lake Huron in the 1900s and the species became extirpated in the main basin in the 1950s (Eshenroder et al. 1995). Stocking began in the 1970s. Abundance of stocked fish in southern Michigan waters increased steadily during the 1970s and 1980s then decreased in response to reduced stocking. Abundance in northern Michigan waters increased briefly during the late 1970s and early 1980s, but decreased slowly after that because of excessive sea lamprey predation and fishery exploitation.

Natural reproduction occurred in Thunder Bay, Michigan, and South Bay, Ontario, but self-sustaining populations have not developed at either location. Restoration efforts now focus on reducing the number of sea lampreys and stocking a variety of lake trout strains on off-shore reefs and in a refuge. The refuge, located in the northern part of the lake, may provide protection from fishery exploitation, and thereby may allow a self-sustaining population to become established.

Lake Erie

Wild lake trout populations collapsed in Lake Erie during the 1920s (Cornelius et al. 1995). Stocking began in the 1980s. Abundance of stocked lake trout increased steadily following initial chemical treatment of sea lampreys in 1986–1987, although abundance of stocked lake trout decreased after 1990 for unexplained reasons. Current restoration efforts focus on controlling sea lampreys and stocking yearling lake trout. Research efforts focus on identifying causes of declining abundance of stocked fish and determining whether adult lake trout will aggregate at suitable spawning locations and reproduce successfully.

Lake Ontario

Wild lake trout populations collapsed in Lake Ontario between 1930 and 1960 (Elrod et al. 1995). Stocking began in the 1970s. Stocked lake trout subsequently survived to maturity, spawned, and deposited eggs that hatched into juveniles. These juveniles, however, evidently did not survive to later ages because fishery biologists have not yet discovered any older, wild-origin lake trout. Current restoration efforts focus on stock strains of lake trout that reproduce more successfully. Research focuses on evaluating factors that limit survival of the fry, such as predation and contaminants.

WETLANDS IN REGULATED GREAT LAKES

Wilcox and Meeker (1995) point out water levels in the Great Lakes are affected by variations in precipitation, evaporation, ice buildup, internal waves (seiches), and human alterations that include modifying the connecting channels between lakes and regulating the water levels of Lake Superior and Lake Ontario. Fluctuations in the water level promote the interaction of aquatic and terrestrial systems, thereby resulting in higher quality habitat and increased productivity. When the fluctuations in water levels are reduced through stabilization, shifting of vegetation types decreases, more stable plant communities develop, and species diversity and habitat value decrease (Wilcox and Meeker 1991, 1992). Although water levels in Lake Superior are regulated by structures at the outlet, water-level cycles and patterns remain fairly similar to natural conditions. Lake Ontario water levels are also regulated, but high and low water extremes have been eliminated since the mid-1970s. The effects of water-level history on wetland plant communities under the two regulation regimes were investigated by studying wetlands on each lake.

Seventeen sites on Lake Ontario and 18 on Lake Superior were sampled. Vegetation was mapped and then sampled along transects that followed elevation contours with specific water-level histories (number of years since last flooded or last dry). The histories and elevations differed between lakes. Correlations between specific elevations and accompanying plant communities were assessed across all wetlands sampled in each lake to determine the range of elevations in which the most diverse plant communities occur; these data were used to create schematic cross-sections depicting the structural habitat provided by the plant communities characteristic of each lake (Wilcox and Meeker 1995).

Vegetation and Water Level

At study sites on both Lakes Ontario and Superior, wetland plant communities differed at different elevations; these plant communities developed as a result of the water-level history of each elevation that was sampled. In general, plant communities at elevations that had not been flooded for many years were dominated by shrubs, grasses, and old-field plants. If flooding was more recent, small shrubs that became established after flooding were present, as were grasses, sedges, and other nonwoody plants.

The plant communities at elevations that were flooded periodically at 10- to 20-year intervals and dewatered for successive years between floods had the greatest diversity of wetland vegetation. Dominants included grasses, sedges, rushes, short emergent plants, and submerged aquatic vegetation. At elevations that were rarely or never dewatered, submerged and floating plants were dominant, with emergent plants also occurring at some sites (Wilcox and Meeker 1995).

Lake Superior

Water levels on Lake Superior have been regulated for much of the past century, although the range of fluctuations and the cyclic nature of high and low lake levels have not been altered substantially. More than 275 taxa were recorded in a sampling of 18 wetlands along the U.S. shoreline, 216 of which were *obligate* (i.e., restricted to one particularly characteristic mode of life) or *facultative* (i.e., capable of living under more than one set of environmental conditions) wetland species. Vegetation mapping showed the most prevalent vegetation types to be those dominated by submerged aquatic vegetation or shrubs, both of which were present in all sites and averaged about 25 percent of the cover. Vegetation types dominated by cattails occurred in about half the sites but averaged only about 6 percent of the cover. Across all sites, 27 different vegetation types were mapped (Wilcox and Meeker 1995).

Lake Ontario

Water levels on Lake Ontario have been regulated since 1960, when the St. Lawrence Seaway began operation. Before regulation, the range of fluctuations during the 20th century was about 6.6 feet. After regulation, the range was reduced slightly between 1960 and 1976, but low water-supply conditions in the mid-1960s and high supplies in the mid-1970s maintained much of the range. Regulation reduced the range to about 2.9 feet in the years after 1976.

The lack of alternating flooded and dewatered conditions at the upper and lower edges of the wetlands resulted in establishment of extensive stands of cattail (see figure 8.5) and domination of other areas by purple loosestrife, reed canary grass, and various shrubs. Although more than 250 taxa were recorded in a sampling of 17 wetlands along the U.S. shoreline, only 151 were obligate or facultative wetland plants. Vegetation mapping showed the cattail-dominated vegetation type

Figure 8.5.   Cattails. Abbot Lake, Peaks of Otter, Virginia
Photograph by Frank R. Spellman

to be most prevalent, occurring at all sites and averaging about 32 percent of the cover. The submerged aquatic vegetation type occurred at 75 percent of the sites and averaged about 30 percent of the cover. Across all sites, 20 different vegetation types were mapped.

Habitat Structure

Differences in the species and structural types of plants at different elevations in wetlands of regulated Lakes Superior and Ontario result in different habitats for faunal organisms because the greater diversity of taxa and vegetation types in Lake Superior wetlands provides more niches for fauna than in Lake Ontario wetlands (Engle 1985; Wilcox and Meeker 1992). The prevalence of dominant cattail stands in Lake Ontario wetlands reduces habitat value there (Weller and Spatcher 1965).

Periodic high waters are necessary to reduce dominant emergent vegetation in Great Lakes wetlands; low waters are necessary to reduce dominant submerged vegetation. High waters followed by low-water years allow a diversity of plants to grow from seed on the exposed sediments, reproduce, and replenish and seed banks. Although competitive species such as cattails will again become dominant, the next high-water-level fluctuations will be reduced by regulation, and therefore the processes for rejuvenating wetland plant communities will be lost and habitat values will decrease (Wilcox and Meeker 1995).

Gastropod Fauna in the Mobile Bay Basin

The historical freshwater gastropod fauna of the Mobile Bay Basin in Alabama, Georgia, Mississippi, and Tennessee was the most diverse in the world, comparable only to the diversity reported for the Mekong River in Southeast Asia. This fauna was represented by nine families and about 118 species. Several families have genera endemic to the Mobile Bay Basin (Bogan, Pierson, and Hartfield 1995).

Although this extremely diverse aquatic gastropod fauna has received little attention in the past 50 years, it was actively studied during the second quarter of the 20th century (Goodrich 1922, 1924, 1936, 1944a, 1944b). During the past 60 years, this unique gastropod fauna has declined precipitously (Athearn 1970; Heard 1970; Stansbery 1971). More recent documentation of the decimation of this fauna was presented by Stein (1976) and Palmer (1986). The endemic genus *Tulotoma*, formerly widespread in the main channel of the Alabama and Coosa rivers, was presumed extinct until recently rediscovered (Hershler, Pierson, and Krotzer 1990). The pleurocerid genus *Gyrotoma*, restricted primarily to the shoals of the Coosa River, contained six recognized species, all of which are now presumed extinct.

STATUS AND TRENDS

Bogan et al. (1995) point out that literature records were compiled to document the gastropod species present historically. Recent surveys of the aquatic gastropod fauna of the Coosa and Cahaba river drainages in Alabama have been conducted by using standard field techniques (Bogan and Pierson 1993a, 1993b).

Recent surveys of the aquatic gastropod fauna at about 800 sites have documented population declines, decreases in species' ranges, and the loss of a major portion of the gastropod diversity, especially in the Coosa River. The Coosa River drainage had at least 82 species historically; today, 26 species are presumed extinct.

The fauna of the Cahaba River drainage has fared much better. The Cahaba River drainage does not suffer from the numerous dams, acid mine drainage, pollution from wastewater treatment plants, and water drawn for domestic water use. Species such as *Lepyrium showwalteri* and *Lioplax cyclostomaformis*, formerly much more widespread in the basin, are now apparently restricted to one or two shoal areas in the Cahaba River main channel. The status of the pebblesnails is uncertain. The former diversity of the genus *Somatogyrus* in the Coosa River has probably suffered the same fate as most of the main channel shoal-dwelling pleurocerid species: extinction. Detailed information on the distribution of the freshwater limpets is not available, but they appear to have suffered similar range restrictions.

Declining species diversity can be directly linked to the inundation of the shoal areas of the rivers of the Mobile Bay Basin by impoundment and siltation resulting from a variety of watershed disturbances, including 33 major dams for hydroelectric generation, locks and flood control on the major rivers of the Mobile Bay Basin, and numerous smaller impoundments on tributary rivers and steams. Most gastropods inhabiting shoal areas are gill-breathing species, typically grazing on the plant life growing on the rock substrate in shallow riffle and shoal areas. They formerly lived on rocks in the shallow shoal areas with highly oxygenated water. The pleurocerid gastropod fauna represented a significant portion of the invertebrate biomass living on these shoal areas.

When this habitat was impounded, the snails were not able to survive the deep, cold, and often oxygen-depleted water. Many areas not impounded have suffered because of the heavy siltation of shoal areas, smothering the plant life that formed the diet of these gastropods. Major sources of siltation include poor agricultural and silvicultural practices, lack of riparian buffer zones, and generally poor land-use practices. The drastic decline in gastropod diversity is especially evident in the Coosa River main channel where numerous species formerly found on shoals have disappeared after the damming of the river (Bogan and Pierson 1993b). Other species have had their ranges fragmented by the damming of the rivers and have become restricted to the unimpounded area below the dams with clean current-swept gravel and bedrock outcrops.

Tulotoma magnifica is the only aquatic gastropod now federally listed as endangered; none is listed as threatened, although 104 species of aquatic gastropods from Alabama are on the federal candidate list. Most are from the Coosa and the Cahaba rivers. Conservation and recovery of the remaining diversity will require immediate action to prevent further declines and extinctions. This will necessitate action to improve water quality across the basin and to decrease the amount of silt entering the streams and rivers. In addition, the survey of the aquatic gastropod fauna of the Mobile Bay Basin is not complete, and additional fieldwork in the main channels of the larger rivers is needed, especially on the vertical limestone wall habitats.

Protozoa

The protozoa ("first animals") are a large diverse assemblage of eukaryotic organisms (more than 50,000 known species that have adapted a form or cell to serve as the entire body). All protozoa are single-celled organisms. Typically, they lack cell walls but have a plasma membrane that is used to take in food and discharge waste. They can exist as solitary or independent organisms (the stalked ciliates such as *Vorticella* sp., for example) or they can colonize like the sedentary *Carchesium* sp. Protozoa are microscopic and get their name because they employ the same type of feeding strategy as animals. Most are harmless, but some are parasitic. Some forms have two life stages: active trophozoites (capable of feeding) and dormant cysts.

As unicellular eukaryotes, protozoa cannot be easily defined because they are diverse and, in most cases, only distantly related to each other. Protozoa are distinguished from bacteria by their eukaryotic nature and by their usually larger size. Protozoa are distinguished from algae because protozoa obtain energy and nutrients by taking in organic molecules, detritus, or other protists rather than from photosynthesis. Each protozoan is a complete organism and contains the facilities for performing all the body functions for which vertebrates have many organ systems.

Like bacteria, protozoa depend upon environmental conditions (the protozoan community quickly responds to changing physical and chemical characteristics of the environment), reproduction, and availability of food for their existence. Relatively large microorganisms, protozoa range in size from 4 microns to about 500 microns. They can both consume bacteria (limit growth) and feed on organic matter (degrade waste).

Protozoa are divided into four groups based on their method of motility. The Mastigophora are motile by means of one or more flagella; the Ciliophora by means of shortened modified flagella called cilia; the Sarcodina by means of amoeboid movement; and the nonmotile Sporozoa. In table 8.1, all four groups are listed, but in this text, only the first three, Mastigophora, Ciliates, and Sarcodina are discussed.

Mastigophora (flagellates) protozoa are mostly unicellular, lack specific shape (have an extremely flexible plasma membrane that allows for the flowing movement of cytoplasm), and possess whiplike structures called flagella. The flagella, which can move in whiplike motion, are used for locomotion, as sense receptors, and to attract food.

These organisms are common in both fresh and marine waters. The group is subdivided into the Phytomastigophrea, most of which contain chlorophyll and are thus plantlike. A characteristic species of Phytomastigophrea is the *Euglena* sp., often associated with high or increasing levels of nitrogen and phosphate in the treatment process. A second subdivision of Mastigophora is the Zoomastigopherea which are animal-like and nonpigmented.

Ciliophora (ciliates) are the most advanced and structurally complex of all protozoa. Movement and food-getting is accomplished with short hairlike structures called cilia that are present in at least one stage of the organism's life cycle.

Table 8.1. Classification of Protozoans

Classification of Protozoans			
Group	Common name	Movement	Reproduction
Mastigophora	Flagellates	Flagella	Asexual
Ciliophora	Ciliates	Cili	Asexual by transverse fission Sexual by conjugation
Sarcodina	Amoebas	Pseudopodia	Asexual and sexual
Sporozoa	Sporozoans	Nonmotile	Asexual and sexual

Three groups of ciliates exist: (1) free-swimmers, (2) crawlers, and (3) stalked. The majority are free-living. They are usually solitary, but some are colonial and others are sessile. They are unique among protozoa in having two kinds of nuclei: a micronucleus and a macronucleus. The micronucleus is concerned with sexual reproduction. The macronucleus is involved with metabolism and the production of RNA for cell growth and function.

Ciliates are covered by a pellicle, which may act as a thick armor. In other species, the pellicle may be very thin. The cilia are short and usually arranged in rows. Their structure is comparable to flagella except that cilia are shorter. Cilia may cover the surface of the animal or may be restricted to banded regions.

Sarcodina have fewer organelles and are simpler in structure than the ciliates and flagellates. Sarcodina move about by the formation of flowing protoplasmic projections called pseudopodia. The formation of pseudopodia is commonly referred to as amoeboid movement. The amoebae are well known for this mode of action (see figure 8.6). The pseudopodia not only provide a means of locomotion but also serve as a means of feeding; this is accomplished when the organism puts out the pseudopodium to enclose the food. Most amebas feed on algae, bacteria, protozoa, and rotifers (Spellman 1996).

ENVIRONMENTAL QUALITY INDICATORS

Lipscomb (1995) points out polluted waters often have a rich and characteristic protozoan fauna. The relative abundance and diversity of protozoa are used as indicators of organic and toxic pollution (Cairns, Lanza, and Parker 1972; Foissner 1987; Niederlehner, Pontasch, Pratt, and Cairns 1990; Curds 1992). Bick (1972), for example, provided a guide to ciliates that are useful as indicators of environmental quality of European freshwater systems, along with their ecological distribution with respect to parameters such as the amount of organic material and oxygen levels. Foissner (1988) clarified the taxonomy of European ciliates as part of a system for classifying the state of aquatic habitats according to their faunas.

SYMBIOTIC PROTOZOA

Protozoa are infamous for their role in causing disease, and parasitic species are among the best-known protozoa. Nevertheless, our knowledge has large gaps, especially of normally free-living protozoa that may become pathogenic in immunocompromised individuals. For example, microsporidia comprise a unique group of obligate, intracellular parasitic protozoa. Microsporidia are amazingly diverse organisms with more than 700 species and 80 genera that are capable of infecting a variety of plant, animal, and even other protist hosts. They are found worldwide and have the ability to thrive in many ecological conditions. Until the past few years, their ubiquity did not cause a threat to human health, and few systematists worked to describe and classify the species. Since 1985, however, physicians have

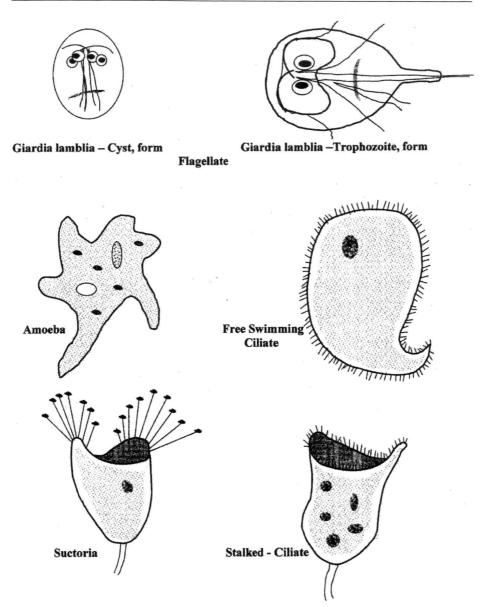

Figure 8.6.   Amoebae and other protozoans

documented an unusual rise in worldwide infections in AIDS patients caused by four different genera of microsporidia. According to the Centers for Disease Control in the United States, difficulties in identifying microsporidian species are impeding diagnosis and effective treatment of AIDS patients (Lipscomb 1995).

Protozoan Reservoirs of Disease

The presence of bacteria in the cytoplasm of protozoa is well known, whereas that of viruses is less frequently reported. Most of these reports simply record the presence of bacteria or viruses and assume some sort of symbiotic relationship between them and the protozoa. Recently, however, certain human pathogens were shown not only to survive but also to multiply in the cytoplasm of free-living, nonpathogenic protozoa (Lipscomb 1995). Indeed, it is now believed that protozoa are the natural habitat for certain pathogenic bacteria. To date, the main focus of attention has been on the bacterium *Legionella pneumophila*, the causative organism of Legionnaires' disease; these bacteria live and reproduce in the cytoplasm of some free-living amoebae (Curds 1992).

SYMBIONTS

Some protozoa are harmless or even beneficial symbionts. A bewildering array of ciliates, for example, inhabit the rumen and reticulum of ruminates and the cecum and colon of equids (e.g., horses, donkey, zebras). Little is known about the relations of the ciliates to their hosts, but a few may aid the animal in digesting cellulose.

ECOLOGICAL ROLE OF PROTOZOA

Although protozoa are frequently overlooked, they play an important role in many communities where they occupy a range of trophic levels. As predators upon unicellular or filamentous algae, bacteria, and microfungi, protozoa play a role both as herbivores and as consumers in the decomposer link of the food chain. As components of the micro- and meiofauna, protozoa are an important food source for microinvertebrates. Thus, the ecological role of protozoa in the transfer of bacterial and algal production to successive trophic levels is important (Lipscomb 1995).

FACTORS AFFECTING GROWTH AND DISTRIBUTION

Most free-living protozoa reproduce by cell division (exchange of genetic material is a separate process and is not involved in reproduction in protozoa). The relative importance for population growth of biotic versus chemical-physical components of the environment is difficult to ascertain from the existing survey data. Protozoa are found living actively in nutrient-poor to organically fresh waters and in freshwater varying between 0C (32°F) and 50C (122°F). Nonetheless, it appears that rates of population growth increase when food is not constrained and temperature is increased (Lee and Fenchel 1972; Fenchel 1974; Montagnes, Lynn, Roff, and Taylor 1988).

Comparison of oxygen consumption in various taxonomic groups show wide variation (Laybourn and Finlay 1976), with some aerobic forms able to function at extremely low oxygen levels, thereby avoiding competition and predation. Many parasitic and a few free-living species are obligatory anaerobes (grow without atmospheric oxygen). Of the free-living forms, the best known are the plagioplids ciliates that live in the anaerobic sulfide-rich sediments of marine wetlands (Fenchel, Perry, and Thane 1977). The importance of plagioplids in recycling nutrients to aerobic zones of wetlands is potentially great.

ECOLOGICAL INTERACTIONS

Because of the small size of protozoa, their short generation time, and (for some species) ease of maintaining them in the laboratory, ecologists have used protozoan populations and communities to investigate competition and predation. The result has been an extensive literature on a few species studied primarily under laboratory conditions. Few studies have been extended to natural habitats with the result that we know relatively little about most protozoa and their roles in natural communities. Intraspecific competition for common resources often results in cannibalism, sometimes with dramatic changes in morphology of the cannibals (Giese 1973). Field studies of interspecific competition are few and most evidence for such species interaction is indirect (Cairns and Yongue 1977).

Aquatic Algae

You don't have to be an environmental specialist to understand that algae can be a nuisance. Many ponds, lakes, rivers, and streams in the United States (and elsewhere) are currently undergoing eutrophication, the enrichment of an environment with inorganic substances (phosphorous and nitrogen). When eutrophication occurs, when filamentous algae such as *Caldophora* break loose in a pond, lake, stream, or river and washes ashore, algae makes its foul, stinking, noxious, decaying presence known.

For environmental specialists, algae are both a nuisance and a valuable ally. In water treatment, for example, although they are not pathogenic, algae are a nuisance. They grow easily on the walls of troughs and basins (treatment unit processes), and heavy growth can cause plugging of intakes and screens. Algae release chemicals that often give off undesirable tastes and odors. In wastewater treatment, on the other hand, controlled algae growth can be valuable in long-term oxidation ponds where the algae aid in the purification process by producing oxygen.

Before beginning a detailed discussion of algae, key terms are defined:

Algae—large and diverse assemblages of eukaryotic organisms that lack roots, stems, and leaves but have chlorophyll and other pigments for carrying out oxygen-producing photosynthesis
Algology or *phycology*—the study of algae
Antheridium—special male reproductive structures where sperm are produced
Aplanospore—nonmotile spores produced by sporangia

Benthic—algae attached and living on the bottom of water bodies

Binary fission—nuclear division followed by division of the cytoplasm

Chloroplasts—packets that contain chlorophyll a and other pigments

Chrysolaminarin—the carbohydrate reserve in organisms of division Chrysophyta

Diatoms—photosynthetic, circular or oblong chrysophyte cells

Dinoflagellates—unicellular, photosynthetic protistan algae

Epitheca—the larger part of the frustule (diatoms)

Euglenoids—contain chlorophylls a and b in their chloroplasts; representative genus is *Euglena*

Fragmentation—a type of asexual algal reproduction in which the thallus breaks up and each fragmented part grows to form a new thallus

Frustule—the distinctive two-piece wall of silica in diatoms

Hypotheca—the small part of the frustule (diatoms)

Neustonic—algae that live at the water-atmosphere interface

Oogonia—vegetative cells that function as female sexual structures in algal reproductive system

Pellicle—a *Euglena* structure that allows for turning and flexing of the cell

Phytoplankton—made up of algae and small plants

Plankton—free-floating, mostly microscopic aquatic organisms

Planktonic—algae suspended in water as opposed to attached and living on the bottom (benthic)

Prototbecosis—a disease in humans and animals caused by the green algae *Prototheca moriformis*

Thallus—the vegetative body of algae

Algae are autotrophic, contain the green pigment chlorophyll, and are a form of diverse aquatic plants. Algae differ from bacteria and fungi in their ability to carry out photosynthesis—the biochemical process requiring sunlight, carbon dioxide, and raw mineral nutrients. Photosynthesis takes place in the chloroplasts. The chloroplasts are usually distinct and visible. They vary in size, shape, distribution, and numbers. In some algal types, the chloroplast may occupy most of the cell space. They usually grow near the surface of water because light cannot penetrate very far through turbid water. Although in mass the unaided eye (multicellular forms such as marine kelp) easily sees them, many of them are microscopic. Algal cells may be nonmotile, motile by one or more flagella, or exhibit gliding motility as in diatoms. They occur most commonly in water (fresh and polluted water, as well as in salt water) in which they may be suspended (planktonic) phytoplanktons or attached and living on the bottom (benthic). A few algae live at the water-atmosphere interface and are termed *neustonic*. Within the fresh and saltwater environments, they are important primary producers (the start of the food chain for other organisms). During their growth phase, they are important oxygen-generating organisms and constitute a significant portion of the plankton in water. Algae also occur in such varied places as the surface layers of soils and porous rocks, on the bark and leaves of trees, in snow, in hot springs, and in symbiotic association with fungi to form lichens (Spellman 1996).

CLASSIFYING ALGAE

According to the five kingdom system of Whittaker, the algae belong to seven divisions distributed between two different kingdoms. Although seven divisions of algae occur, only five divisions are discussed in this text:

- Chlorophyta—green algae
- Euglenophyta—euglenids
- Chrysophyta—golden-brown algae, diatoms
- Phaeophyta—brown algae
- Pyrrophyta—dinoflagellates

The primary classification of algae is based on cellular properties. Several characteristics are used to classify algae, including: (1) cellular organization and cell wall structure; (2) the nature of chlorophyll(s) present; (3) the type of motility, if any; (4) the carbon polymers that are produced and stored; and (5) the reproductive structures and methods.

Algal Cell Wall

Algae show considerable diversity in the chemistry and structure of their cell walls. Some algal cell walls are thin, rigid structures usually composed on cellulose modified by the addition of other polysaccharides. In other algae, the cell wall is strengthened by the deposition of calcium carbonate. Other forms have chitin present in the cell wall. Complicating the classification of algal organisms are the euglenids, which lack cell walls. In diatoms, the cell wall is composed of silica. The frustules (shells) of diatoms have extreme resistance to decay and remain intact for long periods of time, as the fossil records indicate.

Chlorophyll

The principal feature used to distinguish algae from other microorganisms (for example, fungi) is the presence of chlorophyll and other photosynthetic pigments in the algae. All algae contain chlorophyll a. Some, however, contain other types of chlorophylls. The presence of these additional chlorophylls is characteristic of a particular algal group. In addition to chlorophyll, other pigments encountered in algae include fucoxanthin (brown), xanthophylls (yellow), carotenes (orange), phycocyanin (blue), and phycoerythrin (red).

Motility

Many algae have flagella (a threadlike appendage). The flagella are locomotor organelles that may be a single polar or multiple polar type. The *Euglena* is a simple flagellate form with a single polar flagellum. Chlorophyta have either two or four polar flagella. Dinoflagellates have two flagella of different lengths. In some cases, algae are nonmotile until they form motile gametes (a haploid cell or nucleus) during sexual reproduction. Diatoms do not have flagella, but have gliding motility.

Algal Nutrition

Algae can be either autotrophic or heterotrophic. Most are photoautotrophic; they require only carbon dioxide and light as their principal source of energy and carbon. In the presence of light, algae carry out oxygen-evolving photosynthesis; in the absence of light, algae use oxygen. Chlorophyll and other pigments are used to absorb light energy for photosynthetic cell maintenance and reproduction. One of the key characteristics used in the classification of algal groups is the nature of the reserve polymer synthesized as a result of utilizing carbon dioxide present in water.

Algal Reproduction

Algae may reproduce either asexually or sexually. Three types of asexual reproduction occur: binary fission, spores, and fragmentation. In some unicellular algae, binary fission occurs where the division of the cytoplasm forms new individuals such as the parent cell following nuclear division. Some algae reproduce through spores. These spores are unicellular and germinate without fusing with other cells. In fragmentation, the thallus breaks up and each fragment grows to form a new thallus.

Sexual reproduction can involve union of cells where eggs are formed within vegetative cells called *oogonia* (which function as female structures), and sperm are produced in a male reproductive organ called *antheridia*. Algal reproduction can also occur through a reduction of chromosome number or the union of nuclei.

Algal Divisions

1. *Chlorophyta* (green algae). The majority of algae found in ponds belong to this group; they also can be found in salt water and the soil. Several thousand species of green algae are known today. Many are unicellular; others are multicellular filaments or aggregated colonies. The green algae have chlorophylls a and b, along with specific carotenoids, and they store carbohydrates as starch. Few green algae are found at depths greater than seven to ten meters, largely because sunlight does not penetrate to that depth. Some species have a holdfast structure that anchors them to the bottom of the pond and to other submerged inanimate objects. Green algae reproduce by both sexual and asexual means.
2. *Euglenophyta* (euglenoids). Euglenoids are a small group of unicellular microorganisms that have a combination of animal and plant properties. Euglenoids lack a cell wall, possess a gullet, have the ability to ingest food, have the ability to assimilate organic substances, and, in some species, are absent of chloro-

plasts. They occur in fresh, brackish, and salt waters, and on moist soils. A typical *Euglena* cell is elongated and bounded by a plasma membrane; the absence of a cell wall makes them very flexible in movement. Inside the plasma membrane is a structure called the pellicle that gives the organism a definite form and allows the cell to turn and flex. Euglenoids that are photosynthetic contain chlorophylls a and b, and they always have a red eyespot (stigma) that is sensitive to light. Some euglenoids move about by means of flagellum; others move about by means of contracting and expanding motions. The characteristic food supply for euglenoids is a lipopolysaccharide. Reproduction in euglenoids is by simple cell division.

3. *Chrysophyta* (golden brown algae). The Chrysophycophyta group is quite large—several thousand diversified members. They differ from green algae and euglenoids in that: (1) chlorophylls a and c are present; (2) fucoxanthin, a brownish pigment, is present; and (3) they store food in the form of oils and leucosin, a polysaccharide. The combination of yellow pigments, fucoxanthin, and chlorophylls causes most of these algae to appear golden brown. The Chrysophycophyta is also diversified in cell wall chemistry and flagellation. The division is divided into three major classes: golden-brown algae, yellow-brown algae, and diatoms.

   Some Chrysophyta lack cell walls; others have intricately patterned coverings external to the plasma membrane, such as walls, plates, and scales. The diatoms are the only group that has hard cell walls of pectin, cellulose, or silicon, constructed in two halves (the epitheca and the hypotheca) called a frustule. Two anteriorly attached flagella are common among Chrysophyta; others have no flagella.

   Most Chrysophyta are unicellular or colonial. Asexual cell division is the usual method of reproduction in diatoms; other forms of Chrysophyta can reproduce sexually. Diatoms have direct significance for humans. Because they make up most of the phytoplankton of the cooler ocean parts, they are the ultimate source of food for fish. Water and wastewater operators understand the importance of their ability to function as indicators of industrial water pollution. As water quality indicators, their specific tolerances to environmental parameters such as pH, nutrients, nitrogen, concentration of salts, and temperature have been compiled.

4. *Phaeophyta* (brown algae). With the exception of a few freshwater species, all algal species of this division exist in marine environments as seaweed. They are a highly specialized group, consisting of multicellular organisms that are sessile (attached and not free-moving). These algae contain essentially the same pigments seen in the golden-brown algae, but they appear brown because of the predominance of and the masking effect of a greater amount of fucoxanthin. Brown algal cells store food as the carbohydrate laminarin and some lipids. Brown algae reproduce asexually.

5. *Rhodophyta* (dinoflagellates). The principal members of this division are the dinoflagellates. The dinoflagellates comprise a diverse group of biflagellated and nonflagellated unicellular, eukaryotic organisms. The dinoflagellates oc-

cupy a variety of aquatic environments with the majority living in marine habitats. Most of these organisms have a heavy cell wall composed of cellulose-containing plates. They store food as starch, fats, and oils. These algae have chlorophylls a and c and several xanthophylls. The most common form of reproduction in dinoflagellates is by cell division, but sexual reproduction has also been observed (Spellman 1996).

Knowledge of the algae of the United States is not uniform across various groups or environments. Some modern regional floras or lists of plants (e.g., California, southeastern coast, gulf coast) are available for marine benthic macroscopic algae (Dawes 1974; Abbott and Hollenberg 1976; Schneider and Searles 1991), of which there are approximately 900 species on the Pacific Coast and fewer on the Atlantic and Gulf coasts. Local floras are available for many places. Few species are shared between the Atlantic and Pacific coasts. Information about marine microalgae is less accessible.

In general, distribution, status, and trends of algae, even of conspicuous marine algae, are not well established. Floras usually provide ranges, but distribution of many species may be discontinuous, with various causes for the discontinuity. Filling the gaps (or confirming the discontinuities) will require considerable effort.

Although nationwide data on the status and trends of North American algal populations are not readily available, scientists do know that a great deal of formerly aquatic habitat has become unavailable for algae because of landfill, reclamation, and water diversion. In addition, other habitat has been altered through farming and municipal and industrial waste discharge. In the case of reservoirs, however, one kind of aquatic habitat has been replaced by another.

Diatoms

Diatoms are photosynthetic unicellular organisms. They are found in almost all aquatic and semiaquatic habitats and are of great ecological importance because they form an important part of the base of the food web. Although diatoms are widely distributed as a group, most species occur only in habitats with specific physical, chemical, and biological characteristics. Ecologists have long made practical use of this habitat specificity by collecting and analyzing individual species and community data to determine the quality or condition of aquatic habitats. Both long-term monitoring of specific lake and stream habitats and analysis of diatom remains (that become part of the sedimentary record of lakes) allow scientists to obtain a unique long-term historical perspective on these ecosystems. This perspective is especially valuable in assessing the long-term effects of human activities on aquatic and terrestrial ecosystems. Diatoms have been studied throughout the country, but no reasonably complete compilation or summary of these studies exists (Charles and Kociolek 1995).

✔   *Important Point*: The ability to infer ecosystem status and trends from diatoms is largely dependent on the availability of ecological data for the species occurring at study sites.

Diatoms are divided into two groups based on overall symmetry of the cell walls; radially symmetrical forms are informally called "centric" diatoms while bilaterally symmetrical forms are referred to as "pennate" diatoms. One remarkable aspect of these organisms is that they have cell walls made of glass (silicon dioxide). The glass cell walls are ornamented with many holes that are usually arranged in definite patterns. The nature of these perforations as well as their orientation and densities help in the identification of diatom species. Diatom cell walls come in two pieces that fit together the way a Petri dish or pill box does. When these organisms divide, each half reproduces a "daughter" half that, because of the rigidity of the glass walls, must be smaller than the original half.

Despite the important role diatoms play in aquatic ecosystems and their utility in evaluation and monitoring environmental change in these systems, intensive floristic or taxonomic studies on freshwater diatoms in North American have been limited. A two-volume work entitled *The Diatoms of the United States* (Patrick and Reimer 1966, 1975) considered a selected number of genera, and in those genera treated only those species reported from the United States up to 1960. There are only a few regional or statewide taxonomic treatments of diatoms in the United States. The focus has been on specific habitats; areas receiving the most attention have been the Northeast, upper Midwest, the Great Lakes, and isolated areas in the West. Only a few checklists of diatom taxa exist.

Diatom assemblages provide the basis for any important assessments of trends in the status of freshwater ecosystems. These versatile indicators tell us about the acidification of lakes caused by acidic deposition, the eutrophication of lakes caused by human impacts and changing land use, improvements and declines in the quality of our rivers and streams, and changes in climate over the past thousands of years. Because diatoms are important components of the biological community and food web, and are sensitive to changes in water quality, they provide information on both the biological integrity of the ecosystem and those factors likely to be causing any observed changes. Researchers are rapidly developing new techniques for using diatoms to provide even more quantitative and accurate inferences of ecosystem condition and diatoms are being included in a growing number of local and regional-scale monitoring programs.

LAKE ACIDIFICATION

The extent, magnitude, timing, and causes of lake acidification in acid-sensitive regions of the country have been inferred from analysis of diatom assemblages in the stratigraphic record of dated lake sediment cores. These paleolimnological studies show, for example, that about 25–35 percent of the lakes in the Adirondack Mountains with the lowest ability to neutralize acids (acid neutralizing capacity < 400 eq/L) have become more acidic since preindustrial times (Cumming et al. 1992). Lakes in other regions of the country have also acidified but not to the same extent (Charles, Battarbee, Renberg, Van Dam, and Smol 1989). The amount of acidification inferred from diatoms is related to the level of atmospheric loading of strong acid and the ability of watersheds to neutralize those acids. Analysis of diatoms and sedimentary remains of other biological groups (e.g., chrysophytes, chironomids, Cladocera) reveals that acidic deposition has had significant effects on aquatic communities in many lakes. Numbers of taxa are reduced, but some acid-tolerant taxa have significantly increased in abundance.

LAKE EUTROPHICATION

Population estimates of the numbers of lakes in New England and New York that are more eutrophic now than in presettlement times are being obtained from analyses of diatom assemblages from recent and preindustrial levels of sediment cores taken as part of the U.S. Environmental Protection Agency's surface water component of the Environmental Monitoring and Assessment Program (EMAP; Dixit and Smol 1994). The approach of examining lake eutrophication by using diatom assemblages has been widely applied in North America and throughout the world.

RIVERS AND STREAMS

Many long-term diatom data sets exist that can inform us about trends in water quality. The monitoring program conducted by the Federal Water Pollution Control Agency in the 1960s tracked the status of major rivers throughout the country (Williams and Scott 1962). Monitoring of diatom assemblages in rivers and streams is just beginning as part of the U.S. Geological Survey's National Water Quality Assessment (NAWQA) and of the Environmental Monitoring and Assessment Program. The Academy of Natural Sciences of Philadelphia has long-term records for several rivers in the eastern United States. Many of these records show that the quality of water downstream from industrial effluent outfall and sewage treatment plants has improved markedly, but others show worsening conditions, often due to the increased number of sources of stress along the river or in the watershed. Much more could be learned about trends by simply analyzing the immense data that already exist, especially by using new quantitative techniques developed in the past five to ten years.

CLIMATE CHANGE

Diatom assemblage composition is sensitive to changes in water level, salinity, ice cover, wind-mixing patterns, and other characteristics directly and indirectly affected by climate. Paleolimnological studies of sediment cores are providing valuable data on climate change over the past hundreds to thousands of years, which are essential for understanding the nature and magnitude of ecosystem change that can be expected in future years.

Riparian Corridors

Andrews and Auble (1995) note that the strict definition of riparian is "stream-bank" (see figures 8.7A and 8.7B), but riparian ecosystems are often broadly defined to include riverine floodplains. In the broad sense, the riparian zone is both a transition and interface between riverine and upland systems. Functionally and structurally, riparian areas are different from surrounding uplands because of proximity to a water course. In the eastern United States, the upland landscape is generally moist enough to support woody vegetation while the often extensive bottomland forests comprise only those plants able to tolerate flooding and excessive moisture. In much of the West, areas near water courses are often the only places with sufficient moisture for trees. Thus, western riparian ecosystems are often relatively narrow ribbons of trees in a generally unforested landscape.

We lack good estimates of the status of historical changes in area for riparian ecosystems of the West as a whole, although we know that they have always repre-

Figure 8.7A.   Falling spring, Virginia
Photograph by Frank R. Spellman

Figure 8.7B.   Falling spring, Virginia
Photograph by Frank R. Spellman

sented a very small fraction of the land area because of their dependence on water in a dry region. Their importance stems from the unique features that they provide, representing desirable habitat for a variety of species. Many of the same features that make these systems relatively rare and important also make them relatively sensitive.

Western riparian systems have been massively altered in the past 200 years; the history of development in the West is to a large extent one of water development. It is hard to make a hydrologic change without also altering the associated riparian ecosystem. Hydrologic changes can influence the long-term species composition by altering soil salinity and changing the nature of disturbances that create opportunities for regeneration. Some changes on the Middle Rio Grande are relatively straightforward; riparian vegetation is inundated by a reservoir or a channel narrows with lower stream flow. Other effects of hydrologic alteration are more complex and are played out over many decades. The absence of a change in net area may mask dramatic shifts in the location of different vegetation types.

Although hydrology is the dominant factor shaping these ecosystems, it is not the only one. As a case in point, for example, invasions of non-native plants have changed the composition of the communities and the way the systems will likely respond in the future. Moreover, timber clearing, overgrazing by livestock, agricultural conversion, and urban growth are other important causes of change in these ecosystems (Andrews and Auble 1995).

WESTERN RIPARIAN ECOSYSTEMS

In much of western North America, riparian environments are the only part of the landscape moist enough to allow survival of trees. Riparian landscapes are usually defined as ecotones or corridors between terrestrial and aquatic realms (Malanson 1993). In spite of their limited areal extent, riparian ecosystems are essential habitat for many vertebrate species and provide critical physical and biological linkages between terrestrial and aquatic environments (Gregory, Swanson, McKee, and Cummins 1991).

Because of their association with scarce surface water resources, western riparian ecosystems have long been influenced by human activities. Human-caused perturbations can alter energy and material flow in riverine ecosystems, thus modifying riparian plant communities (Brinson 1990). Among the most serious impacts to riparian ecosystems are water impoundment and diversion, groundwater impoundment and diversion, groundwater pumping from alluvial aquifers, livestock grazing, land clearing for agriculture or to increase water yield, mining, road development, heavy recreational demand, fire, the elimination of native organisms (e.g., beaver) or the introduction of exotics, and overall watershed degradation (Stromberg 1993).

Riparian ecosystems along most major western rivers have changed as the result of water development and flood control. Losses of riparian forest downstream of dams have been reported throughout western North America (Rood and Mahoney 1990). In contrast, woodland expansion in other dam-regulated riparian ecosystems provides evidence that the interrelationships between plant communities and hydrogeomorphic processes are complex (Johnson 1994). As the result of widespread, human-induced changes in hydrology and land use, native cottonwood-willow stands are being replaced by non-native woody species such as Russian olive and tamarisk throughout the West (Olson and Knopf 1986; Knopf and Scott 1990; Stromberg 1993).

Most of the Missouri River through the Dakotas to its confluence with the Mississippi River is controlled by a series of large dams and reservoirs constructed between the 1930s and 1950s. These dams radically altered the magnitude, timing, and frequency of flood flows that formerly promoted regeneration and maintenance of extensive riparian cottonwood forests (Johnson, Burgess, and Keammerer 1976; Johnson 1992).

The lower Colorado River riparian ecosystem (Nevada, California, and Arizona) has also been affected by hydrologic change resulting form human activities. Declines in riparian forest dominated by cottonwood and willow have been attributed to change in the physical environment and to the extensive invasion of tamarisk (Busch and Scott 1995).

HYDROLOGY AND RIPARIAN ECOSYSTEM DYNAMICS

Reproduction and growth of riparian plant species are closely associated with peak flows and related channel processes such as meandering. Successful establishment of such plants typically occurs only in channel positions that are moist, bare, and protected from removal by subsequent disturbance (Sigafoos 1964; Everitt 1968; Noble 1979; Bradley and Smith 1986; Stromberg, Pattern, and Richter 1991; Sacchi and Price 1992; Johnson 1994). If stream flow is diverted, young trees may die (Smith, Wellington, Nachlinger, and Fox 1991). Studies of plant water uptake in floodplain ecosystems indicate that maintenance of cottonwood and willow populations depends on groundwater moisture sources that, in turn, are closely linked to in-stream flows (Busch, Ingraham, and Smith 1992). Thus, the establishment and maintenance of riparian plant communities are a function of the interplay among surface water dynamics, groundwater, and river channel processes.

Maps and notes from the journals of Lewis and Clark (1804–1806) suggest that the present distribution and abundance of cottonwoods along the Missouri River within the study reach are generally similar to presettlement conditions. Although flows through this reach are influenced by the Canyon Ferry Dam on the main stem and the Tiber Dam on the Marias River, the gross seasonal timing of flows and the magnitude and frequency of daily maximum flows have not been greatly altered by dam operations. This is due in part to the dam's relatively small storage capacity and the presence of a number of unregulated tributaries below the dams.

In the Colorado River, the link of floodplain groundwater with in-stream flows is illustrated by the association of river discharge and fluctuations in water table depth in the adjacent floodplain. Further evidence for this linkage comes from daily fluctuations in water table depth, which correlated closely with the Colorado River hydroperiod (Busch, unpublished data). Colorado River floodplain soils were dry. Volumetric soil moisture in the upper 3.3 feet of the Colorado River soil profile average 13 percent. Incision of stream channels, through either natural or human-induced causes, can lead to the depression of floodplain water tables (Williams and Wolman 1984). Channelization of the lower Colorado River appears to have led to floodplain groundwater declines, and this has tended to isolate riparian vegetation from its principal moisture source at or near the water table (Busch et al. 1992).

SALINITY AND ALTERATION OF RIPARIAN ECOSYSTEM PROCESSES

In regulated rivers, a lack of flooding or infrequent groundwater incursion into surface soils can result in altered nutrient dynamics. The lack of an aqueous medium for salt dispersal may result in the elevation of soil salinity to levels that are stressful to some of the trees and shrubs native to southwestern riparian ecosystems (Busch and Smith 1995). Colorado River soils were significantly more saline than soils in the adjacent Bill Williams River floodplain. Salinities in Colorado River soils exceeded levels shown to inhibit germination, reduce vigor, and induce mortality in seedling cottonwood and willow (Jackson et al. 1990). Salt-tolerant species could thus benefit from elevated alluvium salinity. Evidence for salinity tolerance in both native and exotic halophytes (plants growing in salty soils or a saltwater environment) shows that arrowweed and tamarisk had significantly higher leaf tissue sodium concentrations than did cottonwood and willow (Busch and Scott 1995).

ESTABLISHMENT PATTERNS OF RIPARIAN TREE POPULATIONS

The structural diversity of riparian cottonwood and willow stands is a function of spatial and temporal patterns of occurrence. These patterns are largely determined by events during the establishment phase (Stromberg et al. 1991; Scott, Wondzell, and Auble 1993). Where stream regulation limits flooding and channel movement (e.g., the lower Colorado River), opportunities for seed germination are limited. In such systems, community structure may become less dynamic unless novel forms of disturbance such as fire increase in importance relative to the nature disturbance regime.

The magnitudes of flows associated with cottonwood establishment are influenced by local channel processes. Along the upper Missouri River, sections of meandering channel alternate with sections where lateral migration does not occur. In meandering sections, successful establishment occurs at relatively low elevations above the channel, producing several bands of even-aged trees (Bradley and Smith 1986).

If, however, lateral movement of the channel is constrained by a narrow valley, successful establishment occurs only at high elevations, often producing a single, narrow band of trees; seedlings initially established at lower positions are removed by water or ice scour. Where the channel is free to move, plant establishment occurs relatively frequently in association with both moderate and high river flows, but where the channel is constrained, plant establishment is associated with infrequent high flows in excess of 50,000 cubic feet per second. Elimination of such high flows would largely eliminate cottonwood and willow stands from the constrained reaches of the upper Missouri River and decrease the frequency of stand establishment in the meandering reaches. From a water-management perspective, then, it is important to recognize how flow variability, including infrequent large flows, shapes the distribution and abundance of riparian tree populations.

DISTURBANCE REGIMES AND THE INVASION OF NON-NATIVE SPECIES

Riparian ecosystems are dependent upon disturbance caused by occasional high flows. Along rivers where these flows have been reduced in frequency and magnitude, natural riparian ecosystems are being lost along with associated invertebrate and vertebrate species. Resource managers concerned with maintaining floodplain ecosystems need to consider ways of preserving flows that produce establishment, growth, and survival of native riparian species. If not, species such as tamarisk can exploit resources more efficiently than native riparian species, thereby altering whole ecosystem properties (Vitousek 1990). Thus, as Hobbs and Huenneke (1992) suggest, modification of the historical disturbance regimes will result in a decline in native species diversity. Although successful plant invasions are often associated with increased disturbance (Hobbs 1989; Rejmanek 1989; Hobbs and Huenneke 1992: Parker, Mertens, and Schemske 1993), in situations where the frequency or intensity of a natural disturbance is decreased, the invasion of competitively superior non-natives may be promoted (Hobbs and Huenneke 1992).

Although most riparian plants are adapted to flooding, the frequency, timing, and duration of floods may be highly altered or regulated in importance relative to flooding as a form of disturbance affecting regulated southwestern rivers, including the Colorado. Colorado River cottonwood and willow canopy cover decreased only slightly following fire, but burned cottonwood-willow stands had significantly greater cover of both arrowweed and tamarisk. Efficiency in water uptake, transport, and use are among the mechanisms responsible for superior postfire recovery of halophyte shrubs compared with trees native to the Colorado River ecosystem (Busch and Smith 1993).

As a result of ecosystem change over the past century, cottonwoods have become rare along the lower Colorado River, and most remaining stands are dominated by senescent (i.e., in decline) individuals. Although a senescent segment was also a substantial portion of the willow population, this species is still relatively abundant in stands classified as cottonwood-willow habitat. Even so, salt-tolerant or water stress-tolerant shrubs such as tamarisk and arrowweed now dominate these habitats.

Similar to tamarisk, the non-native Russian olive is a shrubby tree that has become naturalized throughout the western United States (Olson and Knopf 1986), forming extensive stands in some areas (Knopf and Olson 1984; Brown 1990), particularly where historical river flow patterns have been altered by water development, such as along the Platte River in Nebraska (Currier 1982) and the Bighorn River in Wyoming (Akashi 1988). Such conversion of riparian vegetation from native to non-native species may have profound wildlife management implications. Bird species' richness and density, for example, are higher in native riparian vegetation than in habitats dominated by tamarisk or Russian-olive (Knopf and Olson 1984; Brown 1990; Rosenberg, Ohmart, Hunter, and Anderson 1991).

THE BOTTOM LINE

The health of natural riparian ecosystems is linked to the periodic occurrence of flood flows, associated channel dynamics, and the preservation of base flows capable of sustaining high floodplain water tables. The establishment of native riparian vegetation is diminished when the frequency and magnitude of peak river flows are reduced. Water uptake and water-use patterns indicate that native trees are replaced by non-native species in riparian ecosystems where stream flows are highly modified. Although riparian ecosystems are most directly affected by altered stream flow, additional factors threaten their integrity, including groundwater pumping (Stromberg, Tress, Wilkins, and Clark 1992), grazing (Armour, Duff, and Elmore 1991), timber harvest, land clearing (Brinson, Swift, Plantico, and Barclay 1981), and fire (Busch and Smith 1993). Studies are underway to evaluate whether exotic plants will encroach further into riparian ecosystems, given conditions predicted under global climate change scenarios.

Coastal and Marine Ecosystems

McIvor (1995) notes that the quantity and health of the nation's coastal and marine resources (see figure 8.8) have declined over historical time at the species, community, and ecosystem levels. Human activities contribute to these declines. Human impacts on the coastal and near-shore marine zone include urbanization (direct loss of habitat, lowered water quality), shoreline modification (dredging and filling, diking, and impoundments), overfishing, and high-density recreational use.

Some portion of the overall downward trend is directly attributable to natu-

Figure 8.8.   Sunset in Thailand
Photograph by Revonna Bieber

ral processes. Hurricanes and coastal storms can have significant negative impacts on both barrier islands and seagrass beds. Rising sea level and coastal subsidence—natural processes that are likely being accelerated by anthropogenic (human-caused) activities—are responsible for coastal wetland loss in Louisiana. Rising sea level is also implicated in the erosion of barrier islands. The inescapable conclusion is, however, that even where natural processes play a role, human impact is of equal or greater importance to the long-term health of these resources.

Despite overall declines in coastal and marine resources, there is some room for cautious optimism. Some coral reefs are far enough from human habitation that they are probably stable and not declining. Despite changes in the relative abundances of native fish species and the introduction of exotic species in the tidal portion of the Hudson River, no native fish species have been extirpated within the period of record (1936 to 1990). The population trend for manatee in Florida appears stable and perhaps slightly increasing. Recent local reversals in the decline of seagrasses have occurred in Chesapeake Bay and in lower Tampa and Little Sarasota bays. These successes, however, are tempered by the realization that human populations in coastal states is projected to substantially increase soon.

FISH OF THE TIDAL HUDSON RIVER

The Hudson River drains over about 17,370 square miles, most of it in eastern New York. Although this is a young river with a relatively small watershed at higher latitudes, the Hudson and its tributaries support a rich fish fauna of more than 200 species (Smith and Lake 1990). This fauna is a diverse mixture of native and exotic freshwater species, diadromous (migratory between fresh and salt waters) fish, and marine strays (Barnhouse, Klauda, Vaughan, and Kendall 1988). More than 150 of these species are reported from the tidal portion of the river that extends 151 miles from the battery on Manhattan Island to the Troy Lock; of these, about 80 species are freshwater or diadromous forms and 50 species occur regularly in near-shore areas (Smith 1985). During the past half-century, the near-shore fauna of the tidal portion of the river has undergone two types of changes: Species have been added to and deleted from the fauna, and the relative abundances of the dominant species have changed (Daniels 1995).

Change in the near-shore fish assemblage of the tidal portion of the Hudson River is continuous. To identify trends in the abundance of an assemblage made up of resident freshwater and estuarine species, diadromous fishes, and marine strays, data must be collected in ways that account for the dynamic qualities of the species involved. Although the Hudson River is among the most-studied aquatic systems in North America, data necessary to confirm population trends in its fish assemblage are scant. Abundance data are best for some commercially important and protected fishes. Data on other species are often inadequate, rare, or nonexistent. Early or baseline data are often incompatible with modern surveys, and long-term databases, although growing, are still in their early years.

Some changes appear to be trends. First, the number of fish species in the

Hudson River appears to be increasing. The presence of recent entrants into the river—such as gizzard shad, rudd, grass carp, central mudminnow, white bass, and freshwater drum—may create management concerns in the future.

Second, another group of fish appears to be declining, although it seems that only a few species, if any, have been extirpated. This group consists of fish that were common in the 1936 survey of the river but rare in all recent collections, including the bridle shiner, common shiner, comely shiner, spotfin shiner, creek chub, northern hog sucker, and creek chubsucker. These fish remain common, or at least present, in tributaries to the lower Hudson River. Their absence from the main channel may result from increasing development and loss of riparian vegetation at the mouths of many tributaries, which may isolate tributary populations from those of the main channel and lead to the creation of sub- or new populations.

The third apparent trend is that, although richness is increasing, diversity in the near-shore fish assemblage has declined because of the increase in population size of the dominant species.

Studies that allow a better assessment of trends in the Hudson River fish assemblage will provide broad-based benefits. Management agencies, commercial fishing operations, and individual anglers, for example, all have an interest in the fisheries and fish of the river. Other river users, such as municipal planners and utility companies, also will gain from increased knowledge of the population trends of river-dwelling organisms because the trends reflect changes in water-quality conditions (Daniels 1995).

CHESAPEAKE BAY WATERSHED

The Chesapeake Bay is the nation's largest estuary; its watershed covers 64,000 square miles and is occupied by 13 million people. By the 1980s, the bay's waters were enriched with nutrients from agriculture and loaded with pollutants from urban and suburban areas. The bay's submerged grasses were disappearing, fisheries two centuries old were in serious decline, and wetlands and other natural habitats were under continuing threat of development (Flemer, Mackiernan, Nehlsen and Tippe 1983).

In 1983 the federal government, Virginia, Maryland, Pennsylvania, the District of Columbia, and the Chesapeake Bay Commission formally declared their intent to work cooperatively to restore the natural resources of the bay. Their partnership, known as the Chesapeake Bay Program, attacked water-quality problems by adopting measures to reduce input of nitrogen and phosphorus from urban, industrial, and agricultural sources and to increase levels of dissolved oxygen in bay waters. Simultaneously, scientists and managers determined the status of bay species and natural habitats and began to track historical and ongoing trends.

Status and trends assumed special relevance as they were incorporated into managerial objectives and goals or as indexes of the success of programs and polices

(Chesapeake Bay Implementation Committee 1988). Trends for three habitats—submersed aquatic vegetation of beds, wetlands, and forest; four key aquatic species of oysters, blue crabs, striped bass, and American shad; and waterfowl—are summarized below. These trends represent a mixture of moderate successes and continuing challenges for managers of the bay (Pendleton 1995).

Submerged Aquatic Vegetation (SAV)

Beginning in the late 1960s and continuing into the 1970s, the distribution and abundance of a community of 20 species of submerged grasses declined throughout the bay because of nutrient enrichment, increased loads of suspended sediments, and other factors (Stevenson and Confer 1978; Orth and Moore 1983). In 1978, the first aerial survey estimated 40,700 acres of SAV in the bay (Anderson and Macomber 1980). The next year, 38,000 acres were documented (Orth et al. 1985); since that time, annual surveys have shown modest but continual increases in SAV coverage to an estimated 70,600 acres (Orth, Nowak, Anderson, and Whiting 1993). Recent increases represent gains in brackish mid-bay regions and are tempered somewhat by slow or no SAV recovery in freshwater areas in the upper bay and by the spread of the exotic species hydrilla in the tidal freshwater portions of the Potomac River.

Wetlands

The status and trends for more than a million acres of wetlands in the Chesapeake Bay watershed have been estimated over two time periods, from the mid-1950s to the late 1970s and early 1980s (Tiner and Finn 1986), and from this period to 1989 (Tiner et al., USFWS, unpublished data). Dominant wetland types include nontidal forested wetlands (70 percent of total wetlands), nontidal shrub-scrub wetlands (10 percent), and salt and freshwater marshes (10 percent each).

Losses occurred in all of these wetland types during the period from the mid-1950s to late 1970s and early 1980s. About 9 percent of the watershed's salt marshes were lost to dredging, impoundment, and filling. Nontidal wetlands declined by nearly 6 percent as a result of being drained and converted to agriculture or impounded to form ponds, lakes, and reservoirs. During the 1980s, losses continued; the rate of marsh loss declined, while forested wetland losses increased. Overall, there was an estimated net loss of 0.5 percent of estuarine wetlands and a net loss of 2.0 percent of palustrine ("palus" or marsh; i.e., any inland wetland lacking flowing water) or wetlands (roughly equal to tidal and nontidal wetlands) during the 1980s. These trends mirror historical losses over the past 200 years (Dahl 1990).

Forests

An estimated 95 percent of the Chesapeake Bay was forested before European settlement; around 58 percent remains today (Chesapeake Bay Program 1993).

This percentage is declining for the first time in over a century because of recent forest clearing for urban and suburban development. Forest clearing has proceeded unevenly over the watershed, with some drainage intact and others as much as 85 percent cleared.

Oysters

Oyster landings in Chesapeake Bay have experienced a 95 percent decline since 1980 and are estimated to be at their lowest recorded level (Kennedy 1991; National Marine Fisheries Service, Annapolis, Maryland, unpublished data). Although reproductive success of the oyster remains high (as measured by larval oyster, or spat, set on oyster reefs and other suitable substrates; Maryland Department of Natural Resources, Oxford, Maryland, unpublished data), populations have suffered from harvest to lower levels, two parasitic disease (Dermo and MSX), habitat loss (including decreased water quality), and predation.

Blue Crabs

Blue crab populations in the Chesapeake Bay, as indicated by commercial landings data, vary from year to year, making trends less apparent than those of other bay species (Lipcius and Van Engel 1990; National Marine Fisheries Service, Annapolis, Maryland, unpublished data). Populations appear to follow a 7–12-year cycle and may be in the "trough" of this cycle at present. This perception and increasing annual harvests as fishery efforts shift to crabs from other species have prompted Maryland and Virginia to begin to regulate the blue crab fishery.

Striped Bass

Probably the most monitored fish species in the bay, striped bass populations have increased about 25 percent a year since 1984, after falling to low levels in the early 1980s (Gibson 1993). Increases are at least partially attributed to a moratorium on harvest from 1985 to 1989 to allow improvement of the age and sex structure of the spawning stock. The 1993 young-of-the-year index, a measure of the number of juvenile fish entering the population, is the highest on record (National Marine Fisheries Service, Annapolis, Maryland, unpublished data) and may be related to the timing of high freshwater flows, nutrient inputs, and increases in planktonic prey (Blankenship 1994), which may interact to allow large numbers of young fish to survive after hatching.

American Shad

Like striped bass, American shad have declined in Chesapeake Bay in recent decades; unlike the stripers, this species has not shown a strongly positive population response despite moratoria on fishing in Maryland and Virginia. Long-term trends show a drastic decline in fishery landings to the point of almost total disappearance in the bay (National Marine Fisheries Service, Annapolis, Maryland, unpublished data). This decline has been related to blockages of spawning streams by dams, overharvest, and pollution (Blankenship 1993). Population estimates in 1992 and 1993 for the upper bay, where shad are counted during their upstream migration to the Susquehanna River, show a reversal of a recent positive trend, for reasons yet unknown.

Waterfowl

Midwinter surveys estimate an average of more than 1 million waterfowl along the Atlantic flyway winter in Chesapeake Bay each year (USFWS, Chesapeake Bay Field Office, Annapolis, Maryland, unpublished data). Of the 28 species of ducks, geese, and swans represented in this total, some are declining in abundance, where others show increasing or variable trends in abundance. In general, duck numbers declined and goose populations increased since the late 1950s as submerged aquatic vegetation and other duck foods dwindled and changing farming practices left more grain in fields for geese. Recently, geese have also declined as excessive harvest and poor production on northern breeding grounds reduced their numbers. Their distribution along the Atlantic flyway has also shifted to the north. Mallards and introduced mute swans have shown moderate increases, but many other species, including American black duck, wigeon (common and widespread duck), northern pintail, canvasback, and redhead, have declined or stabilized at population levels substantially lower than in the 1950s.

FLORIDA MANATEES

Lefebvre and O'Shea (1995) point out that the endangered Florida manatee is a survivor. It is one of only three living species of manatees that, along with their closest living relative, the dugong, make up the order Sirenia. This taxonomic distinctiveness reflects their evolutionary and genetic uniqueness. Sirenians are the only herbivorous marine mammals; manatees feed on seagrasses; freshwater plants, including nuisance species such as hydrilla and water hyacinth; and even some shoreline vegetation. Because manatees depend on marine, estuarine, and freshwater ecosystems, our efforts to protect them necessitate protection of aquatic resources.

GULF OF MEXICO COASTAL WETLANDS

The Gulf of Mexico's coastal wetlands are of special interest because the gulf is an exceptionally productive sea that yields more than 2.5 billion pounds of fish and shellfish annually and contains four of the top five fishery ports in the nation by weight (USEPA 1988). The volume of commercial shrimp landings in the gulf has been statistically related to the areal coverage of gulf coastal wetlands (and seagrass beds) that provide crucial nursery habitat to the young (Turner 1977). Coastal wetlands (particularly salt marshes and mangroves) and associated shallow waters function similarly in support of many fish species of commercial interest (Seaman 1985). The gulf wetlands are also well known for their large populations of wild-life, including shorebirds, colonial nesting birds, and 75 percent of the migratory waterfowl traversing the United States (Duke and Kruczynski 1992). The extensive coastal wetlands that remain along the gulf make up about half of the nation's total wetland area (National Oceanic and Atmospheric Administration [NOAA] 1991; Johnston, Watzin, Barras, and Handley 1995).

Galveston Bay

White, Tremblay, Wermund, and Handley (1993) reported both gains and losses in Galveston Bay wetlands from the 1950s to 1989, but the net trend was one of wetland loss, going form 171,000 acres in the 1950s to 138,000 acres in 1989. In general, freshwater scrub-shrub habitats decreased in areas from the 1950s to 1979 and 1989, while forested wetlands increased.

   The five key factors contributing most to wetlands decline in the Galveston Bay since the 1950s are (1) industrial development; (2) urbanization; (3) navigation channels; (4) flood control and multipurpose water projects to meet Houston's future water demand, especially upstream impoundments on the Trinity and San Jacinto rivers; and (5) pollution due to agricultural runoff despite the diminished acreage lost to agricultural expansion. It should be noted that human-induced subsidence due to industrial development (oil and gas activities) and urbanization (groundwater withdrawals) is considered in this analysis (Johnston et al. 1995).

Coastal Louisiana

Coastal wetland losses for Louisiana represent 67 percent of the nation's total loss. Although much of this loss is only indirectly linked to human activities, most of the net current, catastrophic wetland loss is primarily the result of altered hydrology stemming from navigation, flood control, and mineral extraction and transport projects (Sasser, Dozier, Gosselink, and Hill 1986; Louisiana Wetland Protection Panel 1987; Turner and Cahoon 1988). These operations do not always destroy wetlands directly, but they do amplify tidal forces in historically low-energy systems, which upsets the balance of subsidence and accretion, reduces nutrient and sediment influx, decreases freshwater retention, and increases the levels of salt, sulfate, and other substances potentially toxic to indigenous plant species (Good 1993).

   Current wetland losses are concentrated in the southern Deltaic Plain (78 percent). In this region, losses are especially severe in the fringing marshes of the Terrebonne and Barataria basins. Previous losses in the Deltaic Plain occurred primarily in large areas of interior lands. In the Chenier Plain, loss rates were more constant (22 percent); many of the larger areas of loss there seem related to impounded areas with managed water levels.

Mobile Bay

Nonfreshwater marshes surrounding Mobile Bay declined by more than 10,000 acres from 1955 to 1979, representing a loss of 35 percent (Roach, Watzin, and Scurry 1987). When comparing these data to a 1988 wetland habitat map prepared for upper Mobile Bay, it appears that in this portion of the bay no additional net loss of nonfreshwater marsh has occurred since 1979. Some marsh has obviously continued to be lost in certain areas, primarily because of dredge disposal associated with navigation and industry. These losses, though, seem to have been offset by the growth of emergent marsh in existing spoil sites (Watzin, Tucker, and South 1994).

The Southern Science Center's 1988 areal estimates show a substantial increase of 467 acres in freshwater marsh form 1979 to 1988 in upper Mobile Bay. Further investigation revealed that some of this gain was the result of the growth of emergent vegetation in existing disposal areas and in ditches along railroads and highways. Because of disparities in photo interpretation between dates, it is also quite likely that some of these differences are simply due to mapping errors and differences in mapping technique (Watzin et al. 1994).

As a result of mapping errors associated with interpreting forested and scrub-shrub wetlands in the 1956 photographs, Roach et al. (1987) had little faith in the quantitative estimate of change between 1956 and 1979 for these wetland types. The Southern Science Center's 1988 wetland area figures for forested wetlands appear relatively accurate; they indicate that about 1,201 acres of forested wetlands (2.7 percent) were lost in upper Mobile Bay between 1979 and 1988. These losses can be attributed to conversion of forested habitats to scrub-shrub areas (e.g., clear-cutting associated with timber harvest), small impoundments, and commercial and residential development (Watzin et al. 1994).

Tampa Bay

Haddad (1989) reported that between the 1950s and 1980s emergent wetlands decreased 18 percent. During the same time frame, mangroves decreased about 7 percent, salt marshes declined 30 percent, and freshwater wetlands decreased 21 percent. Lewis, Durako, Moffler, and Phillips (1985) estimate that 44 percent of the salt marsh and mangroves has been lost in Tampa Bay since the late 1800s. Although their numbers and those of Haddad (1989) are not readily comparable because of differences in time frame, methodology, vegetation classification, and area mapped, the results taken together confirm that significant losses of wetland habitat have occurred. Marsh and mangrove losses are the product of dredge and fill activities that are now under strict regulatory control; although permitted dredging continues, protective measures exist to minimize loss that is not for public benefit.

The Bottom Line

To protect the future of Gulf Coast wetlands, status and trends over time must be continually recorded and noted in the scientific and public literature. Preliminary data from selected coastal areas studied in the 1980s show a reduced rate of wetland loss compared with earlier decades. While this is good news, the pressures of a continuously expanding human population make it unclear whether this trend will continue in the 21st century. Only additional monitoring data can answer this question (Johnston et al. 1995).

SEAGRASS DISTRIBUTION IN THE NORTHERN GULF OF MEXICO

Seagrass ecosystems are widely recognized as some of the most productive benthic habitats in estuarine and near-shore waters of the gulf coast. Seagrass meadows provide food for wintering habitat for several species of commercially important finfish and shellfish. Physical structure provided by seagrasses affords juveniles refuge from predation and allows for attachment of epiphytes and benthic organism. Seagrass communities also support several endangered and threatened species, including some sea turtles and manatees. Changes in seagrass distribution can reflect the health of a water body, and losses of seagrasses may signal water-quality problem in coastal waters. Losses of seagrasses in the northern Gulf of Mexico over the past five decades have been extensive—from 20 percent to 100 percent for most estuaries, with only a few areas experiencing increases in seagrasses.

Although often considered continuous around the entire periphery of the gulf, seagrasses exist only in isolated patches and narrow bands form Mobile Bay, Alabama, to Aransas Bay, Texas. This pattern of occurrence results from a combination of low salinities, high turbidity, and high wave energy in shallow waters. Seagrasses are more extensively developed from Mobile Bay to Florida Bay.

Seagrass habitats in the Gulf of Mexico have declined dramatically during the past 50 years, mostly because of coastal population growth and accompanying municipal, industrial, and agricultural development. Although proximate causes of local declines can sometimes be identified, most habitat loss has resulted from widespread deterioration of water quality (Neckles 1993).

Five species of seagrasses occur in the Gulf of Mexico: turtle grass, shoal grass, manatee grass, star grass, and widgeon grass. The latter has a distribution in water with lower salinity, but is commonly reported in association with the seagrasses throughout the Gulf Coast.

Seagrass Meadows of the Laguna Madre of Texas

Onuf (1995) notes that a series of lagoons forms an almost continuous fringe of water behind coastal barriers for 310 miles from Galveston Bay, Texas, to the Mexican border. At the northeastern end, where river discharge and precipitation

greatly exceed evaporation from the embayments, fringing marshes are the dominant wetland type. Toward the southwest, freshwater inputs decrease, fringing
marshes are replaced by wind-tidal flats that support highly productive algal mats
during period inundation, and seagrasses dominate the shallow waters of the embayments.

Seagrasses are so prevalent in Laguna Madre that they define the structure of
the physical environment, and are the sources of biological production for the
ecosystem. Consequently, seagrass meadows serve a critical nursery function in
support of the region's rich fisheries, and one waterfowl species has established an
exclusive dependence on Laguna Madre and its most common seagrass. More than
75 percent of the world population of redhead ducks winters in the greater Laguna
Madre ecosystem (inclusive of the Laguna Madre de Tamaulipas, immediately
south of the delta of the Rio Grande in Mexico; Weller 1964) and feeds almost
exclusively on one species of seagrass while in residence (shoal grass). Because of
the degree of dependence of the redhead population on the Laguna Madre and
reports of major disruptions to Laguna Madre's seagrass community, the National
Biological Service began a research program in coastal Texas.

Coastal Barrier Erosion

The conterminous United States has nearly 88,182 miles of tidal shoreline that
exists in a delicate balance with the forces of nature (Culliton et al. 1990). Much
of this shoreline and the coastal barriers in particular are experiencing greatly increased pressures as a result of rapid population growth and accompanying development. Although coastal areas are highly desirable for their abundant natural
resources and habitability, they are also extremely dynamic environments in which
conditions hazardous to humans (e.g., erosion, flooding, pollution) may be present. In many regions, these hazards, which threaten not only humans but also
valuable marine resources and even entire ecosystems, are increasing at alarming
rates as coastal development, recreation, and waste disposal increase, often in direct
conflict with long-term natural coastal processes (Williams and Johnston 1995).

Coastal barriers are geologically recent depositional sand bodies that are
highly variable in shape, size, and their response to natural processes and human
alterations. They may stretch many kilometers in length and contain high sand
dunes—such as the Outer Banks of North Carolina—or they may be small and
isolated islands, so low in relief that they are routinely overwashed by spring tides
and minor storms. Their dynamic nature means coastal barriers are constantly
shifting and being modified by winds and waves, but scientific field investigations
over the past several decades are revealing some disturbing trends.

Long-term survey data by the U.S. Geological Survey and others, based on
analyses of archive maps, reports, and aerial photographs, demonstrate that coastal
erosion is affecting each of the 30 coastal states (Williams, Dodd, and Gohn
1991). About 80 percent of U.S. coastal barriers are undergoing net long-term
erosion at rates of less than 3.3 feet to as much as 65.6 feet per year. Natural
processes such as storms, rises in relative sea level, and sediment starvation (a re-

duction in volume of sediment transported by rivers reaching the coast), which may also be a result of human interference, are responsible for most of their erosion; but human factors such as mineral extraction, emplacement of hard coastal-engineering structures, and dredging of sand from navigation channels are now recognized as having major effects on shoreline stability.

Reef Fishes of the Florida Keys

The Florida Keys is a chain of islands extending 199 miles along the southern edge of the Florida Plateau form Biscayne Bay to the Dry Tortugas 63 miles west of Key West. The Florida Reef tract, a band of living coral reefs paralleling the Keys, extends from Fowey Rocks to the Marquesas and includes about 81 miles of bank reefs and 6,000 patch reefs. For convenience, the Keys can be divided into the Upper, Middle, and Lower Keys (Smith-Vaniz, Bohnsack, and Williams 1995).

The environmental and economic importance of the Florida Keys is indicated by the many protected or regulated areas, which include several national wildlife refuges, national parks, marine sanctuaries, and state-protected areas. Because many recreational and commercial activities occur in near-shore habitats, these areas have high potential for environmental damage.

Relatively high rates of human population increase (28–44 percent) are predicted over the next 20 years in some parts of the Keys; Monroe County, which includes all of the Keys, had a population growth of 160 percent during the past 40 years. Human activities associated with this growth increased the area's overall economy. In recognition of this possibility, the Florida Keys National Marine Sanctuary was designed in 1990 under the Marine Protection, Research, and Sanctuaries Act, U.S. Public Law 101–605. The sanctuary includes 3,673 square miles of coastal waters around the Florida Keys. The Sanctuaries and Reserves Division of the National Oceanic and Atmospheric Administration was charged with developing a comprehensive management plan and regulations to protect sanctuary resources (NOAA 1995).

The diversity of and richness of fishes in the Florida Keys are unparalleled in shelf waters of the continental United States and reflect the mixing of dissimilar faunal components (Gilbert 1973) and the variety of habitats. More fish species have been reported from Alligator Reef in the Upper Keys than at any single location in the Western Hemisphere (Starck 1968). These fishes consist primarily of continental, warm-temperate species characteristic of the northern Gulf of Mexico, and tropical Caribbean species, especially on the Atlantic side of the Florida Keys. Mixing of the warm-temperate and tropical Caribbean species occurs from north to south with distribution limits of individual species determined by seasonal temperature variations and the exchange of Gulf of Mexico and Atlantic Ocean waters in near-shore habitats in the Middle to Lower Keys. The key silverside is the only fish confined to the Florida Keys. It is not as rare as had previously been thought, and a recommendation has been made to change its official state listing from "threatened" to "special concern" (Gilbert 1992).

Two studies of single sites indicate the total diversity of Florida Keys fishes.

Longley and Hildebrand (1941) listed 442 species from the Dry Tortugas, 300 of which are closely associated with coral reefs. Starck (1968) recorded 517 fish species from Alligator Reef, including 389 considered members of the reef community. The category "coral reef fish" is arbitrary, however, because a continuum exists from obligate species that spend their entire adult lives largely hidden within recesses of the reef, to opportunistic species that use many habitats. Also, most economically important reef fish are dependent on seagrasses and mangroves along the Keys and in Florida Bay for critically important nursery habitat. The availability of such habitats permits a higher density of organisms and a more complex reef community (Parrish 1989).

As more researchers, anglers, recreational scuba divers, and snorkelers have visited the Keys, an appreciation of the complex nature of reef fish communities has increased (Sale 1991). Research that uses visual census techniques has focused on the more common and readily observable reef fish. The most comprehensive census study to date (Bohnsack, Harper, McClellan, Sutherland, and White 1987) provided a detailed quantitative description of the fish fauna of Looe Key National Marine Sanctuary for depths less that 43 feet. Quantitative studies of this kind serve as essential baseline references required for monitoring and detecting future changes in reef fish abundances and distributions. That study, with additional data from Key Largo, showed that fish faunas of the outer reefs in the Keys are diverse and complex, and their community structures are similar to well-developed reefs throughout the Caribbean (Smith-Vaniz et al. 1995).

Summary of Key Terms

Aquatic organisms—classified according to their mode of life. There are five classes: benthos, periphytons, plankton, pelagic (nektons), and neustons.
  • Benthos (mud dwellers) live within or on bottom sediments.
  • Periphytons (Aufwucks) live attached to plants or rocks.
  • Planktons (drifters) are small microscopic plants (phytoplankton) and animals (zooplankton) that move about freely with the current.
  • Pelagic (nektons) are organisms swimming freely.
  • Neustons are organisms that live on the surface of the water.
High turbidity—turbulence in water that can reduce light penetration, which can limit photosynthesis.
Lentic class (calm zone)—a class of freshwater ecology consisting of lakes, ponds and swamps. These are composed of five zones: littoral, limnetic, profundal, and benthic zones.
  • Littoral zone is the outermost shallow region of the lentic class, which has light penetration to the bottom.
  • Limnetic zone is the open water zone of the lentic class to a depth of effective light penetration.
  • Euphotic refers to all lighted regions (light penetration) formed of the littoral and limnetic zones.

- Profundal zone is a deep-water region beyond light penetration of the lentic class.
- Benthic zone is the bottom region of a lake.

Limiting Factors—factors in natural waters such as temperature, light, turbidity, dissolved gases, biogenic salts, and water movements that limit the existence, growth, abundance, or distribution of organisms.

Limnology—the study of freshwater ecology, which is divided into two classes: lentic and lotic

Lotic (washed)—class of freshwater ecology consisting of rivers and streams and is composed of two zones: rapids and pools.

- In the rapids zone, the stream velocity prevents sedimentation, with a firm bottom provided for organisms specifically adapted to live attached to the substrate.
- The pool area is a deeper region with a slow enough velocity to allow sedimentation. The bottom is soft due to silts and settleable solids that cause lowered DO due to decomposition.

Nutrients—in natural waters are usually in the form of biogenic salts, which are essential for the synthesis of protoplasm. In streams and rivers, the three primary sources of basic nutrients are runoff, dissolution of rocks, and sewage discharge.

Photosynthetic rate—the rate of photosynthesis depending on the intensity of light and the photoperiod.

Temperature-caused density differences—can produce stratification especially in lakes. A stratified lake can be divided into three horizontal layers: epilimnion (upper, usually oxygenated layer); mesolimnion or thermocline (middle layer of rapidly changing temperature); and hypolimnion (lowest layer, which is subject to deoxygenation).

Water movements—process to mix oxygen into the water, distribute nutrients, and affect the type of bottom.

Chapter Review Questions

8.1   Explain the "balanced aquarium" concept.

8.2   The major difference between land and freshwater habitats is in the _____ in which they both exist.

8.3   Atmospheric air contains at least _____ times more oxygen than water.

8.4   Limnology divides freshwater ecosystems into two groups or classes: _____ and _____ habitats.

8.5   A sessile pond is a _____ pond.

8.6   The topmost zone near the shores of a lake is known as the _____ zone.

8.7   _____ are small organisms that can feed and reproduce on their own and serve as food for small chains.

8.8   Zone of a lake not penetrated by light is called the _____ zone.

8.9   The _____ zone supports scavengers and decomposers.

8.10 Name the three zones of a lotic habitat.

8.11 _____ float or rest on the surface of water.

8.12 Define limiting factor.

8.13 List at least twelve factors that affect biological communities in streams.

8.14 What causes winter kill of fish in winter?

8.15 Oxygen and _____ are often limiting factors in ecosystems.

8.16 The amount of DO is affected inversely by the amount of _____ in the water.

8.17 Explain eutrophication.

8.18 Lake Ontario has reached this stage: _____.

8.19 Small particles of limestone are _____ and, when formed in abundance, cause high turbidity in lakes.

8.20 A lotic system is a _____ system, while a lentic system is a _____ system.

Cited References and Recommended Reading

Abbott, I. A., and Hollenberg, G. J. 1976. *Marine algae of California*. Palo Alto, CA: Stanford University Press.

Akashi, Y. 1988. *Riparian vegetation dynamics along the Bighorn River*. M.S. thesis, University of Wyoming, Laramie.

Allen, J. D. 1996. *Stream ecology: Structure and function of running waters*. London: Chapman & Hall.

American Society for Testing and Materials (ASTM). 1969. *Manual on water*. Philadelphia: American Society for Testing and Materials.

Anderson, B. W., and Ohmart, R. D. 1984. *Vegetation community type maps lower Colorado River*. Boulder City, NV: U.S. Bureau of Reclamation.

Anderson, R. R., and Macomber, R. T. 1980. *Distribution of submerged vascular plants, Chesapeake Bay, Maryland*. Annapolis, MD: Environmental Protection Agency, Chesapeake Bay Program.

Andrews, A. K., and Auble, G. T. 1995. Riparian ecosystems. In *Our living resources*. Washington, DC: U.S. Department of the Interior, National Biological Service.

Armour, C. L., Duff, D. F., and Elmore, W. 1991. The effects of livestock grazing on riparian and stream ecosystems. *Fisheries* 16:7–11.

Athearn, H. 1970. Discussion of Dr. Heard's paper. *Malacologia* 10, no. 1:28–31.

Aulerich, R. J., and Ringer, R. K. 1977. Current status of PCB toxicity to mink, and effect on their reproduction. *Archives of Environmental Contamination and Toxicology* 6:279–92.

Barnhouse, L. W., Klauda, R. J., Vaughan, D. S., and Kendall, R. I., eds. 1988. *Science, law, and Hudson River power plants: A case study in environmental impact assessment*. American Fisheries Society Monograph 4.

Baumann, P. C., Mac, J. J., Smith, S. B., and Harshbarger, H. C. 1991. Tumor frequencies in walleye and brown bullhead and sediment contaminants in tributaries of the Laurentian Great Lakes. *Canadian Journal of Fisheries and Aquatic Sciences* 48:1804–1810.

Baumann, P. C., and Whittle, D. M. 1988. The status of selected organics in the Laurentian

Great Lakes: An overview of DDT, PCBs, dioxins, furans, and aromatic hydrocarbons. *Aquatic Toxicology* 11:241–57.

Bick, H. 1972. *Ciliated protozoa: An illustrated guide to the species used as biological indicators in freshwater biology*. Geneva: World Health Organization.

Blankenship, K. 1993. Biologists puzzled by sudden decline of East Coast shad. *Bay Journal* 3, no. 9:1,6.

Blankenship, K. 1994. Bay bounces back from record-setting spring "freshet." *Bay Journal* 3, no. 10:1,7.

Bogan, A. E., and Pierson, J. M. 1993a. *Survey of the aquatic gastropods of the Cahaba River Basin, Alabama: 1992*. Montgomery: Alabama Natural Heritage Program.

Bogan, A. E., and Pierson, J. M. 1993b. *Survey of the aquatic gastropods of the Coosa River Basin, Alabama: 1992*. Montgomery: Alabama Natural Heritage Program.

Bogan, A. E., Pierson, J. M., and Hartfield, P. 1995. Decline of the freshwater gastropod fauna in the Mobile Bay Basin. In *Our living resources*. Washington, DC: U.S. Department of the Interior, National Biological Service.

Bohnsack, J. A., Harper, D. E., McClellan, D. B., Sutherland, D. L., and White, M. W. 1987. Resource survey of fishes within Looe Key National Marine Sanctuary. *National Oceanic and Atmospheric Administration Technical Memorandum NOS MEMD* 5:1–108.

Bradley C. E., and Smith, D. G. 1986. Plains cottonwood recruitment and survival in a prairie meandering river floodplain, Milk River, southern Alberta and northern Montana. *Canadian Journal of Botany* 64:1433–42.

Brewer, S. K. 1992. *Community structure of benthic macroinvertebrates in Navigation Pool no. 8, Upper Mississippi River: Comparisons between 1975 and 1990*. M.S. thesis, University of Wisconsin, La Crosse.

Brinson, M. M. 1990. Riverine forests. In *Ecosystems of the world*. Vol. 14. *Forested ecosystems*, ed. A. E. Lugo, M. Brinson, and S. Brown, 87–141. Amsterdam: Elsevier.

Brinson, M. M., Swift, B. L., Plantico, R. C., and Barclay, J. S. 1981. *Riparian ecosystems: Their ecology and status*. U.S. Fish and Wildlife Service Biological Report 81/17.

Brown, C. R. 1990. *Avian use of native and exotic riparian habitats on the Snake River, Idaho*. M.S. thesis, Colorado State University, Fort Collins.

Burch, J. B. 1989. *North American freshwater snails*. Hamburg, MI: Malacological.

Busch, D. E., Ingraham, N. L., and Smith, S. D. 1992. Water uptake in woody riparian phreatophytes of the southwestern United States: A stable isotope study. *Ecological Applications* 2:450–59.

Busch, D. E., and Scott, M. L. 1995. Western riparian ecosystems. In *Our living resources*. Washington, DC: U.S. Department of the Interior, National Biological Service.

Busch, D. E., and Smith, S. D. 1993. Effects of fire on water and salinity relations of riparian woody taxa. *Oecologia* 94:186–94.

Busch, D. E., and Smith, S. D. 1995. Mechanisms associated with the decline and invasion of weedy species in two riparian ecosystems of the southwestern *U.S. Ecological Monographs*.

Cairns, J., Lanza, G. R., and Parker, B. C. 1972. Pollution related structural and functional changes in aquatic communities with emphasis on freshwater alga and protozoa. *Proceedings of the National Academy of Sciences* 124:79–127.

Cairns, J., and Ruthven, J. A. 1972. A test of the cosmopolitan distribution of fresh-water protozoans. *Hydrobiologia* 39:405–27.

Cairns, J., and Yongue, W. H. 1977. Factors affecting the number of species of freshwater protozoan communities. In *Aquatic microbial communities*, ed. J. Cairns, 257–303. New York: Garland.

Charles, D. F., Battarbee, R. W., Renberg, I., Van Dam, H., and Smol, J. P. 1989. Paleoecological analysis of lake acidification trends in North America and Europe using diatoms and chrysophytes. In *Acidic prescription: Soils, aquatic processes, and lake acidification*, Vol. 4, ed. S. A. Norton, S. E. Lindberg, and A. L. Page, 207–76. New York: Springer-Verlag.

Charles, D. F., and Kociolek, P. 1995. Freshwater diatoms: Indicators of ecosystem change. In *Our living resources*. Washington, DC: U.S. Department of the Interior, National Biological Service.

Chesapeake Bay Implementation Committee. 1988. *The Chesapeake Bay Program: A commitment renewed*. Annapolis, MD: U.S. Environmental Protection Agency. Chesapeake Bay Program Office.

Chesapeake Bay Program. 1993. *Environmental indicators: Measuring our progress*. Annapolis, MD: U.S. Environmental Protection Agency, Chesapeake Bay Program Office.

Coon, T. G., Eckblad, J. W., and Trygstad, P. M. 1977. Relative abundance and growth of mussels (*Mollusca Eulamellibrachia*) in Pools 8, 9, and 10 of the Mississippi River. *Freshwater Biology* 7:279–85.

Cornelius, F. C., Muth, K. M., and Kenyon, R. 1995. Lake trout rehabilitation in Lake Erie: A case history. *Journal of Great Lakes Research*.

Culliton, T. M., Warren, M. A., Goodspeed, T. R., Remer, D. G., Blackwell, C. M., and McDonough III, J. J. 1990. *Fifty years of population change along the nation's coasts, 1960–2010*. National Oceanic and Atmospheric Administration Coastal Trend Series, 2nd report.

Cumming, B. F., Smol, J. P., Kingston, J. C., Charles, D. F., Birks, H. J. B., Camburn, K. E., Dixit, S. S., Uutala, A. J., and Selle, A. R.1992. How much acidification has occurred in Adirondack region lakes (New York, USA) since preindustrial times? *Canadian Journal of Fisheries and Aquatic Sciences* 49:128–41.

Curds, C. R. 1992. *Protozoa and the water industry*. Cambridge: Cambridge University Press.

Currier, P. J. 1982. *The floodplain vegetation of the Platte River: Phytosociology, forest development, and seedling establishment*. Ph.D. dissertation, Iowa State University, Ames.

Dahl, T. E. 1990. *Wetlands losses in the United States 1780s to 1980s*. Washington, DC: U.S. Department of the Interior, Fish and Wildlife Service.

Dahlgren, R. B. 1990. Fifty years of the harvest on the Upper Mississippi River National Wildlife and Fish Refuge: Consistencies, anomalies, and economics. In *Proceedings of the 46th Annual Meeting of the Upper Mississippi River Conservation Committee*, 142–60, Rock Island, IL.

Daniels, R. A. 1995. Near-shore fish assemblage of the tidal Hudson River. In *Our living resources*. Washington, DC: U.S. Department of the Interior, National Biological Service.

Dawes, C. J. 1974. *Marine algae of the west coast of Florida*. Coral Gables, FL: University of Miami Press.

Dawley, C. 1947. Distribution of aquatic mollusks in Minnesota. *American Midland Naturalist* 38:671–97.

Dixit, S. S., and Smol, J. P. 1994. Diatoms and indicators in the Environmental Monitoring and Assessment Program—Surface Waters (EMAP—SW). *Environmental Monitoring and Assessment* 31:275–306.

Duke, T. W., and Kruczynski, W. L., eds. 1992. *Status and trends of emergent and submerged vegetated habitats of the Gulf of Mexico, USA*. Gulf of Mexico Program, U.S. Environmental Protection Agency, John C. Stennis Space Center, MS.

Duncan, R., and Thiel, P. A. 1983. *A survey of the mussel densities in Pool 10 of the Upper Mississippi River*. Wisconsin Department of Natural Resources Technical Bulletin 139.

Ecological Analysis, Inc. 1981. *Survey of freshwater mussels at selected sites in Pools 11 through 24 of the Mississippi River*. Rock Island, IL: U.S. Army Corps of Engineers, Final Report 9031.

Ellis, M. M. 1931a. A survey of conditions affecting fisheries in the Upper Mississippi River. *U.S. Bureau of Fisheries Circular* 5:1–18.

Ellis, M. M. 1931b. Some factors affecting the replacement of the commercial fresh-water mussels. *U.S. Bureau of Fisheries Circular* 7:1–10.

Elrod, J. H., Gorman, R. O., Schneider, C. P., Eckert, T. H., Schaner, T., Bowlby, J. N., and Schleen, L. P. 1995. Lake trout rehabilitation in Lake Ontario. *Journal of Great Lakes Research.*

Engle, S. 1985. *Aquatic community interactions of submerged macrophytes.* Wisconsin Department of Natural Resources Technical Bulletin 158.

Enger, E., Kormelink, J. R., Smith, B. F., and Smith, R. J. 1989. *Environmental science: An introduction.* Dubuque, IA: William C. Brown.

Ensor, K. L., Pitt, W. C., and Helwig, D. D. 1993. *Contaminants in Minnesota wildlife 1989–1991.* St. Paul: Minnesota Pollution Control Agency.

Eschmeyer, P. 1968. *The lake trout.* U.S. Fish and Wildlife Service. Fishery Leaflet 555.

Eshenroder, R. L., Payne, N. R., Johnson, J. E., Bowen II, C. A., and Ebener, M. P. 1995. Lake trout rehabilitation in Lake Huron. *Journal of Great Lake Research.*

Everitt, B. L. 1968. Use of the cottonwood in an investigation of the recent history of a floodplain. *American Journal of Science* 266:417–39.

Fenchel, T. 1974. Intrinsic rate increase: The relationship with body size. *Oecologia* 14:317–26.

Fenchel, T., Perry, T., and Thane, A. 1977. Anaerobiosis and symbiosis with bacteria in freeliving ciliates. *Journal of Protozoology* 24:154–63.

Finke, A. H. 1966. *Report of a mussel survey in Pools 4 (Lake Pepin), 5, 6, 7, and 9 of the Upper Mississippi River during 1965.* LaCrosse: Wisconsin Department of Natural Resources.

Flemer, D.A., Mackiernan, G., Nehlsen, W., and Tippe, V. 1983. *Chesapeake Bay: A profile of environmental change.* Annapolis, MD: U.S. Environmental Protection Agency, Chesapeake Bay Program Office.

Foissner, W. 1987. Soil protozoa: Fundamental problems, ecological significance, adaptations in ciliates and testaceans, bioindicators, and guide to the literature. *Progress in Protisology* 2:69–212.

Foissner, W. 1988. Taxonomic and nomenclatural revision of Stadecek's list of ciliates as indicators of water quality. *Hydrobiologia* 166:1–64.

Fuller, S. L. H. 1978. *Freshwater mussels of the Upper Mississippi River: Observation at selected sites within the 9-foot channel navigation project on behalf of the U.S. Army Cops of Engineers.* Final Rep. 78–33. Philadelphia, PA: Academy of Natural Sciences.

Fuller, S. L. H. 1979. Historical and current distribution of fresh-water mussels in the Upper Mississippi River. In *Proceedings of the UMRCC Symposium on Upper Mississippi River Bivalve Mollusk,* ed. J. L. Rasmussen, 71–119. Rock Island, IL: Upper River Conservation Committee.

Gibson, M. R. 1993. Historical estimates of fishing mortality on the Chesapeake Bay striped bass stock using separate virtual population analysis applied to market class catch data. *Report to the Atlantic States Marine Fisheries Commission Striped Bass Technical Committee.* Washington, DC: Atlantic States Marine Fisheries Commission.

Giese, A. C. 1973. *Blepharisma.* Palo Alto, CA: Stanford University Press.

Giesy, J. P., Ludwig, J. P., and Tillitt, D. E. 1994. Deformities in birds of the Great Lakes region: Assigning causality. *Environmental Science and Technology* 28:128A–135A.

Gilbert, C. R. 1973. Characteristics of the western Atlantic reef-fish fauna. *Quarterly Journal of the Florida Academy of Sciences (1972)* 35, no. 2–3:130–44.

Gilbert, C. R. 1992. Key silverside. In *Rare and endangered biota of Florida.* Vol. 2. *Fishes,* ed. C. R. Gilbert, 213–17. Gainesville: University Press of Florida.

Good, B. 1993. Louisiana's wetlands: Combating erosion and revitalizing native ecosystems. *Restoration and Management Note* 11:125–33.

Goodrich, C. 1922. The anculosae of the Alabama River drainage. *University of Michigan, Museum of Zoology Miscellaneous Publication* 7:1–57.

Goodrich, C. 1924. The genus Gyrotoma. *University of Michigan Museum of Zoology Miscellaneous Publication* 12:1–29.

Goodrich, C. 1936. Goniobasis of the Coosa River, Alabama. *University of Michigan Museum of Zoology Miscellaneous Publication* 31:1–60.

Goodrich, C. 1944a. Certain operculates of the Coosa River. *The Nautilus* 58, no. 1:1–10.

Goodrich, C. 1944b. Pulmonates of the Coosa River. *The Nautilus* 58, no. 10:11–15.

Gregory, S. V., Swanson, F. J., McKee, W. A., and Cummins, K. W. 1991. An ecosystem perspective of riparian zones. *Bioscience* 41, no. 8:540–51.

Grier, N. M., and Mueller, J. F. 1922. Notes on the naiad fauna of the Upper Mississippi River. II. The naiads of the Upper Mississippi drainage. *Nautilus* 36:46–49, 96–103.

Haddad, K. D. 1989. Habitat trends and fisheries in Tampa and Sarasota bays. In *Tampa and Sarasota bays: Issues, resources, status, and management,* 113–28. National Oceanic and Atmospheric Administration, Estuary-of-the Month Seminar Series 11.

Handley, L. R. 1995. Seagrass distribution in the northern Gulf of Mexico. In *Our living resources.* Washington, DC: U.S. Department of the Interior, National Biological Service.

Hansen, M. J., and Peck, J. W. 1995. Lake trout in the Great Lakes. In *Our living resources.* Washington, DC: U.S. Department of the Interior, National Biological Service.

Hansen, M. J., Peck, J. W., Schorfhaar, R. G., Selgeby, J. H., Schreiner, D. R., Schram, S. T., Swanson, B. L., MacCallum, W. R., Burnham-Curtis, M. K., Heinrich, J. W., and Young, R. J. 1995. Lake trout restoration in Lake Superior. *Journal of Great Lakes Research.*

Hartig, J., and Mikol, G.. 1992. "How clean is clean?" An operational definition for degraded areas in the Great Lakes. *Journal of Environmental Engineering and Management* 2:15–23.

Heard, W. H. 1970. Eastern freshwater mollusks (II). The south Atlantic and gulf drainages. *Malacologia* 10, no. 1:23–27.

Hershler, R., Pierson, J. M., and Krotzer, R. S. 1990. Rediscovery of *Tulotoma magnifica* (Conrad) (*Gastropoda Viviparidae*). *Proceedings of the Biological Society of Washington* 103, no. 4:815–24.

Hesselberg, R. J., and Gannon, J. E. 1995. Contaminant trends in Great lakes fish. In *Our living resources.* Washington, DC: U.S. Department of the Interior, National Biological Service.

Hesselberg, R. J., Hickey, J. P., Northrup, D. A., and Wilford, W. A. 1990. Contaminant residues in the bloater of Lake Michigan, 1969–1986. *Journal of Great Lakes Research* 16:121–29.

Hile, R., Eschmeyer, P. H., and Lunger, G. F. 1951. Status of the lake trout fishery in Lake Superior. *Transactions of the American Fisheries Society* 80:278–312.

Hobbs, R. J. 1989. The nature and effects of disturbance relative to invasions. In *Biological invasions: A global perspective,* ed. J. A. Drake, H. A. Mooney, F. di Castri, R. H. Groves, F. J. Druger, M. Rejmanek, and M. Williamson, 389–405. Chichester, UK: Wiley.

Hobbs, R. J., and Huenneke, I. F. 1992. Disturbance, diversity, and invasion: Implications for conservation. *Conservation Biology* 6:324–37.

Holey, M. E., Rybicki, R. W., Eck, G. W., Brown, Jr. E. H., Marsden, J. E., Lavis, D. S., Toneys, M. L., Trudeau, T. N., and Horrall, R. M. 1995. Progress toward lake trout restoration in Lake Michigan. *Journal of Great Lakes Research.*

Holland-Bartels, L. E. 1990. Physical factors and their influence on the mussel fauna of a main channel border habitat of the Upper Mississippi River. *Journal of the North American Benthological Society* 9:327–35.

Hora, M. E. 1984. Polychlorinated biphenyls (PCBs) in common carp of the Upper Mississippi River. In *Contaminants in the Upper Mississippi River,* ed. J. G Wiener, R. V. Anderson, and D. R. McConville, 231–39. Stoneham, MA: Butterworths.

Jackson, J., Ball, J. T., and Rose, M. R. 1990. *Assessment of the salinity tolerance of eight Sonoran Desert riparian trees and shrubs.* Yuma, AZ: U.S. Bureau of Reclamation.

Johnson, W. C. 1992. Dams and riparian forests: Case study from the upper Missouri River. *Rivers* 3, no. 4:229–42.

Johnson, W. C. 1994. Woodland expansion in the Platt River, Nebraska: Patterns and causes. *Ecological Monographs* 64:45–84.

Johnson, W. F., Burgess, R. I., and Keammerer, W. R. 1976. Forest overstory vegetation and environment on the Missouri River floodplain in North Dakota. *Ecological Monographs* 46:59–84.

Johnston, J. B., Watzin, M. C., Barras, J. A., and Handley, L. R. 1995. Gulf of Mexico coastal wetlands: Case studies of loss trends. In *Our living resources.* Washington, DC: U.S. Department of the Interior, National Biological Service.

Kennedy, V. S. 1991. Eastern oyster. In *Habitat requirements for Chesapeake Bay living resources,* 2nd ed., 3-1–3-20. Annapolis, MD: Chesapeake Bay Program, Living Resources Subcommittee, U.S. Fish and Wildlife Service.

Kevern, N. R., King, D. L., and Ring, R. 1999. Lake classification systems, part I. *The Michigan Riparian,* 1, December.

Knopf, F. L., and Olson, T. E. 1984. Naturalization of Russian-olive implications to Rocky Mountain wildlife. *Wildlife Society Bulletin* 12:289–98.

Knopf, F. L., and Scott, M. L. 1990. Altered flows and created landscapes in the Platte River headwaters, 1840–1990. In *Management of dynamic ecosystems,* ed. J. M. Sweeney, 47–70. West Lafayette, IN: The Wildlife Society, North Central Section.

Korschgen, C. E. 1989. Riverine and deepwater habitats for diving ducks. In *Habitat management for migrating and wintering waterfowl in North America,* ed. L. M. Smith, R. L. Person, and R. M. Kaminski, 157–80. Lubbock: Texas Tech University Press.

Kreier, J. P., and Baker, J. R. 1987. *Parasitic protozoa.* Boston, MA: Allen & Unwin.

Laws, E. A. 1993. *Aquatic pollution: An introductory text.* New York: John Wiley & Sons.

Laybourn, J., and Finlay, B. J. 1976. Respiratory energy losses related to cell weight and temperature in ciliated protozoa. *Oecologia* 44:165–74.

Lee, C. C., and Fenchel, T. 1972. Studies on ciliates associated with sea ice from Antarctica. II. Temperature. Responses and tolerances in ciliates form Antarctica temperate and tropical habitats. *Archive fur Protistenkunde* 114:237–44.

Lefebvre, L. W., and O'Shea, T. J. 1995. Florida manatees. In *Our living resources: Coastal and marine ecosystems.* Washington, DC: U.S. Department of the Interior, National Biological Service, 267–69.

Lerczak, T. V., and Sparks, R. E. 1995. Fish populations in the Illinois River. In *Our living resources.* Washington, DC: U.S. Department of the Interior, National Biological Service.

Lewis, R. R., III, Durako, M. J., Moffler, M. D., and Phillips, R. C. 1985. Seagrass meadows of Tampa Bay. In *Proceedings, Tampa Bay Area Scientific Information Symposium,* ed. S. Treat, J. Simon, R. Lewis III, and R. Whitman Jr., 210–46. Florida Sea Grant Report 65.

Lipcius, R. M., and Van Engel, W. A. 1990. Blue crab population dynamics in Chesapeake Bay: Variation in abundance (York River, 1972–1988) and stock-recruit functions. *Bulletin of Marine Science* 46:180–94.

Lipscomb, D. 1995. Protozoa. In *Our living resources.* Washington, DC: U.S. Department of the Interior, National Biological Service.

Longley, W. H., and Hildebrand, S. F. 1941. Systematic catalogue of the fishes of Tortugas, Florida, with observations on color, habits, and local distribution. *Papers from Tortugas Laboratory No. 34 (Carnegie Institution of Washington Publication 535).*

Louisiana Wetland Protection Panel. 1987. *Saving Louisiana's coastal wetlands: The need for a long-term plan of action.* U.S. Environmental Protection Agency, EPA-230–02–87–026.

Mac, M. J. 1995. Aquatic ecosystems. In *Our living resources.* Washington, DC: U.S. Department of the Interior, National Biological Service.

Mac, M. J., and Edsall, C. C. 1991. Environmental contaminants and the reproductive success of lake tout in the Great Lakes: An epidemiological approach. *Journal of Toxicology and Environmental Health* 33:375–94.

Mac, M. J., Schwartz, T. R., Edsall, C. C., and Frank, A. M. 1993. PCBs in Great Lakes lake trout and their eggs: Relations to survival and congener composition 1979–1988. *Journal of Great Lakes Research* 19:752–65.

Malanson, G. P. 1993. *Riparian landscapes.* Cambridge: Cambridge University Press.

Mathiak, H. A. 1979. *A river survey of the unionid mussels of Wisconsin, 1973–1977.* Horicon, WI: Sand Shell Press.

McHenry, J. R., Ritchie, J. C., Cooper, C. M., and Verdon, J. 1984. Recent rates of sedimentation in the Upper Mississippi River. In *Contaminants in the Upper Mississippi River,* ed. J. G. Wiener, R. V. Anderson, and D. R. McConville, 99–118. Boston, MA: Butterworth.

McIvor, C. C. 1995. Coastal and marine ecosystems. In *Our living resources.* Washington, DC: U.S. Department of the Interior, National Biological Service.

Miller, G. T. 1988. *Environmental science: An introduction.* Belmont, CA: Wadsworth.

Mills, H. B., Starrett, W. C., and Bellrose, F. C. 1966. *Man's effect on the fish and wildlife of the Illinois River.* Illinois Natural History Survey Biological Notes 57.

Moe, R. 1995. Marine and freshwater algae. In *Our living resources.* Washington, DC: U.S. Department of the Interior, National Biological Service.

Montagnes, D. J. S., Lynn, D. H., Roff, J. C., and Taylor, W. D. 1988. The annual cycle of heterotrophic planktonic ciliates it the waters surrounding the Isles of Shoals, Gulf of Maine: An assessment of their trophic role. *Marine Biology* 99:21–30.

Narr, J. 1990. *Design for a livable planet.* New York: Harper & Row.

National Oceanic and Atmospheric Administration (NOAA). 1991. *Coastal wetlands of the United States: An accounting of a national resource base.* National Oceanic and Atmospheric Administration Report 91–3.

National Oceanic and Atmospheric Administration (NOAA). 1995. *Florida Keys National Marine Sanctuary draft management plan/environmental impact statement.* Vols. 1–3. Silver Spring, MD: National Oceanic and Atmospheric Administration.

Neckles, H.A., ed. 1993. *Seagrass monitoring and research in the Gulf of Mexico: Draft report of a workshop held at Mote Marine Laboratory in Sarasota, Florida, January 28–29, 1992.* Lafayette, LA: National Biological Survey, National Wetlands Research Center.

Niederlehner, B. R., Pontasch, K. W., Pratt, J. R., and Cairns, J. 1990. Field evaluation of predictions of environmental effects from multispecies microcosm toxicity test. *Archives of Environmental Contamination and Toxicology* 19:62–71.

Noble, M. G. 1979. The origin of *Populus deltoids* and Salix interior zones on point bars along the Minnesota River. *American Midland Naturalist* 2, no. 1:59–67.

Northington, D. K., and Goodin, J. R. 1984. *The botanical world.* St. Louis: Times Mirror/Mosby College Press.

Odum, E. P. 1971. *Fundamentals of ecology.* Philadelphia: Saunders College.

Olson, T. E., and Knopf, F. L. 1986. Naturalization of Russian-olive in the western United States. *Western Journal of Applied Forestry* 1:65–69.

Onuf, C. P. 1995. Seagrass meadows of the Laguna Madre of Texas. In *Our living resources.* Washington, DC: U.S. Department of the Interior, National Biological Service.

Orth, R. J., and Moore, K. A. 1983. Chesapeake Bay: An unprecedented decline in submerged aquatic vegetation. *Science* 222:51–53.

Orth, R. J., Nowak, J. F., Anderson, G. F., and Whiting, J. R. 1993. *Distribution of submerged aquatic vegetation in the Chesapeake Bay and tributaries and Chincoteague Bay—1992.* Annapolis, MD: U.S. Environmental Protection Agency, Chesapeake Bay Program Office.

Orth, R. J., Simons, J., Allaire, R., Carter, V., Hindman, L., Moore, K., and Rybick, N. 1985. *Distribution of submerged aquatic vegetation in the Chesapeake Bay and tributaries—1984: Final Report.* Annapolis, MD: U.S. Environmental Protection Agency, Chesapeake Bay Program.

Palmer, A. R. 1986. Quantum changes in gastropod shell morphology need not reflect speciation. *Evolution* 39:699–705.

Parker, I. M., Mertens, S. K., and Schemske, D. W. 1993. Distribution of seven native and two exotic plants in a tallgrass prairie in southeastern Wisconsin: The importance of human disturbance. *American Midland Naturalist* 130:43–55.

Parrish, J. D. 1989. Fish communities of interacting shallow-water habitats in tropical oceanic regions. *Marine Ecology Progress Series* 58:143–60.

Patrick, R., and Reimer, C. W. 1966/1975. *The diatoms of the United States.* Vols. 1 and 2. Philadelphia, PA: The Academy of Natural Science.

Pendleton, E. 1995. Natural resources in the Chesapeake Bay watershed. In *Our living resources.* Washington, DC: U.S. Department of the Interior, National Biological Service.

Perry, E. W. 1979. A survey of Upper Mississippi River mussels. In *A compendium of fishery information on the Upper Mississippi river* (2nd ed.), ed. J. L. Rasmussen, 118–39. Rock Island, IL: Upper Mississippi River Conservation Committee.

Platnow, N. S., and Karstad, L. H. 1973. Dietary effect of polychlorinated biphenyls on mink. *Canadian Journal of Comparative Medicine* 37:391–400.

Price, P. W. 1984. *Insect ecology.* New York: John Wiley & Sons.

Pycha, R. L., and King, G. R. 1975. *Changes in the lake trout population of southern Lake Superior in relation to the fishery, the sea lamprey and stocking, 1950–70.* Great Lakes Fishery Commission Technical Report 28, Ann Arbor, MI.

Rejmanek, M. 1989. Invasibility of plant communities. In *Biological invasions: A global perspective,* ed. J. A. Drake, H. A. Mooney, F. di Castri, R. H. Groves, F. J. Kruger, M. Rejmanek, and M. Williamson, 369–88. Chichester, UK: Wiley.

Roach, E. R., Watzin, M. C., and Scurry, J. D. 1987. Wetland changes in coastal Alabama. In *Symposium on the Natural Resources of the Mobile Bay Estuary,* ed. T. A. Lowery, 92–101. Mobile: Alabama Sea Grant Extension Service, A. MASGP-87-007.

Rood, S. B., and Mahoney, J. M. 1990. Collapse of riparian poplar forests downstream from dams in western prairies: Probable causes and prospects for mitigation. *Environmental Management* 14:451–64.

Rosenberg, K. V., Ohmart, R. D., Hunter, W. C., and Anderson, B. W. 1991. *Birds of the lower Colorado River Valley.* Tucson: University of Arizona Press.

Sacchi, C. F., and Price, P. W. 1992. The relative roles of abiotic and biotic factors in seedling demography of arroyo willow. *American Journal of Botany* 79, no. 4:395–405.

Sale, P. F., ed. 1991. *The ecology of fishes on coral reefs.* New York: Academic Press.

Sasser, C. E., Dozier, M. D., Gosselink, J. G., and Hill, J. M. 1986. Spatial and temporal changes in Louisiana's Barataria Basin marshes, 1945–1980. *Environmental Management* 10:671–80.

Schneider, C. W., and Searles, R. B. 1991. *Seaweeds of the southeastern United States: Cape Hatteras in Cape Canaveral.* Durham, NC: Duke University Press.

Scott, M. L., Friedman, J. M., and Auble, G. T. 1994. Fluvial process and the establishment of bottomland trees. *Geomorphology.*

Scott, M. L., Wondzell, M. A., and Auble, G. T. 1993. Hydrograph characteristics relevant to the establishment and growth to western riparian vegetation. In *Proceedings of the 13th annual American Geophysical Union Hydrology Days*, ed. H. J. Morel-Seytoux, 237–46. Atherton, CA: Hydrology Days.

Seaman, W., Jr. 1985. *Florida aquatic habitat and fishery resources*. Kissimmee: Florida Chapter of the American Fisheries Society.

Shimek, B. 1921. Mollusks of the McGregor, Iowa, Region I. *Iowa Conservationist* 5:1.

Sigafoos, R. S. 1964. *Botanical evidence of floods and flood-pain deposition*. U.S. Geological Survey Professional Paper 485A.

Smith, B. R. 1971. Sea lampreys in the Great Lakes of North America. In *The biology of lampreys*, ed. M. W. Hardisty and L. C. Potter, 207–47. Vol. 1. New York: Academic Press.

Smith, C. L. 1985. *Inland fishes of New York State*. Albany, NY: Department of Environmental Conservation.

Smith, C. L., and Lake, T. R. 1990. Documentation of the Hudson River fish fauna. *American Museum Novitiates* 2981.

Smith, R. L. 1974. *Ecology and field biology*. New York: Harper & Row.

Smith, S. B., Bleuin, M. A., and Mac, J. J. 1994. Ecological comparisons of Lake Erie tributaries with elevated incidence of fish tumors. *Journal of Great Lakes Research* 20:701–16.

Smith, S. D., Wellington, A. B., Nachlinger, J. L., and Fox, C.A. 1991. Functional responses of riparian vegetation to streamflow diversion in the eastern Sierra Nevada. *Ecological Applications* 1:89–97.

Smith, T. M., and Smith, R. L. 2006. *Elements of ecology* (6th ed.). New York: Benjamin Cummings.

Smith-Vaniz, W. F., Bohnsack, J. A., and Williams, J. D. 1995. Reef fishes of the Florida Keys. In *Our living resources*. Washington, DC: U.S. Department of the Interior, National Biological Service.

Spellman, F. R. 1996. *Stream ecology and self-purification*. Lancaster, PA: Technomic.

Stansbery, D. H. 1971. Rare and endangered mollusks in eastern United States. In *Proceedings of a symposium on Rare and Endangered Mollusks (Naiads)*, ed. S. E. Jorgenson and R. E. Sharp, 5–18. Washington, DC: U.S. Bureau of Sport Fisheries and Wildlife.

Starck, W. A., II. 1968. A list of fishes of Alligator Reef, Florida, with comments on the nature of the Florida reef fish fauna. *Undersea Biology* 1, no. 1:5–36.

Stein, C. B. 1976. Gastropods. Endangered and threatened species of Alabama, ed. H. Boschung. *Bulletin of Alabama Museum of Natural History* 2:21–41.

Steingraeber, M. T., Schwartz, T. R., Wiener, J. G., and Lebo, J. A. 1994. Polychlorinated biphenyl congeners in emergent mayflies from the Upper Mississippi River. *Environmental Science and Technology* 28:707–14.

Stevenson, J. C., and Confer, N. M. 1978. *Summary of available information on Chesapeake Bay submerged vegetation*. U.S. Fish and Wildlife Service Biological Report FWS/OBS-78/66.

Stromberg, J. C. 1993. Fremont cottonwood-Gooding willow riparian forests: A review of their ecology, threats, and recovery potential. *Journal of the Arizona-Nevada Academy of Science* 26:97–110.

Stromberg, J. C., Pattern, D. T., and Richter, B. D. 1991. Flood flows and dynamics of Sonoran riparian forests. *Rivers* 2, no. 3:221–35.

Stromberg, J. C., Tress, J. A., Wilkins, S. D., and Clark, S. D. 1992. Response of velvet mesquite to groundwater decline. *Journal of Arid Environments* 23:45–58.

Taylor, W., and Sanders, R. 1991. Protozoa. In *Ecology and classification of North American freshwater invertebrates*, ed. J. H. Thorp and A. P. Covich, 37–93. New York: Academic Press.

Tchobanoglous, G., and Schroeder, E. D. 1985. *Water supply*. Reading, MA: Addison-Wesley.

Thiel, P. A. 1981. *Survey of unionid mussels in the Upper Mississippi River (Pools 1-11)*. Madison: Wisconsin Department of Natural Resources, Technical Bulletin 124.

Thiel, P. A., Talbot, M., and Holzer, J. 1979. Survey of mussel in the Upper Mississippi River, Pools 3–8. In *Proceedings of the UMRCC Symposium on Upper Mississippi River Bivalve Mollusks*, ed. J. L. Rasmussen, 148–56. Rock Island, IL: Upper Mississippi River Conservation Committee.

Thompson, D. H. 1928. The "knothead" carp of the Illinois River. *Illinois Natural History Survey Bulletin* 17, no. 8:285–320.

Tiner, R. W. and Finn, J. T. 1986. *Status and recent trends of wetlands in five Mid-Atlantic states: Delaware, Maryland, Pennsylvania, Virginia, and West Virginia*. Newton Corner, MA: U.S. Fish and Wildlife Service, Region 5, and U.S. Environmental Protection Agency, region 3, Philadelphia, PA. Cooperative Technical Publication.

Turner, R. E. 1977. *Intertidal vegetation and commercial yields of penaeid shrimp*. Kissimmee, FL: Transactions of the American Fisheries Society.

Turner, R. E., and Cahoon, D., eds. 1988. *Causes of wetlands loss in the coastal Gulf of Mexico*. Vol. I, Executive summary. Minerals Management Service OCS Study/MMS 87–0119.

U.S. Environmental Protection Agency (USEPA). 1988. *The gulf initiative: Protecting the Gulf of Mexico*. John C. Stennis Space Center, MS: USEPA.

U.S. Geological Surveys. 1995. *Our living environment*. Washington, DC: Author.

Vitousek, P. M. 1990. Biological invasions and ecosystem processes: towards an integration of population biology and ecosystem studies. *Oikos* 57:7–13.

Watzin, M. C., Tucker, S., and South, C. 1994. *Environmental problems in the Mobile Bay ecosystem: The cumulative effects of human activities*. U.S. Environmental Protection Agency Technical Report.

Welch, P. S. 1963. *Limnology*. New York: McGraw-Hill.

Weller, M. W. 1964. Distribution and migration of the redhead. *Journal of Wildlife Management* 28:64–103.

Weller, M. W., and Spatcher, C. S. 1965. *Role of habitat in the distribution and abundance of marsh birds*. Ames: Iowa Agricultural and Home Economics experiment Station, Special Report 43.

Wetzel, R. G. 1983. *Limnology*. New York: Harcourt Brace Jovanovich.

White, W. A., Tremblay, T. A., Wermund, E. G., and Handley, L. R. 1993. *Trends and status of wetland and aquatic habitats in the Galveston Bay system*. The Galveston Bay (Texas) National Estuary Program GBNEP-31.

Wiener, J., Korschgen, C., Dahlgren, R., Sauer, J., Lubinski, K., Rogers, S., and Brewer, S. 1995. Biota of the upper Mississippi River ecosystem. In *Our living resources*. Washington, DC: U.S. Department of the Interior, National Biological Service.

Wilcox, D. A., and Meeker, J. E. 1991. Disturbance effects on aquatic vegetation in regulated lakes in northern Minnesota. *Canadian Journal of Botany* 69:1542–51.

Wilcox, D. A., and Meeker, J. E. 1992. Implications for faunal habitat related to altered macrophytes structure in regulated lakes in northern Minnesota. *Wetlands* 12:192–203.

Wilcox, D. A., and Meeker, J. E. 1995. Wetlands in regulated Great Lakes. In *Our living resources*. Washington, DC: U.S. Department of the Interior, National Biological Service.

Williams, G. P., and Wolman, M. G. 1984. *Downstream effects of dams on alluvial rivers*. U.S. Geological Survey Professional Paper 1286.

Williams, L. G., and Scott, C. 1962. Principal diatoms of major waterways of the United States. *Ecology* 7:365–79.

Williams, J. D., Warren Jr., M. L., Cummings, K. S., Harris, J. L., and Neves, R. J. 1993. Conservation status of freshwater mussels of the United States and Canada. *Fisheries* 18:6–22.

Williams, S. J., Dodd, K., and Gohn, K. K. 1991. *Coasts in crisis*. Reston, VA: U.S. Geological Survey Circular 1075.

Williams, S. J., and Johnston, J. B. 1995. Coastal barrier erosion: Loss of valuable coastal ecosystems. In *Our living resources*. Washington, DC: U.S. Department of the Interior, National Biological Service.

Williams, T .T. 1995. *Desert quartet*. New York: Pantheon Books.

Williams, T. T. 2002. *Red: Passion and patience in the desert*. New York: Vintage Books.

Wilson, D. M., Naimo, T. J., Wiener, J. G., Anderson, R.V., Sandheinrich, M. B., and Sparks, R. E. 1994. Declining populations of the fingernail clam in the Upper Mississippi River. *Hydrobiologia*.

Wlosinski, J. H., Olsen, D. A., Lowenberg, C., Owens, T. W., Rogala, J., and Laustrup, M. 1995. Habitat changes in the upper Mississippi River floodplain. In *Our living resources*. Washington, DC: U.S. Department of the Interior, National Biological Service.

Wren, C. D. 1991. Cause-effect linkages between chemicals and populations of mink and otter in the Great Lakes basin. *Journal of Toxicology and Environmental Health* 33:549–85.

Younker, G. L., and Andersen, C. W. 1986. *Mapping methods and vegetation changes along the lower Colorado River between Davis Dam and the border with Mexico*. Boulder City, NV: U.S. Bureau of Reclamation.

Butterfly.
Photograph by Revonna Bieber

CHAPTER 9

Current Issues in Ecology

Listen. Below us. Above us. Inside us. Come. This is all there is.

—Terry Tempest Williams, 2002

Topics

Ozone: The Jekyll and Hyde of Chemicals
Is Pollution a Real Concern or Just Hype?
Global Climate Change and Pollution
Global Warming
Global Warming and Sea-Level Rise
Human Influences
Agricultural Ecology
Non-native Species
Habitat Assessment
The Bottom Line on Our Environment and Future
Summary of Key Terms
Chapter Review Questions

Ecology is a thriving scientific discipline. However, as a scientific discipline, ecology is under constant pressure, partly caused by increased awareness of the general public to current environmental issues. Current environmental issues include climate change, environmental pollution, land use, and others. This chapter presents information on many of these current issues and themes in ecology. Specifically, the topics discussed are:

- methods man uses that alter Earth's environment in dangerous ways
- nature and causes of global warming
- changes in the greenhouse effect
- how global warming and the greenhouse effect are related and the current counter-arguments
- factors at work in changes in Earth's average surface temperature
- global signs of advancing long-term surface temperature change
- probable consequences of global warming—sea-level rise
- how sea-level rise would affect coastal areas: the Bruun rule

- how a rising sea level would affect human populations
- causes and effects of acid precipitation
- history of photochemical smog and how the internal combustion engine creates the conditions that provide it
- how man's activities on Earth are affecting stratospheric ozone, and what steps for change are in progress
- examples of human influences
- non-native species
- habitat assessments

Note: Portions of chapter are based on information from the U.S. Department of the Interior's (1995) *Our Living Resources* and Spellman and Whiting's (2006) *Environmental Science and Technology: Concepts and Applications*, 2nd ed.

Ozone: The Jekyll and Hyde of Chemicals

In Robert Lewis Stevenson's classic horror novel *Dr. Jekyll and Mr. Hyde*, Jekyll and Hyde are different aspects of the same person. Dr. Jekyll's kind, compassionate character is countered by Mr. Hyde's evil, dispassionate nature. Ozone has the same potential for good and evil within a single entity.

Ozone (O_3) is a molecule containing three atoms of oxygen. In the Earth's stratosphere, about 50,000 to 120,000 feet high, ozone molecules band together to form a protective layer that shields the Earth from some of the sun's potentially destructive ultraviolet (UV) radiation. Stratospheric ozone (ozone in its kindly Dr. Jekyll incarnation), formed in the atmosphere by radiation from the sun, provides us with an enormously beneficial function. Life as we know it on Earth could have evolved only with the protective ozone shield in place.

The Centers for Disease Control in Atlanta look at ozone more critically, however. They point out that ozone (in its evil Mr. Hyde form) is an extraordinarily dangerous pollutant. Only two-hundredths of a gram of ozone is a lethal dose. A single 14-ounce aerosol can filled with ozone could kill 14,000 people. Ozone is nearly as effective at destroying lung tissue as mustard gas. Not only is ozone a poisonous gas for us on Earth, it is a main contributor to air pollution, especially smog.

Is Pollution a Real Concern or Just Hype?

How serious a matter is environmental air pollution? Simply answered: You can literally bet your life on its seriousness. And that is exactly what we are doing— betting our lives, all of us. Remember, environmental pollution transcends national boundaries, and threatens the global ecosystem. We should be concerned about environmental problems. The problem is very serious. But should we panic? Not exactly.

Humans have altered the environment in dramatic fashions, especially over the past 200 years. In this chapter, we focus on human activities that profoundly affect the environment. These activities are not secret or mysterious; in fact, they are obvious—and most of us take part in some of these activities somehow, daily. As Graedel and Crutzen (1989) put it, the activities of man are changing our atmosphere. To summarize these activities: (1) man's industrial activities emit a variety of atmospheric pollutants; (2) man's practice of burning large quantities of fossil fuel introduces pollutants into the atmosphere; (3) man's transportation practices emit pollutants into the atmosphere; (4) man's mismanagement and alteration of land surfaces (deforestation) lead to atmospheric problems; (5) man's practice of clearing and burning massive tracts of vegetation produces atmospheric contaminants; and (6) man's agricultural practices, which produce chemicals such as methane, have an impact on the atmosphere. These man-made alterations to the Earth's atmosphere have produced profound effects, including increased acid precipitation, localized smog events, greenhouse gases, ozone depletion, and increased corrosion of materials induced by atmospheric pollutants.

What exactly should we do?

We should understand the man-made mechanisms at work destroying our environment, and what we are collectively doing to our environment—and we must be aware that our environment is finite, not inexhaustible or indestructible. Our environment can be destroyed. We must also clearly identify and understand both the causal and the remedial factors involved. Recognizing one particular salient point is absolutely essential: Life on Earth and the nature of Earth's atmosphere are connected—literally chained together. The atmosphere drives Earth's climate and ultimately determines its suitability for life. We must work to preserve the quality of our atmosphere.

Only through a cool-headed, scientific, intellectual, informed mind-set will we be able to solve our environmental dilemma. To save our environment (and ourselves) we must develop a vision of an environmentally healthy world. And this is something we can accomplish.

In this chapter, we discuss many current environmental issues relevant to the maintenance of our atmosphere, hydrosphere, and lithosphere. We discuss global warming, acid precipitation, photochemical smog, stratospheric ozone depletion, introduction of non-native species, and habitat assessment.

Global Climate Change and Pollution

As Edward T. LaRoe (1995) puts it, scientists have long recognized climate, especially temperature and precipitation, as one of the major ecological forces affecting the abundance, location, and ecological health of living organisms. This relationship is so strong that in many cases, if biologists know what plants and animals are present in an area, they can approximate the climate of the area. Quantifying these relationships will allow scientists to predict the ecological consequences of global climate change.

Recently, the scientific community reached a remarkable consensus on the likelihood and magnitude of global climate change, describing a likely scenario of a 3C (5.4F) average global warming, significant changes in the patterns and abundance of precipitation, and a 1.9-foot sea-level rise in the next 60 years (Houghton, Jenkins, and Ephramus 1990; LaRoe 1991). These changes will occur faster than previous change in geologic history and are therefore expected to have greater ecological impact.

Because of the strong relation between climate and ecosystem health and distribution, the U.S. Global Change Research Program has as a major component the monitoring of plants and animals to detect, understand, and ultimately predict the effects of global climate change on living resources (Committee on Earth and Environmental Sciences [CEES] 1990). The National Biological Service's research includes several projects to monitor the effects of climate change on animal and plant populations and ecosystems. Not only will the results of these projects allow a better understanding of the ecological effects of climate change, but they will also give an early, clear indication of the onset and magnitude of climate change because living resources may be sensitive indicators of global change.

Determining if long-term change in a species' population abundance or distribution was caused by specific climate changes is an extremely difficult scientific problem for two reasons: first, both climate and biological factors vary greatly from year to year, and these annual variations often mask long-term trends, making them difficult to detect. Second, several factors such as habitat loss, hunting pressure, competition with other native species and non-native species, and contaminants are simultaneously affecting species' population size and distribution along with climate change so that it is difficult to determine definitively the effect of any one cause.

Some species of plants and animals already may be affected by one type of global climate change: global warming. Much of the evidence for this is anecdotal or poorly documented. For example, some cold-intolerant species such as opossums and armadillos have expanded their range significantly northward during the past 50 years, and some heat-sensitive species, such as white birch, have receded northward during the same period. Data from some recent studies also suggest that global warming may be influencing the distribution or physiology of other plants and animals. Although these data are not sufficient to determine cause-and-effect relationships, they are intriguing enough to identify further research needs.

Keep in mind that the data are subject to the complexities common to most work on global change: that all the trends show dramatic year-to-year variation in response to short-term temperature changes and all have multiple possible explanations; and while all show intriguing statistical correlations, none demonstrate a cause-and-effect relationship. Moreover, these trends do not affect all species, because different species have different sensitivities to temperature and because global climate change is not the only factor affecting species. A number of competing hypotheses can be used to explain these changes. Nonetheless, together these articles suggest that global warming should be considered as a contributing factor.

In short, the jury is still out on whether the warming trend we are now experiencing is anything more than a natural cycle that has repeatedly occurred since time immemorial. However, because of the importance and currency of the phenomenon, the data that follow investigate interesting trends between one aspect of climate change—global warming—and plant and animal behavior. First, however, let's begin the following discussion with some background information and facts (questions and answers) about global warming.

Global Warming

Humanity is conducting an unintended, uncontrolled, globally pervasive experiment whose ultimate consequences could be second only to nuclear war. The Earth's atmosphere is being changed at an unprecedented rate by pollutants resulting from human activities, inefficient and wasteful fossil fuel use and the effects of rapid population growth in many regions. These changes are already having harmful consequences over many parts of the globe. (Toronto Conference Proceedings, June 1988)

The preceding quotation clearly states the issue. But what is global warming? It is a long-term rise in the average temperature of Earth. This appears to be the case, even though the geological record shows abrupt climate changes occur from time to time (Crowley and North 1988). Here's a second question, one many people use to question the validity of the concept of global warming as an environmental hazard. Is global warming actually occurring? The answer to this accompanying question is of enormous importance to all life on Earth—and is the subject of intense debate throughout the globe. Again, all the debate for the occurrence of global warming can't dispute the historical record that points out that measurements made in central England, Geneva, and Paris from about 1700 until the present indicate a general upward trend in surface temperature (Thompson 1995).

For the sake of discussion, let's assume that global warming is occurring. With this assumption in place, we must ask other questions, ones that deal with why, how, what, and what. Why is global warming occurring? How can we be sure it is occurring? What will be the ultimate effects? What can and are we going to do about it? These questions are difficult to answer. The real danger is that we may not be able to definitively answer these questions before it is too late—when we've reached the point that the process has progressed beyond the power of man to effect prevention or mitigation. This situation raises a red flag—a huge red flag—and additional questions. Are we to stand by and do nothing? Are we to simply ignore the potential impact of this problem? Are we to take the consequences of global warming lightly? Are we not to take precautionary actions now and wait until later—much later, when it is too late? Indeed, a red flag has been raised (a cause-and-effect relationship to the greenhouse effect), but there is still time before it begins to wave in the climate change that is inevitable—when mitigation becomes harder, more expensive, and impossible to effect.

Exactly what is the nature of the problem of global warming? All the answers are not provided here (or anywhere else), but a discussion of the phenomena and its potential impact on Earth is provided.

GREENHOUSE EFFECT

To understand Earth's greenhouse effect, here's an explanation most people (especially gardeners) are familiar with. In a garden greenhouse, the glass walls and ceilings are largely transparent to shortwave radiation from the sun, which is absorbed by the surfaces and objects inside the greenhouse. Once absorbed, the radiation is transformed into long-wave (infrared) radiation (heat), which is radiated back from the interior of the greenhouse. But the glass does not allow the long-wave radiation to escape, instead absorbing the warm rays. With the heat trapped inside, the interior of the greenhouse becomes much warmer than the air outside.

A similar greenhouse effect takes place with the Earth and its atmosphere (see figure 9.1). The shortwave and visible radiation that reaches Earth is absorbed by the surface as heat. The long heat waves are then radiated back out toward space, but the atmosphere instead absorbs many of them. This is a natural and balanced process, and indeed is essential to our life on Earth. The problem comes when changes in the atmosphere radically change the amount of absorption, and therefore the amount of heat retained. Scientists, in recent decades, speculate that this may have been happening as various air pollutants have caused the atmosphere to absorb more heat.

That this phenomenon takes place at the local level with air pollution, causing heat islands in and around urban centers, is not questioned. The main contributors to this effect are the greenhouse gases: water vapor, carbon dioxide, carbon monoxide, methane, volatile organic compounds (VOCs), nitrogen oxides, chlorofluorocarbons (CFCs), and surface ozone. These gases delay the escape of infrared radiation from the Earth into space, causing a general climatic warming. Note that scientists stress that this a natural process—indeed, the Earth would be 33C cooler than it is presently if the "normal" greenhouse effect did not exist (Hansen et al. 1986).

The problem with Earth's greenhouse effect is that human activities are now rapidly intensifying this natural phenomenon, which may lead to global warming. There is much debate, confusion, and speculation about this potential consequence. Scientists are not entirely sure of, or agree about, whether the recently perceived worldwide warming trend is because of greenhouse gases or from some other cause, or whether it is simply a wider variation in the normal heating and cooling trends they have been studying. But if it continues unchecked, the process may lead to significant global warming, with profound effects. Human impact on

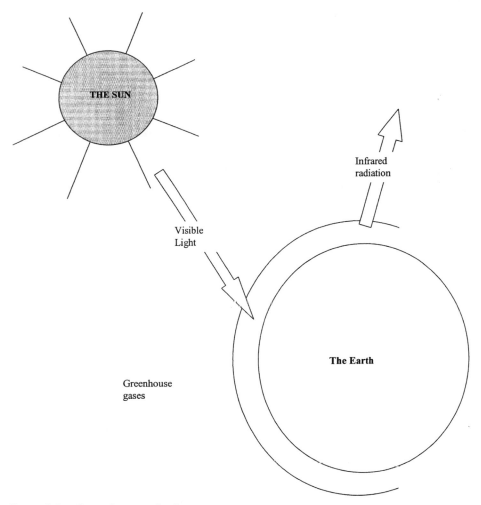

Figure 9.1. Greenhouse effect

greenhouse effect is real; it has been measured and detected. The rate at which the greenhouse effect is intensifying is now more than five times what it was during the past century (Hansen et al. 1989).

GREENHOUSE EFFECT AND GLOBAL WARMING

Those who support the theory of global warming base their assumptions on man's altering Earth's normal greenhouse effect, which provides necessary warmth for life. They blame human activities (burning of fossil fuels, deforestation, and the use of certain aerosols and refrigerants) for the increased quantities of greenhouses gases. These gases have increased the amounts of heat trapped in the Earth's atmosphere, gradually increasing the temperature of the whole globe.

Many scientists note that (based on recent or short-term observation) the past decade has been the warmest since temperature recordings began in the late 19th century and that the more general rise in temperature in the past century has coincided with the Industrial Revolution, with its accompanying increase in the use of fossil fuels. Other evidence supports the global warming theory. For example, in the Arctic and Antarctica, places synonymous with ice and snow, we see evidence of receding ice and snow cover.

Taking a long-term view, scientists look at temperature variations over thousands or even millions of years. Having done this, they cannot definitively show that global warming is anything more than a short-term variation in Earth's climate. They base this assumption on historical records that show the Earth's temperature does vary widely, growing colder with ice ages and then warming again. On another side of the argument, some people point out that the 1980s saw 9 of the 12 warmest temperatures ever recorded, and the Earth's average surface temperature has risen approximately 0.6C (1F) in the past century (U.S. Environmental Protection Agency [USEPA] 1995). But at the same time, still others offer as evidence that the same decade also saw three of the coldest years: 1984, 1985, and 1986. So what is really going on? We are not certain. But let's assume that we are indeed seeing long-term global warming. If this is the case, we must determine what is causing it. But here, we face a problem. Scientists cannot be sure of greenhouse effect's causes. The global warming may simply be part of a much longer trend of warming since the last ice age. Though much has been learned in the past two centuries of science, little is actually known about the causes of the worldwide global cooling and warming that have sent the Earth through a succession of major ice ages and smaller ones. We simply don't have the enormously long-term data to support our theories. In short, we do not know what we do not know about global climate change . . . period.

FACTORS INVOLVED WITH GLOBAL WARMING/ COOLING

Right now, scientists are able to point to six factors that could be involved in long-term global warming and cooling.

1. Long-term global warming and cooling could result if changes in the Earth's position relative to the sun (the Earth's orbit around the sun) occur, with higher temperatures when the two are closer together and lower when farther apart.
2. Long-term global warming and cooling could result if major catastrophes (meteor impacts or massive volcanic eruptions) that throw pollutants into the atmosphere, which can block out solar radiation, occur.
3. Long-term global warming and cooling could result if changes in albedo (reflectivity of the Earth's surface) occur. If the Earth's surface were more reflec-

tive, for example, the amount of solar radiation radiated back toward space instead of absorbed would increase, lowering temperatures on Earth.

4. Long-term global warming and cooling could result if the amount of radiation emitted by the sun changes.
5. Long-term global warming and cooling could result if the shape and relationship of the land and oceans change.
6. Long-term global warming and cooling could result if the composition of the atmosphere changes.

"If the composition of the atmosphere changes"—this possibility, of course, relates directly to our present concern: Have human activities had a cumulative impact large enough to affect the total temperature and climate of Earth? We are not certain, right now. We are somewhat concerned and alert to the problem, but we are not certain. So the question is, What are we doing about global warming? The answer to this pertinent question is provided in the next section.

HOW IS CLIMATE CHANGE MEASURED?

Worldwide, scientists are trying to establish ways to test or measure whether or not greenhouse-induced global warming is occurring. Scientists are currently looking for signs that collectively are called a greenhouse "signature" or "footprint." If it is occurring, eventually it will be obvious to everyone—but what we really want is clear advance warning. Thus, scientists are currently attempting to collect and then decipher a mass of scientific evidence to find those signs to give us clear advance warning. According to Franck and Brownstone (1992), these signs are currently believed to include changes in:

- global temperature patterns, with continents being warmer than oceans; lands near the Arctic warming more than the tropics; and the lower atmosphere warming, while the higher stratosphere becomes cooler;
- atmospheric water vapor, with increasing amounts of water evaporating into the air as a result of the warming, more in the tropics than in the higher latitudes. Since water vapor is a "greenhouse gas," this would intensify the warming process;
- sea surface temperature, with a fairly uniform rise in the temperature of oceans at their surface and an increase in the temperature differences among oceans around the globe;
- seasonality, with changes in the relative intensity of the seasons, with the warming effects especially noticeable during the winter and in higher latitudes.

In a measured, scientific way, these signs give a general overview of some of the changes that would be expected to occur with global warming. Note, however, that from a viewpoint of life on Earth, changes resulting from long-term global warming would be drastic—profoundly serious. Probably the most dramatic—and the effect with the most far-reaching results—would be sea-level rise.

GLOBAL CLIMATE CHANGE EFFECT ON SELECTED BIRDS, 1901–1989

Root and Weckstein (1995) point out that over time the ranges of species expand and contract, and abundance patterns shift. Ranges can expand when population pressure forces migration to new areas. Contractions can occur when populations decline and individuals abandon less-than-ideal habitats, which are often along the edges of species' ranges.

The data collected by volunteers for the National Audubon Society's Christmas Bird Counts provide excellent information to examine ranges and abundance patterns of wintering North American birds over both a very broad spatial scale and a long temporal scale. The changes that Root and Weckstein (1995) found were primarily due to human activity, both purposeful (e.g., management of game species) and accidental. Some of these changes could be viewed as being beneficial (e.g., water management programs increasing bald eagle numbers), while others could be viewed as negative (e.g., logging allowed barred owls to invade spotted owl territories).

CLIMATE CHANGE IN THE NORTHEAST

Climate is a principal determinant of biological distributions and of patterns that characterize the seasonal physiology and behavior of many organisms (Gates 1993). Consequently, a changing climate should elicit response in these biological properties. Detecting and characterizing such changes are logical early steps in assessing the significance of climate change to species and ecosystems (Schwartz 1990). Most published work on this subject involves species and ecosystem modeling based on known physiological and behavioral traits of selected species. Oglesby and Smith (1995) present evidence from an array of phenological data suggesting that climate change is occurring and that biological effects may already be of considerable magnitude. (Phenological data are those associated with the relationship between climate and periodic phenomena such as bird migration and flowering).

As Oglesby and Smith (1995) put it, various research efforts report trends toward earlier arrival dates for migratory birds and earlier blooming dates for spring wildflowers are concurrent with patterns of climate warming and consistent with what might be expected in the context of global warming. At the same time, local changes in land cover, with the forested area of the region increasing by more than 30 percent since 1900, may provide greater amounts of suitable habitat for attracting and holding migrating land birds, thereby contributing to observed patterns of change in migratory behavior.

The bottom line: climate change is the one variable affecting diverse groups of organisms that offers a rational and parsimonious explanation for the observed changes in the timing of migration in birds and blooming in plants that we and others have observed. Research either planned or in progress includes analyzing additional data sets as well as more sophisticated statistical analysis; determining the species most appropriate for monitoring climate change; finding and analyzing data sets that describe the phenology of other taxa; and possibly extending the study to other locations (Oglesby and Smith 1995).

POTENTIAL IMPACTS OF CLIMATE CHANGE ON NORTH AMERICAN FLORA

Morse, Kutner, and Kartesz (1995) remind us that climate change is a natural phenomenon that has occurred throughout the history of the Earth. The frequency and magnitude of climate change have varied substantially during and between glacial periods, and temperatures on both global and local scales have been both substantially warmer and colder than present-day averages (Ruddiman and Wright 1987; Pielou 1991; Peters and Lovejoy 1992). While potential magnitudes of local and global climate change are of concern, it is the predicted rate of temperature change that poses the greatest threat to biodiversity. The ability of species to survive rapid climate changes may partially depend on the rate at which they can migrate to newly suitable areas.

In the next few centuries, climate may change rapidly because of human influences. The concentrations of "greenhouse" gases in the atmosphere are being altered by activities such as carbon dioxide emission from burning fossil fuels. Models of climate change (Intergovernmental Panel on Climate Change [IPCC] 1990, 1992) predict an increase in mean global temperature of about 1.5–4.5°C (2.7–8.1°C) in the next century. Temperature changes suggested by general circulation models would present natural systems with a warmer climate than has been experienced during the past 100,000 years. While this would be a substantial change from the current climate, the rate of climate change is the greatest determinant of the impact on biological diversity. Future climate change due to human influences could occur many times faster than any past episode of global climate change (IPCC 1990, 1992; Schneider, Mearns, and Gleick 1992).

The strong association between distributions of plant species and climate suggests that rapid global climatic changes could alter plant distributions, resulting in extensive reorganization of natural communities (Graham and Grimm 1990). Climate changes could also lead to local extirpations of plant populations and species extinctions. The effects of global climate change are likely to vary regionally, depending on factors such as proximity to oceans and mountain ranges. Alteration of the amount and timing of precipitation and evaporation would affect soils and habitats; freshwater ecosystems are likely to be vulnerable to these changes in hydrology (Carpenter, Fisher, Grimm, and Kitchell 1992). Even minor fluctuations in the availability of water can radically affect habitat suitability for many wetland plant species. Rapid, large-scale shifts in temperature, precipitation, and other climate patterns could have broad ecological effects, presenting major challenges to the conservation of biodiversity.

Global Warming and Sea-Level Rise

In the past few decades, human activities (burning fossil fuels, leveling forests, and producing synthetic chemicals such as CFCs) have released into the atmosphere huge quantities of carbon dioxide and other greenhouse gases. These gases are warming the Earth at an unprecedented rate. If current trends continue, they are expected to raise Earth's average surface temperature by at least 1.5C to 4.5C (or more) in the next century—with warming at the poles perhaps two to three times as high as warming at the middle latitudes (Wigley, Jones, and Kelly 1986).

If we assume global warming is inevitable and/or is already underway, what, then, must we do? Obviously we cannot jump off the planet and head for greener pastures. We live on Earth and are stuck here (we have no effective method or technology to allow us to leave, or a convenient place to go if we tried). If this is the case, understanding the dynamics of change that are evolving around us and taking whatever prudent actions we can to mitigate the situation makes good sense.

We must also take this attitude and approach with the effect global warming is having on the rise in sea level. This rise is already underway, and with it will come increased storm damage, pollution, and subsidence of coastal lands.

Consider the following information taken from USEPA's 1995 report *The Probability of Sea Level Rise*.

1. Global warming is most likely to raise sea levels 15 centimeters (cm) by the year 2050 and 34 cm by the year 2100. There is also a 10 percent chance that climate change will contribute 30 cm by 2050 and 65 cm by 2100. These estimates do not include sea-level rise caused by factors other than greenhouse warming.

2. There is a 1 percent chance that global warming will raise the sea level 1 meter in the next 100 years and 4 meters in the next 200 years. By the year 2200, there is also a 10 percent chance of a 2-meter contribution. Such a large rise in sea level could occur either if Antarctic Ocean temperature warms 5°C, if Antarctic ice streams respond more rapidly than most glaciologists expect, or if Greenland temperatures warm by more than 10C. None of these scenarios is likely.

3. By the year 2100, climate change is likely to increase the rate of sea-level rise by 4.1 millimeters per year (mm/yr). There is also a 1-in-10 chance that the contribution will be greater than 10 mm/yr, as well as a 1-in-10 chance that it will be less than 1 mm/yr.

4. Stabilizing global emissions in the year 2050 would be likely to reduce the rate of sea-level rise by 28 percent by the year 2100, compared with what it would be otherwise. These calculations assume that we are uncertain about the future trajectory of greenhouse gas emissions.

5. Stabilizing emissions by the year 2025 could cut the rate of sea-level rise in half. If a high global rate of emissions growth occurs in the next century, sea level is likely to rise 6.2 mm/yr by 2100; freezing emissions in 2025 would prevent the rate from exceeding 3.2 mm/yr. If less emissions growth were expected, freezing emissions in 2025 would cut the eventual rate of sea-level rise by one-third.

6. Along most coasts, factors other than anthropogenic climate change will cause the sea to rise more than the rise resulting from climate change alone. These factors include compaction and subsidence of land, groundwater depletion, and natural climate variations. If these factors do not change, global sea level is likely to rise 45 cm by the year 2100, with a 1 percent chance of a 112 cm rise. Along the coast of New York, which typifies the United States, sea level is likely to rise 26 cm by 2050 and 55 cm by 2100. There is also a 1 percent chance of a 55 cm rise by 2050 and a 120 cm rise by 2100.

Along with the EPA's findings reported above, additional lines of evidence corroborate that global mean sea level has been rising during at least the past 100 years. According to Broecker (1987), this evidence is apparent in tide gauge records; erosion of 70 percent of the world's sandy coasts and 90 percent of America's sandy beaches; and the melting and retreat of mountain glaciers. Edgerton (1991) points out that the correspondence between the two curves of rising global temperatures and rising sea levels during the past century appears to be more than coincidental.

Major uncertainties are present in estimates of future sea-level rise. The problem is further complicated by our lack of understanding of the mechanisms contributing to relatively recent rises in sea level. In addition, different outlooks for climatic warming dramatically affect estimates. In all this uncertainty, one thing is sure: Estimates of sea-level rise will undergo continual revision and refinement as time passes and more data is collected.

MAJOR PHYSICAL EFFECTS OF SEA-LEVEL RISE

With increased global temperatures, global sea-level rise will occur at a rate unprecedented in human history (Edgerton 1991). Changes in temperature and sea level will be accompanied by changes in salinity levels. For example, a coastal freshwater aquifer is influenced by two factors: pumping and mean sea level. In pumping, if withdrawals exceed recharge, the water table is drawn down and saltwater penetrates inland. With mean sea level, the problem occurs if the sea level rises and the coastline moves inland, reducing aquifer area. Additional problems brought about by changes in temperature and sea level are seen in tidal flooding, oceanic currents, biological processes of marine creatures, in runoff and landmass erosion patterns, and saltwater intrusion.

Two major factors contribute to beach erosion. First, deeper coastal waters enhance wave generation, thus increasing their potential for overtopping barrier islands. Second, shorelines and beaches will attempt to establish new equilibrium positions according to what is known as the Bruun rule; these adjustments will include a recession of shoreline and a decrease in shore slope (Bruun 1962, 1986).

MAJOR DIRECT HUMAN EFFECTS OF SEA-LEVEL RISE

Along with the physical effects of sea-level rise, in one way or another, directly or indirectly, accompanying effects have a direct human side, especially concerning human settlements and the infrastructure that accompanies them: highways, airports, waterways, water supply and wastewater treatment facilities, landfills, hazardous waste storage areas, bridges, and associated maintenance systems. Sea-level rise could also cause intrusion of saltwater into groundwater supplies (Edgerton 1991).

To point out that this infrastructure will be placed under tremendous strain by a rising sea level coupled with other climatic change is to understate the possible consequences. Indeed, the impact on infrastructure is only part of the direct human impact. For example, there is widespread agreement among scientists that any significant change in world climate resulting from warming or cooling will (1) disrupt world food production for many years, (2) lead to a sharp increase in food prices, and (3) cause considerable economic damage.

Just how much of a rise in sea level are we talking about? According to USEPA (1995), "if the experts on whom we relied fairly represent the breadth of scientific opinion, the odds are fifty-fifty that greenhouse gases will raise sea level at least 15 cm by the year 2050, 25 cm by 2100, and 80 cm by 2200" (123).

Human Influences

The following sections are directed at neither a specific species nor an ecosystem but at human activities that affect living resources nationally and internationally. These broad-scale effects on, and changes in, ecosystem health are frequently the result of local or regional actions and land-use practices that collectively have effects across the nation (Willford 1995; Spellman and Whiting 2006).

ACID PRECIPITATION

In the evening, when you stand on your porch during a light rain and look out on your terraced lawn and that flourishing garden of perennials, you probably feel a sense of calm and relaxation hard to describe—but not hard to accept. The sound of raindrops against the roof of the house and porch, against the foliage and lawn, the sidewalk, the street, and that light wind through the boughs of the evergreens soothes you. Whatever it is that makes you feel this way, rainfall is a major ingredient.

But someone knowledgeable and/or trained in environmental science might take another view of such a seemingly welcome and peaceful event. We might wonder to ourselves whether the rainfall is as clean and pure as it should be. Is this actually rainfall—or is it rain carrying acids as strong as lemon juice or vinegar with it—capable of harming both living and nonliving things such as trees, lakes, and manmade structures? This may seem strange to some folks who might wonder why anyone would be concerned about such an off-the-wall matter.

Maybe such an interest was off-the-wall before the Industrial Revolution, but today the purity of rainfall is a major concern for many people, especially the levels of acidity. Acidic deposition, or "acid rain," describes any form of precipitation, including rain, snow, and fog, with a pH of 5.5 or below. Most rainfall is slightly acidic because of decomposing organic matter, the movement of the sea and volcanic eruptions, but the principal factor is atmospheric carbon dioxide, which causes carbonic acid to form. Acid rain (pH <5.5; in the pollution sense) is produced by the conversion of the primary pollutants sulfur dioxide and nitrogen oxides to sulfuric acid and nitric acid, respectively. These processes are complex, and depend on the physical dispersion processes and the rates of the chemical conversions. The basic cycle is shown in figure 9.2.

Contrary to popular belief, acid rain is not a new phenomenon, nor does it

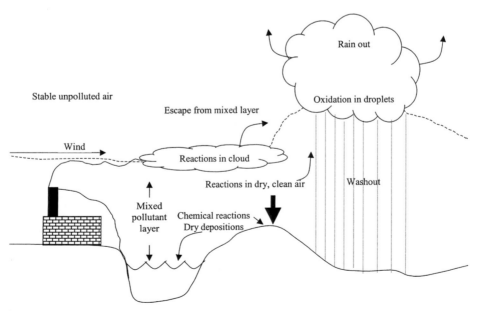

Figure 9.2. Acid rain cycle

result solely from industrial pollution. Natural processes such as volcanic eruptions and forest fires produce and release acid particles into the air. The burning of forest areas to clear land in Brazil, Africa, and other countries also contributes to acid rain. However, the rise in manufacturing that began with the Industrial Revolution literally dwarfs all other contributions to the problem.

The main culprits are emissions of sulfur dioxide from the burning of fossil fuels, such as oil and coal, and nitrogen oxide, formed mostly from internal combustion engine emissions, which is readily transformed into nitrogen dioxide. These mix in the atmosphere to form sulfuric acid and nitric acid.

In dealing with atmospheric acid deposition, the Earth's ecosystems are not completely defenseless; they can deal with a certain amount of acid through natural alkaline substances in soil or rocks that buffer and neutralize acid. The American Midwest and southern England are areas with highly alkaline soil (limestone and sandstone) that provide some natural neutralization. Areas with thin soil and those with soil laid on granite bedrock, however, have little ability to neutralize acid rain.

Scientists continue to study how living beings are damaged and/or killed by acid rain. This complex subject has many variables. We know from various episodes of acid rain that pollution can travel over very long distances. Lakes in Canada, Maine, and New York feel the effects of coal-burning in the Ohio Valley. For this and other reasons, the lakes of the world are where most of the scientific studies have taken place. In lakes, the smaller organisms often die off first, leaving the larger animals to starve to death. Sometimes the larger animals (fish) are killed directly; as lake water becomes more acidic, it dissolves heavy metals leading to concentrations at toxic and often lethal levels. Have you ever wandered up to the local lakeshore and observed thousands of fish belly-up? Not a pleasant sight or smell, is it? Loss of life in lakes also disrupts the system of life on the land and the air around the lakes.

In some parts of the United States, the pH of rainfall has fallen well below 5.6. In the northeastern United States, for example, the average pH of rainfall is 4.6, and rainfalls with a pH of 4.0, which is 100 times more acidic than distilled water, are not unusual. Despite intensive research into most aspects of acid rain, scientists still have many areas of uncertainty and disagreement. That is why the progressive, forward-thinking countries emphasize the importance of further research into acid rain.

PHOTOCHEMICAL SMOG

When various hydrocarbons, oxides of nitrogen, and sunlight come together, they can initiate a complex set of reactions that produce a number of secondary pollutants known as photochemical oxidants or photochemical smog. Photochemical smog of the type most people are familiar with was first noticed in Los Angeles in the early 1940s. Determining its true cause has taken many years. According to Black-Covilli (1992), at first it was thought to arise from dust and smoke emitted from factories and incinerators. Accordingly, Los Angeles County officials issued a ban on all outdoor burning of trash and initiated steps toward control of industrial smoke emission. Before long, though, county authorities determined that their initial efforts were not working; the smog continued unabated. Then they went after another suspected culprit, sulfur dioxide (SO_2) given off by oil refineries and by combustion of sulfur-bearing coal. So they placed controls on sulfur dioxide emissions—but still gained no benefit.

The biochemist Dr. Arie Haagen-Smit, as a result of chance during research aimed at finding the compounds responsible for the pleasant tastes and odors of fruit, found the cause of the smog problem, which showed beyond a doubt that the internal combustion engine was the principal source.

How do internal combustion engines produce smog? A few of the finer details are yet unclear, but the following, in simplified form, appears to be what happens. Smog begins with the high temperatures in the internal combustion engine, which cause atmospheric oxygen and nitrogen to react, producing nitric oxide. At the same time, varying quantities of fuel in the engine fail to burn completely. This results in a mixture of aldehydes, ketones, olefins, and aromatic hydrocarbons that is expelled in the exhaust. The exhaust enters the atmosphere, where ultraviolet radiation from the sun causes a complex series of reactions to take place. These reactions involve atmospheric oxygen, nitric oxide, and organic compounds. The result is that nitrogen dioxide (NO_2) and ozone (O_3) are formed, both of which are highly toxic and irritating. In addition, this reaction also causes the formation of other constituents of photochemical smog, including formaldehyde, peroxybenzohl nitrate, peroxyacetyl nitrate (PAN), and acrolein.

Photochemical smog is known to cause many annoying respiratory effects—coughing, shortness of breath, airway constriction, headache, chest tightness, and eye, nose, and throat irritation (Masters 1991).

STRATOSPHERIC OZONE DEPLETION

Ozone is formed in the stratosphere by radiation from the sun, and helps to shield life on Earth from some of the sun's potentially destructive ultraviolet (UV) radiation.

In the early 1970s, scientists suspected that the ozone layer was being depleted. By the 1980s, it became clear that the ozone shield was indeed thinning in some places, and at times, even has a seasonal hole in it, notably over Antarctica. The exact causes and actual extent of the depletion are not yet fully known, but most scientists believe that various chemicals in the air are responsible.

Most scientists identify the family of chlorine-based compounds, most notably chlorofluorocarbons (CFCs) and chlorinated solvents (carbon tetrachloride and methyl chloroform) as the primary culprits involved in ozone depletion. In 1974, Molina and Rowland hypothesized that the CFCs, containing chlorine, were responsible for ozone depletion. They pointed out that chlorine molecules are highly active and readily and continually break apart the three-atom ozone into the two-atom form of oxygen generally found close to Earth in the lower atmosphere.

The Interdepartmental Committee for Atmospheric Sciences (1975) estimates that a 5 percent reduction in ozone could result in nearly a 10 percent increase in cancer. This already frightening scenario was made even more frightening in 1987 when evidence showed that CFCs destroy ozone in the stratosphere above Antarctica every spring. The ozone hole had become larger, with more than half of the total ozone column wiped out and essentially all ozone disappeared from some regions of the stratosphere (Davis and Cornwell 1991).

In 1988, Zurer reported that, on a worldwide basis, the ozone layer shrank approximately 2.5 percent in the preceding decade. This obvious thinning of the ozone layer, with its increased chances of skin cancer and cataracts, is also implicated in suppression of the human immune system, and damage to other animals and plants, especially aquatic life and soybean crops. The urgency of the problem spurred the 1987 signing of the Montreal Protocol by 24 countries, which required signatory countries to reduce their consumption of CFCs by 20 percent by 1993, and by 50 percent by 1998, marking a significant achievement in solving a global environmental problem.

SIGNIFICANCE OF FEDERAL LANDS FOR ENDANGERED SPECIES

The federal government has overall trust responsibilities for species listed as threatened or endangered under the Endangered Species Act (ESA). The options available for managing and protecting these species, however, are directly related to the ownership of the lands on which the species are found (Stein, Breen, and Warner 1995).

National Heritage Programs—a partnership between state and federal agencies and the Nature Conservancy—gather and manage a variety of information linking both biological and nonbiological factors of relevance to biodiversity conservation. Central to this effort is the inventory of all known occurrences for species of conservation concern, including all federally listed endangered or threatened species. What constitutes an occurrence depends on the biology of the particular species, but most often reflects a "mapable" and geographically distinct population or subpopulation. Pertinent information is documented for each occurrence, such as the biological health and population trends of the occurrence, habitat quality, protection or management status, and land ownership.

Heritage Programs in all 50 states queried their databases for all documented occurrences of federally listed species in their jurisdiction and reported the class of landowner or type of managing agency. (Note: "Species" under the ESA includes subspecies as well as full species; in the strictest taxonomic sense, these collectively would be referred to as "taxa.")

While the Heritage Programs are the most comprehensive source for such local information on rate species and reviewed about 350,000 occurrence records for this analysis, this information is incomplete for four reasons: (1) Heritage Programs may not be aware of all occurrences, and indeed, many populations for species of concern may yet be discovered; (2) most programs have a data-entry backlog; (3) not all data centers have completely recorded the land ownerships for all their occurrence records; and (4) species occurrences to lakes and rivers are generally not recorded as under the jurisdiction of a federal agency except where they are entirely included in such areas as national parks or wildlife refuges. On the other hand, in many states more is known about the status of listed species on federal lands than on state or private lands. This imbalance in the available data, largely the result of federally funded inventories on federal land, will tend to overstate the proportion of a species' range of population on federal lands.

INCREASED AVIAN DISEASES WITH HABITAT CHANGE

Changes in disease patterns and trends reflect changing relationships between the affected species (host) and the causes of disease (agent). Host-agent interactions are closely linked to environmental factors that either enhance or reduce the potential or disease to occur. As a result, wildlife disease patterns and trends are, to a substantial extent, indicators of environmental quality and changing host-agent interactions within the environment being evaluated. The types, distribution, and frequency of diseases causing major avian die-offs have changed greatly during the 20th century. Too little is known to assess the changes of most avian diseases that result in chronic attrition rather than major die-offs, or about those that affect reproductive success, reduce body condition, or affect survival in other indirect ways. Nevertheless, the changing patterns and trends in highly visible avian diseases provide notice of problems needing attention (Friend 1995).

Information on the status of disease in wild birds was obtained from National Wildlife Health Center (NWHC) evaluations of the cause of death for more than 30,000 carcasses from across the United States during the past two decades, reports of avian mortality received from collaborators, the scientific literature, and NWHC field investigations of bird mortality. Comprehensive assessments of causes of wild bird mortality, magnitude of losses, and geographic distribution of specific disease are not possible from these data, although we can identify general relationships for waterfowl and some other species.

✔ *Important Point*: The occurrence of disease involves three factors: a susceptible host, presence of an agent capable of causing disease, and suitable environmental conditions for contact between the host and agent in a manner that results in disease. Environment is often the dominant factor in this relationship.

The most dramatic example of geographic expansion of a noninfectious indigenous disease is avian botulism, caused by the bacterium *Clostridium botulinum*. In 1914 a Bureau of Biological Survey researcher began investigating catastrophic die-offs that had begun in 1901 and in which millions of waterbirds along the Great Salt Lake, Utah, had died. Later studies revealed that avian botulism was responsible for those die-offs. Historically, avian botulism was referred to as "western duck disease" because of its rather limited geographical distribution of occurrence (Kalmbach and Gunderson 1934).

Avian botulism now occurs all over the United States and in many other countries as well. Because of the visibility of massive die-offs, avian botulism is probably the best-documented nonhunting waterfowl mortality (Stout and Cornwell 1976). The continued reporting of avian botulism die-offs since the early 1900s makes researchers suspect that much of the disease's geographic expansion is of recent origin. Also, most (15 of 21) initial outbreaks of avian botulism in countries other than in North America have occurred since 1970.

Avian cholera, caused by the bacterium *Pasteurella multocida*, has been recognized as an important infectious disease of domestic poultry in the United States since at least 1867 (Rhoades and Rimler 1991). Therefore, it is noteworthy that a 1930 evaluation of the status of waterfowl commented on the lack of documentation of avian cholera in wild waterfowl (Phillips and Lincoln 1930). In 1944, however, the disease was documented in wild waterfowl in the United States (Quortrup, Queen, and Merovka 1946). Limited geographical expansion of avian cholera in wild waterfowl occurred during the 1940s and 1950s, and sporadic occurrences were documented at a few new locations during the 1960s. By the end of the 1960s, though, avian cholera was reported as established in the Central and Pacific flyways. Outbreaks in the Mississippi flyway were unusual, and only two outbreaks had occurred in the Atlantic flyway. With the exception of a single instance during the breeding season, outbreaks occurred in winter (Stout and Cornwell 1976). During the 1970s, avian cholera became established as a major cause of waterfowl mortality in all four flyways within the United States and as a recurring cause of waterfowl mortality in Canada. Geographic expansion of die-off locations continues, and outbreaks now occur during all seasons of the year (Friend 1987).

Duck plague is another emerging disease of North American waterfowl. This herpes virus infection first appeared on the North American continent in 1967 when it caused large-scale losses in the domestic duck industry and losses of a small number of wild waterfowl (Leibovitz and Hwang 1968). The first major die-off involving wild waterfowl occurred during January 1973 at the Lake Andes National Wildlife Refuge in South Dakota (Friend and Pearson 1973). Duck plague has expanded throughout North American since the initial outbreak, along with an increasing number of outbreaks in each decade. Nearly all occurrences of duck plague have involved nonmigratory movements. A February 1994 outbreak in the Finger Lakes region of New York State involving mallards and American black ducks is the first major outbreak involving migratory waterfowl since the January 1973 Lake Andes outbreak.

Other diseases affecting wild birds are newly recognized, are occurring with increasing frequency, or have expanded their geographic occurrence during the 20th century. Changes in disease patterns in wild birds are consistent with such changes in other species, including humans, and reflect environmental changes that foster the eruption of disease and the spread of infectious agents.

Agricultural Ecology

Approximately 45 percent of the U.S. land area is used for agricultural purposes, with 472 million acres in cropland and 587 million acres in range or pasture (Knutson, Penn, and Boehm 1990). American agriculture has become the most productive in the world based on technology and increased specialization. Energy, machinery, agrochemicals, and irrigation are principal components of modern American agriculture, all of which potentially affect farm and off-farm environmental quality. In addition, government policies have pervasively affected U.S. agriculture, often precluding producers from responding to changing market conditions or affecting adoption of farm practices that potentially improve environmental quality (National Research Council 1989; Reichelderfer 1990).

Energy and technology have propelled American agriculture from pioneering conversion of the landscape to intensive, high-yield, monocultural production. The composition of agriculture in terms of farm numbers, size, and methods of production have changed dramatically throughout this century. The effects of the agricultural industry on the diversity, distribution, and abundance of wildlife continue to be profound (Allen 1995).

Larger, more economically efficient producers that could tolerate smaller profit margins have absorbed the assets of smaller, less successful operations. In 1991, the U.S. human population on farms was less than one-tenth of what it was in 1920 (Haynes 1991). As the number of farms decreased by two-thirds during this same period, farm size increased. In response to fewer farms and the need to increase production efficiency, fields have become larger, crop diversity has decreased, crop rotation patterns have become simpler and less frequent, and agrochemicals play a major role in crop production. Over the past 30 years, these elements have had significant effects on environmental quality within agricultural ecosystems (Allen 1995).

The Conservation Title of the Food Security Act of 1985, commonly referred to as the Farm Bill, was formulated in a time of commodity surpluses, economic stress within the agricultural community, and increasing public concern about environment quality. The Conservation Reserve Program (CRP), a cornerstone of the 1985 Farm Bill, was enacted to remove highly erosive cropland from production. This legislation reflects an effective integration of economic support to the agricultural community with environmental polices advocated by a strong coalition of organizations representing a wide spectrum of the American public. The CRP has provided substantial benefits to wildlife populations across the nation. To appreciate the CRP's significance to wildlife, we must remember that tremendous changes in agriculture have influenced the abundance and quality of habitat in this century (Soil and Water Conservation Society 1994).

World War II, for example, brought an increased demand for American agricultural products. New technologies adopted in the postwar period reduced production costs and further escalated farm output. Tractors and farm machinery became more powerful and efficient. Time and energy savings decreased the amount of human labor needed to work larger fields. Advances in biological and chemical technologies further increased agricultural efficiency and crop yields. The use of nitrogen fertilizer increased from 217 tons in 1940 to 7,459 million tons in 1970 (Haynes 1991). By the early 1970s, crop yields had skyrocketed to new records.

American agriculture entered the world market in the 1970s in response to increased global demands for agricultural products. American farmers expanded production by cultivating existing croplands more intensively and bringing new, less fertile, and more fragile lands into production. The 1980s arrived with the farm industry in crisis due to overproduction, increased costs for fuels and fertilizers, elevated interest rates, declining land values, and decreased demand for export sales. The agricultural economic predicament, as well as heightened public concern about environmental quality, set the stage for the 1985 Farm Bill and establishment of CRP (Allen 1995).

The effects of modern agriculture on wildlife are indisputable, ranging from habitat elimination to long-term effects of agrochemicals on water quality and reproductive success of ground-nesting birds (Capel, Crawford, Robel, Burger, and Southerton 1993). Habitat diversity in agricultural ecosystems has declined drastically as a consequence of the elimination of hay and pasture needed by draft animals and shifted to crop monocultures. In many regions, wetland drainage, consolidation of fields and farms, and elimination of fencerows and idle areas have reduced habitat diversity even further, thereby diminishing the ability of agricultural ecosystems to sustain viable populations of wildlife. The amount of undisturbed grass-dominated cover and noncropped areas has decreased, resulting in lower availability of habitat and higher losses to predators for many nongame and game species of wildlife. In many agricultural regions, crucial wildlife habitat components such as undisturbed grasslands have become dissected into small, isolated patches, or spatially segregated tracts. Increased agrochemical use has been implicated in the long-term decline of species such as the northern bobwhite.

Monocultures, with minimal rotations between crops, have accelerated soil erosion and led to a greater dependence on chemical fertilizers and pesticides (Bender 1984), resulting in surface and groundwater contamination (Ribaudo 1989). Larger, heavier equipment used for tillage, planting, application of agrochemicals, and harvesting contributes to increased soil compaction and decreased soils tilth (suitability), further contributing to erosion. Agriculture has become the largest single nonpoint source of water pollution, delivering not only soil particles but also absorbed and dissolved nutrients and pesticides (National Research Council 1989).

Non-native Species

The following excerpt is from an article by Scott Harper that appeared in the June 5, 2007, *Virginian-Pilot*, Norfolk, Va.

The Invader: Troublesome Chinese mitten crab infiltrates Chesapeake Bay

It has hairy claws and should not be here. But the Chinese mitten crab—a large, spidery creature classified by the U.S. government as "intrusive wildlife"—keeps popping up in the Chesapeake Bay and, more recently, in the Delaware Bay. While none of the foreign species has been confirmed in Virginia waters—the closest one was found in the bay in southern Maryland—scientists on Friday issued an alert to biologists, fishermen and wildlife managers on the East Coast to be on the lookout for the nuisance crab.

In the Chesapeake Bay, experts worry that the invaders might compete with and damage populations of native blue crabs, which already are stressed by pollution, lost habitat and fishing pressures.

As Hiram W. Li (1995) puts it, introduced species evolved elsewhere and have been transported and purposefully or accidentally disseminated by humans. Many synonyms are used to describe these species: alien, exotic, non-native, and nonindigenous. The spread of non-native species during the past century has been unprecedented in Earth's history, with the speed and scale of these infestations more rapid than natural invasions. The spread of non-native species in human-disturbed habitats reflects a deterioration of the North American landscape.

Introduced species disrupt the functioning of native ecosystems upon which humans depend. Many non-native species become pests by rapidly dispersing into communities in which they have not evolved, and by displacing native species because of evolutionary mismatches. For example, non-native species contributed to 68 percent of the fish extinctions in the past 100 years, and the decline of 70 percent of the fish species listed in the Endangered Species Act (Lassuy 1994).

The economic cost incurred because of non-native species reaches millions, or even billions, of dollars. Non-native species damage agricultural crops and rangelands, contribute to the decline of commercially important fishes, spread diseases that affect domestic animals and humans, and disrupt vital ecosystem functions.

Some species that have become pests were first introduced to "create" a desired landscape; these non-natives include exotic game animals, fish, and decorative plants. Mack and Thompson (1982), for example, traced the widespread dissemination of 139 weedy, non-native plants in the United States to seed catalogues and the commercial seed trade of the 19th century. Similarly, feral (wild) domestic animals such as mustangs are a major problem on public lands, and sound management of such animals has been impeded by romantic images of America's past.

Accidental introductions through human travel is a theme repeated in several articles, indicating that cargo traffic (ship, air, land) is a major vector of non-native species and should be monitored as world trade increases. The zebra mussel is the most notorious hitchhiker, but introductions through ballast water are not isolated to the Laurentian Great Lakes. Li (1995) and his colleagues found that 11 exotic benthic invertebrates have become established in Oregon estuaries. Similarly, dinoflagellates causing red tide toxins have spread into Australian waters through cargo traffic. The importation of raw logs from New Zealand and Siberia endanger Pacific Northwest forests through forest pests hitchhiking in the bark and wood (Li 1995). It is clear that international cargo traffic must be monitored to reduce the spread on non-native species.

Disease may be one of the most important problems caused by non-native species. After Columbus landed in the New World, for example, 95 percent of the Native American tribes became extinct because their people were susceptible to European microbes (Diamond 1992). Likewise, exotic diseases have devastated populations of aquatic organisms worldwide, killed many native trees, and exterminated much of Hawaii's avifauna. Non-native species are the primary vector for theses diseases; for instance, the spread of fish diseases worldwide resulted from the unprecedented transfer of non-native fishes for hatchery production.

It is clear from the research and practical experience that changes caused by non-native species are widespread and profound. New problems continually arise, because humans deliberately and accidentally release non-native species and encourage their invasion through massive disturbances of the landscape, thereby mitigating against native species' resistance to invaders by stressing native populations (Li 1995).

NONINDIGENOUS FISH

Within the United States alone, humans have intentionally or unintentionally introduced more than 4,500 species of terrestrial and aquatic species to areas outside their historical range (U.S. Congress 1993). Although many terrestrial introductions are viewed as beneficial to humans because of economic and social considerations, all but a few intentional aquatic introductions have proven to be mixed blessings (Courtenay and Williams 1992; Steirer 1992; U.S. Congress 1993). No unintentional aquatic introductions have been considered beneficial (Steirer 1992; U.S. Congress 1993).

Both intentional and unintentional introductions have enabled nonindigenous fish to become temporary, and often permanent, residents in nearly every U.S aquatic system. Complete eradication or exclusion is neither economically plausible nor socially justified (U.S. Congress 1993); therefore, nonindigenous fish are and will continue to be components of these aquatic systems. Because nonindigenous fish have the potential to alter significantly the U.S. aquatic ecosystems during the next century and beyond, their interactions within the aquatic community must be monitored and analyzed to ensure that effective management actions are taken before a crisis arises (Boydstun, Fuller, and Williams 1995).

NON-NATIVE REPTILES AND AMPHIBIANS

Interest in established, non-native species of reptiles and amphibians in the United States (including territories and possessions) has been increasing the past 25 years.

Concerns regarding the interactions of introduced and native species have driven this interest (Wilson and Porras 1983). Most successful introductions have taken place in the southern tier of states (California to Florida) and on islands. This success rate is probably due, in part, to favorable environmental conditions (Mc-Coid and Kleberg 1995). Movements by indigenous peoples to islands also may have substantially augmented existing faunas. For example, in American Samoa, virtually the entire terrestrial reptile fauna may have been introduced by the original human colonizers (McCoid and Kleberg 1995).

Of the documented 53 established non-native amphibian and reptile species, at least five—spectacled caiman, marine toad, African clawed frog, bullfrog, and brown tree snake—have been established at least 30 years and have been sufficiently monitored to enable preliminary assessment of impacts on the native biota.

1. *Spectacled caiman*—has been established in southern Florida for about 30 years (Ellis 1980). There are few published accounts of this species in Florida, but one (Ellis 1980) indicated that these animals eat fish, amphibians, and mammals. This information, coupled with the species' ability to tolerate crowding in bodies of water and relatively rapid maturation, suggests that impacts on native alligators might be expected (McCoid and Kleberg 1995). Studies in the species' native range, however, suggest that the spectacled caiman does not co-occur with larger species of crocodilians, perhaps because of their predation on the smaller caimans. Since the American alligator reaches a larger size than the spectacled caiman, it is possible that the American alligator will deter the caiman from substantially expanding its range.

2. *Marine toad*—native to the tropical New World, is widely introduced and now has a virtually circumtropical range (Zug and Zug 1979). Populations were originally established for insect control, but the species itself became a pest. Information from Australia (Tyler 1989) indicates that ingestion of marine toads, because they have highly toxic skin glands, results in deaths of native reptiles, birds, and mammals. Observations on Guam, where the marine toad has been established since 1937 (McCoid 1993), indicate that poisonings of pet dogs and cats by biting or mouthing marine toads are relatively common (McCoid and Kleberg 1995). On Guam, the island-wide decline of a large varanid lizard is attributed to its predation on the introduced toad (McCoid, Hensley, and Witteman 1994). In Florida, where the marine toad has been established since 1955, poisonings of pets (Ashton and Ashton 1988) and declines of native amphibians in areas of co-occurrence with the marine toad are reported. In a laboratory situation, a native toad was behaviorally dominated and excluded from feeding by marine toads (Boice and Boice 1970). There is a literature survey on the marine toad that includes information on extralimital populations (Lawson 1987).

3. *African clawed frog*—despite initial fears of its effect on aquatic California vertebrates (St. Amant 1975), a subsequent study (McCoid and Fritts 1980a) indicated that these fears may be unwarranted because the only vertebrates found in stomach analyses were immature African clawed frogs and an introduced

fish species. Other studies (McCoid and Fritts 1980b, 1993) characterize populations as living primarily in temporary and artificial bodies of water, where most native aquatic vertebrates are expected to be absent. Recently, populations in southern California may have declined because of drought (McCoid, Pregill, and Sullivan 1993). Although African clawed frogs have been established in California since the mid-1960s (McCoid and Fritts 1980b), impacts on native invertebrates, their primary food source, have not been assessed.

4. *Bullfrog*—although precise dates of introductions of the bullfrog into many areas of western North America are not well known (Bury and Whelan 1984), the earliest introduction occurred in 1896 (Hayes and Jennings 1986). Impacts on native ranid frogs, however, are well documented and may account for range restrictions of native ranids (Moyle 1973; Hayes and Jennings 1986; Stuart and Painter 1993). Recent information indicates that the Mexican garter snake is also declining because of predation by bullfrogs.

5. *Brown tree snake*—since the introduction of the brown tree snake on Guam about 40 years ago, the snake has reached enormous densities (Rodda, Fritts, and Conry 1992) and is implicated in the demise of the entire native forest-dwelling bird community (Savidge 1987) and some of the larger lizard species (Rodda and Fritts 1992). Additional impacts include disruption of electrical power (Fritts, Scott, and Savidge 1987), predation on domesticated animals (Fritts and McCoid 1991), and human health risks (Fritts, McCoid, and Haddock 1990; Fritts et al. 1994). There are several overviews of the brown tree snake problem on Guam (Fritts et al. 1987; McCoid 1991).

NON-NATIVE BIRDS

Two of the three most common nesting species in North America today are birds whose ancestors were brought here from Europe. Some non-native birds are more conspicuous than others, so comparisons are only relative, but according to the two largest continental surveys, non-native species (excluding house finches) constitute, on average, about 6 percent of the bird population during summer months (Breeding Bird Survey [BBS]) and about 8 percent in winter (Christmas Bird Count). Percentages vary considerable by habitat and geographic location (Robbins 1995).

Many exotic bird species were introduced to the United States by European colonists who missed the familiar birds of their homeland and tried to establish populations of familiar Old World species. Farmers also saw opportunities for pest control by birds such as starlings and house sparrows, but they did not anticipate the degree to which these exotic species would outcompete native birds for nesting sites. Most introductions, however, were by sporting or hunting organizations and state game departments that wished to provide more hunting opportunities.

Competition between exotic and native species has been particularly severe on islands.

In the Hawaiian Islands, introduced songbird species far exceed native ones.

Visitors to Honolulu, for example, see exotic songbirds only if they hike mountain trails in search of the few remaining endemic species. MacArthur and Wilson (1967) predicted that for every new species colonized or introduced on an island, an average of one species will become extinct. Even Puerto Rico has breeding populations of about 20 kinds of exotic songbirds, far outnumbering the endemics.

The best-known introductions in North America are those that were highly successful: the house sparrow, European starling, rock dove or common pigeon, ring-necked pheasant, mute swan, gray or Hungarian partridge, and the chukar. They readily adapted to their new environments, and most have prospered here for more than 100 years (Robbins 1995).

ZEBRA MUSSELS IN SOUTHWESTERN LAKE MICHIGAN

The zebra mussel is a European species that was accidentally introduced into North America. It has had a tremendous impact on freshwater ecosystems of the United States and Canada. Since the zebra mussel was first discovered in Lake St. Clair in 1988, it has spread to each of the Great Lakes and to the major river systems of central and eastern United States. Communities along the affected lakes and rivers rely on these waters for drinking, industrial waters supplies, transportation, commercial fishing and shelling, and recreation. Rapidly expanding populations of zebra mussels could ultimately affect many of these activities, in addition to changing the structure of the ecosystem (Keniry and Marsden 1995).

By firmly attaching to hard surfaces, zebra mussels have clogged water-intake pipes and fouled hard-shelled animals such as clams and snails. In addition, zebra mussels have reduced plankton populations as colonies of mussels filter large volumes of water for food (e.g., Holland 1993), potentially depleting food resources of larval and planktivorous fishes such as smelt, chub, and alewife. Transfer of suspended material to the lake bottom in mussel waste products also leads to increased water clarity (Reeders, bij de Vaate, and Noordhuis 1992) and increased growth of aquatic plants, a phenomenon already observed in some of the shallower harbors of Chicago. Although clear water is often considered aesthetically pleasing, this clarity indicates that drastic changes have occurred at the base of the food web and that energy flow through the ecosystem has been altered (Keniry and Marsden 1995).

✔   *Important Point*: The first live zebra mussel was discovered in Lake Michigan near Chicago in 1989.

AFRICANIZED BEES IN NORTH AMERICA

Kunzmann, Buchmann, Edwards, Thoenes, and Erickson (1995) point out that the honeybee genus *Apis* likely has the greatest breadth of pollen diet of any insect and, because of its human-caused cosmopolitan distribution, the species directly affects the reproductive biology of about 25 percent of the world's flowering plants (Schmalzel 1980; Buchmann, O'Rourke, Shipman, Thoenes, and Schmidt 1992). This situation has profound consequences for agribusiness, native plants and animals, and ecosystems. In 1956, bee geneticist Warwick E. Kerr imported queen bees of an African race into Brazil to breed a more productive honeybee that was better adapted to the Neotropical climate and vegetation (Kerr 1967). The following year, 26 of Kerr's Africanized honeybee queens were inadvertently released into the surrounding forest (Winston 1987). Since then, the Africanized hybrids have been expanding their range northward, with an average rate of between 200 and 300 miles each year.

The first U.S. Africanized honeybee colony was reported in October 1990, at Hidalgo, Texas, along the international boundary. By fall 1993, Africanized honeybees (AHBs) had extended their territory north and west into numerous counties of Arizona, New Mexico, and Texas. Since the first U.S. AHB swarm was detected, the rate of spread has accelerated to over 375 miles per year in the southwestern United States (Guzman-Novoa and Page 1994).

European honeybees (EHBs) were introduced in North America as early as the 16th century by Spanish conquistadors and missionaries (Brand 1988). Today, one of the three most common subspecies or races of the EHB, the Italian honeybee, is nearly pandemic throughout North America because of its popularity with professional and hobbyist beekeepers. As a consequence, these non-native bees have become naturalized and have been a part of the North American arthropod biota for about 3,500 bee generations, or at least the past 200 years (Buchmann et al. 1992). European honeybees are commonly seen visiting agricultural food crops, cultivated flowers, and roadside wildflowers to gather nectar and pollen. They are even common in areas far from human population centers. These bees are also the preferred, "managed" pollinator for over 100 U.S. agricultural crops (e.g., fruits, vegetables, and some nuts), most of which depend on benefit from insect pollination. The value of these pollination services by EHBs is estimated at $5–$10 billion annually in the United States (Southwick and Southwick 1992).

Africanized and European honeybees represent divergent subspecies within the *mellifera* species of the genus *Apis*. Both have nearly the same biochemistry, morphology, genetics, diet, and reproductive and other behaviors. Their diet includes pollen and spores from most seed plants. Both EHBs and AHBs are social bees living in perennial colonies. They are active on most days collecting nectar, water, pollen, and plant resins for their subsistence. These honeybees "hoard" excess honey as energy-rich carbohydrate reserves in hexagonal wax combs. Energy from honey consumption partially supports brood-rearing and, most importantly, supplies the energy necessary for foraging flights by thousands of adult worker bees.

Africanized and European honeybees exhibit different foraging strategies (largely tropical versus temperate attributes). Africanized honeybee colonies in Africa, and now in much of the Neotropics, are attuned to finding and exploiting isolated mass-flowering tropical trees, and also use pollen and nectar from the nocturnal flowers of bat-pollinated flowering plants. Some tropic *Apis* species even migrate to follow nectar and pollen flows across the floral landscape. Consequently, these bees depend on increased colony mobility (reproductive swarming and abandoning the hive) as behavioral responses to seasonal floral richness or dearth. EHBs are better at hoarding vast amounts of honey and surviving long, cold winters (Kunzmann et al. 1995).

The behavioral ecology of AHBs and their interactions with EHBs and thousands of species of native U.S. bees remain largely unknown. Africanized honeybees have slightly shorter developmental times than do European bees, enabling them to produce more bees per unit time compared with EHBs. Africanized bees will also accept smaller cavities to nest in than European bees. This behavior increases potential competition for nesting sites with birds and other animals and also increases the potential for greater numbers of honeybee colonies in an area. Africanized honeybees commonly abandon their hives, often 15–30 percent annually or even much greater in some localities. Absconding colonies may travel as far as 100 miles before selecting a new nesting site (USDA 1994), thus they have been able to rapidly colonize new areas in the Neotropics.

The most often-discussed characteristic separating the two races is the AHBs' propensity to vigorously defend their colony and nest site. Although all honeybees respond to threats to their colonies, AHBs respond more quickly and in much greater numbers than do EHBs. In comparison to EHBs, greater numbers of AHBs will pursue intruders for much greater distances to defend their colonies. Recent research reported that three to four times as many AHBs responded and left eight to ten times more stings in a black leather measuring target in stinging experiments (USDA 1994).

Biochemical comparisons of AHB and EHB venoms indicate they are nearly identical. Nineteen stings per 1 kg (2.2 lb) of human victim body weight is the predicted median lethal dose (Schumacher, Schmidt, Egen, and Dillion 1992). Massive stinging incidents by AHBs are more likely to result in toxic envenomation. Reported 1993 stinging incidents in Mexico have involved more than 60 human fatalities (one death per 1.4 million). From 1988 to 1992, the Mexican national African Bee Program eliminated 11,000 AHB swarms in densely populated urban areas (Guzman-Novoa and Page 1994). To date, the worst U.S. stinging incident occurred in July 1992, when a 44-year old man mowing his lawn experienced a mass bee attack resulting in 800–1,000 stings (McKenna 1992).

Competition among nectar- and pollen-feeding invertebrate and vertebrate pollinators, resource partitioning, insect and plant community interactions, and ecosystem processes are affected by introduced EHBs and AHBs, with important short- and long-term ecological and perhaps evolutionary consequences. The influence of exotic honeybees on individual species or communities of native tropical (or temperate) plants or animal can only have one of three outcomes: the native species will suffer, benefit, or remain more or less unaffected. The key to understanding these seemingly obvious outcomes is, however, based on obtaining sufficient information to delineate the very complex short- and long-term competitive dynamics between introduced bees, native bees and pollinators, and native plants in diverse, interacting, natural communities.

One observational and manipulative competition study between honeybees, bumblebees, solitary bees, and ants was at midelevations in the Santa Catalina Mountains in the Sonoran Desert near Tucson, Arizona (Schaffer et al. 1983). Dramatic shifts in abundance of ants and bumblebees were detected when honeybees were present (introduced) or sealed inside their hives. The researchers suggested that direct competition between introduced honeybees and native hymenopteran floral visitors was because the honeybees numerically dominated the site. Initial evidence seems to indicate that honeybees seek out and preempt the most profitable habitats and partially exclude native bees indirectly by rapidly reducing the standing crop of plant nectar and pollen (*Agave* in this study; Kunzmann et al. 1995).

Many important nectar- and pollen-producing plants visited by AHBs bloom at night and are pollinated by bats. Africanized honeybees find and exploit these rich flowers at first light, and we predict that saguaros and other columnar cacti will be heavily used as food plants for AHBs in Arizona. Early Arizona data for AHB colonies illustrate that most AHB colonies have been found in the subtropical climate zones in Sonoran Desert scrub.

Determining which plants are used primarily for nectar versus pollen, or both, depends on direct observations of bees on flowers or indirectly by identifying pollen grains in stored nest samples of honey. In Panama, Roubik (1989) found that AHB colonies harvested pollen from at least 142–204 flowering plant species in a forest containing about 800–1,000 species. European honeybees collected pollen or nectar from about 185 plant species from a secondary forest and agricultural area in Mexico (Villanueva 1984). These studies suggest that honeybees are using about 25 percent of the local flora, but intensively use far fewer species at any give time (Roubik 1989). In Arizona, EHBs will often harvest pollen from more than 60 species annually, but of these, only 10–15 are harvested heavily and consistently from year to year (Buchmann et al. 1992). Because of their pollen herbivory and reproductive contact with so many plants, there can be serious long-term ecological and evolutionary consequences of these interactions that we simply do not yet understand.

BULLFROGS: INTRODUCED PREDATORS

In the American Southwest, much of the native fish fauna is facing extinction (Minckley and Deacon 1991); frogs in California (Fellers and Drost 1993) and frogs and garter snakes in Arizona (Schwalbe and Rosen 1988) are also in critical decline. Habitat destruction and introduced predators appear to be primary causes of native frog declines (Jennings and Hayes 1994), and habitat modification often yields ponds and lakes especially suitable for introduced species. Introduced bullfrogs (*Rana catesbeiana*) have been blamed for amphibian declines in much of western North America (e.g., Hayes and Jennings 1986; Leonard, Brown, Jones,

McAllister, and Storm 1993; Vial and Saylor 1993). Extensive cannibalism by bullfrogs renders them especially potent predators at the population level. The tadpoles require only perennial water and grazeable plant material; hence, transforming young can sustain a dense adult bullfrog population even if alternate prey are depleted. This may increase the probability that native species may be extirpated by bullfrog predation (Rosen and Schwalbe 1995).

Introduced predatory fishes are apparently an important cause of frog declines (Hayes and Jennings 1986). They have been strongly implicated in one important case of decline of native ranid frog (family Ranidae, the "true" frogs; Bradford 1989). Some introduced crayfish may also be devastating in some areas (Jennings and Hayes 1994).

WILD HORSES AND BURROS ON PUBLIC LANDS

On December 15, 1971, Congress passed legislation to protect, manage, and control wild horses and burros on public lands. The Wild Free-roaming Horses and Burros Act (Public Law 92–195) described these animals as fast-disappearing symbols of the historic and pioneer spirit of the West. The Bureau of Land Management (BLM) and the U.S. Forest Service are charged with administering the law, which specifies how wild horses and burros are to be managed on the range and how excess animals are to be disposed. Section 3(a) requires the secretary of the interior to manage wild free-roaming horses and burros in a manner designed to achieve and maintain a thriving natural ecological balance on public lands. This section also specifies requirements for inventorying, monitoring, and establishing appropriate management levels, making removals, placing excess animals, and establishing criteria for destruction of animals (Pogacnik 1995).

Although these animals were once considered endangered by the nearly unrestrained onslaught of the mustangers and others, they have thrived under federal protection. With few predators and with protection from humans, wild horse and burro populations on BLM-administered lands (where most of the animals are located) quickly grew until control of the populations and the effect on their habitat became a major concern.

The act requires that BLM maintain a current inventory of wild horses and burros on certain public lands. At present, BLM censuses each of the 196 head-management areas on a rotating basis, usually every three years, using census techniques based on research published by the National Academy of Sciences (1982). Censuses in 1993 identified a nationwide population of 46,500 wild horses and burros. Accuracy for the 1993 census ranged from 85–99 percent for wild horses and 75–88 percent for wild burros.

Annual population growth in wild horse herds varies from 5–25 percent, depending on range and environmental conditions, with 15 percent being a long-term average. At this rate of increase, wild horse populations may double in five years. The annual growth in wild burro populations has not been determined, but their reproductive capacity may be similar to that of wild horse herds.

The act specifies that wild horses and burros may be managed only on lands where they existed on December 15, 1971, the time of the act's passage. The population of wild horses and burros within those 1971 areas of use was estimated at 17,000 animals; however, at that time no formal inventory policies or procedures existed to census populations. The BLM now has 269 herd areas, 196 within which all wild horses and burros will be removed.

Wild horse and burro herd areas occupy almost 43 million acres of public and private land in Arizona (about 4 million acres), California (6 million + acres), Colorado (800,000 acres), Idaho (450,000 + aces), Montana (55,000 + acres), Nevada (nearly 19 million acres), New Mexico (nearly 150,000 acres), Oregon (nearly 4 million acres), Utah (2.5 million acres), and Wyoming (nearly 6 million acres) (BLM 1993).

Within most herd areas, wild horse and burros graze with domestic livestock and a variety of indigenous wildlife species. Because they are generalist species, wild horses and burros inhabit a variety of habitats and vegetative communities.

The BLM's land-use planning process and evaluation of current inventory and monitoring data are used to determine a population level that maintains a thriving natural ecological balance with other uses. The act directs BLM to achieve appropriate population levels by removals, humane destruction, or other options, including antifertility methods.

BLM no longer destroys healthy excess wild horses and burros. Since 1973, when the first removals occurred, BLM has removed 141,762 wild horse and burros from public land and placed 122, 627 animals into private care through the Adopt-A-Horse program.

Removing excess animals from populations that exceed appropriate numbers is expensive, has restricted BLM's attempts to pursue other management alternatives, and therefore has often allowed populations to increase dramatically. When populations reached crisis proportions, funding was increased and large numbers of excess animals were removed form the range and placed with private citizens through the adoption program. The number of animals removed often was greater than the number that could be adopted, resulting in high costs for feeding and veterinary services while animals were held pending adoption.

In June 1992, the director of BLM approved the Strategic Plan for the Management of Wild Horses and Burros on Public Lands (BLM 1992). This plan represents BLM's first comprehensive policy for addressing wild horse and burro management. To reduce the frequency of removals, the plan recommends the use of antifertility management to slow population growth to a level where removals are only required on a cycle of five or more years instead of the current three-year cycle. Pending the availability of practical and cost-effective fertility-control techniques, selective removal of animals based on age or sex is being used to reduce the growth rate in wild horse populations. The negative aspects of selective removal include the difficulty of predicting results through computer modeling and the extensive monitoring needed to ensure that age and sex ratios have not been altered to a level that could threaten the herd. Selective removals for controlling population growth are considered a temporary management option until research on immunocontraception is completed and can be implemented.

The BLM supports research on the use of immunocontraception for controlling wild horse population growth. Successful immunocontraceptive antigens have been developed; researchers are now trying to develop a system that would inhibit reproduction for two to three years (Pogacnik 1995).

Before the passage of the act, wild horses and burros were often captured and destroyed as nuisances or were sold for profit, chiefly for use in commercial products. The methods employed in their capture and destruction were often less than humane. As public awareness of these animals grew, so too did support for federal legislation to protect them from inhumane treatment.

PURPLE LOOSESTRIFE

> Down the green valley where the fallen dew
> Lies thick beneath the elm and count her store,
> Till the brown Satyrs in a jolly crew
> Trample the loosestrife down along the shore,
> And where their horned master sits in state
> Bring strawberries and bloomy plums upon a wicker crate!
>
> —Oscar Wilde, 1913

Purple loosestrife is an exotic wetland perennial herb, with a square, woody stem and opposite or whorled leaves, introduced to North America from Europe in the early 19th century (Stuckey 1980). By the 1930s, the plant was well established along the New England seaboard. The construction of inland canals and waterways in the 1880s favored the expansion of purple loosestrife into interior New York and the St. Lawrence River Valley (Thompson, Stuckey, and Thompson 1987). The continued expansion of loosestrife has coincided with increased development and use of road systems (Thompson et al. 1987), commercial distribution of the plant for horticultural purposes, and regional propagation of seed for bee forage (Pellet 1977). The plant now occurs in dense stands throughout the northeastern United States, southeastern Canada, the Midwest, and in scatted locations in the western United States and southwestern Canada. Newly created irrigation systems in many of the western states have supported its further spread (Malecki 1995).

Purple loosestrife is a classic example of an introduced species whose distribution and spread have been enhanced by the absence of natural enemies and the disturbance of natural systems, primarily by human activity. Although noted for the beauty of its late summer flowers, which also provide a nectar source for bees, loosestrife has few other redeeming qualities. Its invasion into a wetland system results in suppression of the native plant community and the eventual alteration of the wetland's structure and function (Thompson et al. 1987). Large, monotypic stands not only jeopardize various threatened and endangered plants and wildlife, such as Long's bulrush (*Scripus longii*) in Massachusetts (Coddington and Field 1978), small spikerush (*Eleocharis parvula*) in New York (Rawinski 1982), and the bog turtle (*Clemmys muhlenbergii*) in the northeastern Untied States (Bury 1979), but they also eliminate natural foods and cover essential to many wildlife, including waterfowl (Rawinski and Malecki 1984).

Purple loosestrife has many traits that enabled it to become a nuisance in North America. A single, mature plant can produce more than 2.5 million seeds annually; these seeds are long-lived (Welling and Becker 1990) and easily dispersed by water and in mud, adhering to aquatic wildlife, livestock, and people (Thompson et al. 1987). Established plants are tall (about 2 meters or 6.5 feet) with 30–50 stems forming wide-topped crowns that dominate the herbaceous canopy. A strong rootstock serves as a storage organ, providing resources for growth in spring and regrowth if the aboveground shoots are cut, burned, or killed by application of foliar herbicides. No native herbivores or pathogens in North America are known to suppress purple loosestrife (Hight 1990).

No effective method is available to control loosestrife, except in small localized stands that can be intensively managed. In such isolated areas, the plant can be eliminated by uprooting by hand and ensuring that all vegetative parts are removed. Other control techniques include water-level manipulation, mowing or cutting, burning, and herbicide application (Malecki and Rawinski 1985). Although these controls can eliminate small and young stands, they are costly, require continued long-term maintenance, and in the case of herbicides, are nonselective and environmentally degrading.

Purple loosestrife is now a naturalized weed that always will be a part of most North American wetlands. Researchers hope that introducing select insects will result in replacing monotypic stands of loosestrife by native vegetation and an overall decrease in the occurrence of the plant (Malecki 1995).

Habitat Assessment

A systematic approach toward the development of science-based ecological information at multiple scales and across large areas has been lacking from our management of natural resources. Significant gains in achieving an environmentally sustainable society with an acceptable standard of living can be had by addressing this issue (Jennings 1995).

Because the dynamics of larger systems (e.g., landscapes) constrain the behavior and occurrence of the smaller systems that they encompass (e.g., populations or species), by means that are independent of the smaller systems, conservation efforts implemented at the levels of populations or species cannot be effective when systemwide changes are occurring at the landscape level. Environmental changes that were formerly limited to affecting populations and species are now manifest at scales by which natural community and landscape systems function. Therefore, if we are to make significant progress in slowing the loss of our biological heritage, the basis for solving problems and implementing decisions must be predicated on information derived from multiple scales of geographic resolution as well as of biotic organization.

Furthermore, the mechanisms, or the "emergent properties," by which an ecological system operates cannot be identified by a simple aggregation of its smaller components or by a reduction of its larger components (Allen and Starr 1982; O'Neill, DeAngelis, Walde, and Allen 1986). To adequately characterize an ecosystem, it must be observed as a functioning whole rather than only inferred by reducing it to its component parts and then reaggregating the information discovered about the components. For ecosystems that cover large areas, observation is difficult, perhaps impossible, without using aerial photography and satellite imagery along with computerized systems that can handle the large amounts of information for analysis.

There are four requisites to the effective management of biological diversity, soil, water, and natural process across large landscapes: standardized definitions of the resources; replicative scientific methods for inventories that go beyond lists of species to include natural communities and their processes; a high-quality environmental information system with easy access for all; and the expertise to usefully synthesize the information (Jennings and Reganold 1991). The National Wetlands Inventory, Gap Analysis, and the Multi-Resolution Land Characteristics Database are achieving these requisites.

The Bottom Line on Our Environment and Future

People in general often react to difficult-to-deal-with information emotionally, not logically, and news of a pending environmental crisis is no exception. Those reactions cover a wide range of conditions, from denial to hysteria, indifference, obsession, anger, activism, or foolhardiness, and none of those reactions—or the people behind them—does much toward bringing us closer to a solution.

By the time our societal and political systems get through with the information our scientists provide us, public evaluation by anything close to scientific methodology is impossible, because the information the public receives is distorted and incomplete—chopped up and twisted to provide a good sound byte, and tipped to suit the political beliefs of those in power and the financial concerns of the owners.

Little doubt exists that we are putting our environment—and thus ourselves—at serious risk. While sometimes we are not certain (yet) of the exact causes—or all the causes—the changes we observe in our world are measurable. As active members of an increasingly global economy, we increase our risk by pretending it will go away. Sooner or later, we will have to face these problems—and common sense, a coolheaded approach, and developing technology will be how we solve them (Spellman and Whiting 2006).

Summary of Key Terms

Chlorofluorocarbons (CFCs)—synthetic chemicals that are odorless, nontoxic, non-flammable, and chemically inert. CFCs have been used as propellants in aerosol cans, as refrigerants in refrigeration and air conditioners, and in the manufacture of foam packaging. They are partly responsible for the destruction of the ozone layer.

Global warming—the long-term rise in the average temperature on Earth.

Greenhouse effect—the trapping of heat in the atmosphere. Incoming short wavelength solar radiation penetrates the atmosphere, but the longer-wavelength outgoing radiation is absorbed by water vapor, carbon dioxide, ozone, and several other gases in the atmosphere and is reradiated to Earth, causing an increase in atmospheric temperature.

Greenhouse gases—the gases present in the Earth's atmosphere that cause the greenhouse effect.

Heat islands—large metropolitan areas where heat generated has an influence on the ambient temperature (adds heat) in and near the area.

Photochemical smog—a complex mixture of air pollutants produced in atmosphere by the reaction of hydrocarbons and nitrogen oxides under the influence of sunlight.

Sea-level rise—the natural rise of sea level that occurs in cyclical patterns throughout history; may be the result of man's impact on global warning.

Chapter Review Questions

9.1   Why is smog an extremely difficult pollutant to regulate? Explain.

9.2   Several factors are pushing environmental air quality concerns increasingly into the international area. What are some of these concerns? Explain.

9.3   What are the unique factors responsible for the Antarctic ozone hole?

9.4   What effect, if any, does acidic deposition have on forests and crops?

9.5   What are the potential environmental effects associated with global warming?

Cited References and Recommended Reading

Adams, D. D., and Page, W. P., eds. 1985. *Acidic deposition: Environmental economic and policy issues.* New York: Plenum.

Allen, A. W. 1995. Agricultural ecosystems. In *Our living resources.* Washington, DC: U.S. Department of the Interior, National Biological Service.

Allen, T. F., and Starr, T. B. 1982. *Hierarchy: Perspectives for ecological complexity.* Chicago: University of Chicago Press.

Armentrout, P. 1997. *The ozone layer.* New York: Rourke.

Ashton, R. E., Jr., and Ashton, P. S. 1988. *Handbook of reptiles and amphibians of Florida*, Parts 1–3. Miami, FL: Windward.

Bender, M. 1984. Industrial versus biological traction. In *Meeting the expectations of the land: Essays in sustainable agriculture and stewardship*, ed. W. Jackson, W. Berry, and B. Coleman, 87–105. San Francisco: North Point Press.

Black-Covilli, L. L. 1992. Basic air quality. In *Fundamentals of environmental science and technology*, ed. Porter-C. Knowles. Rockville, MD: Government Institutes.

Bureau of Land Management (BLM). 1992. *Strategic plan for the management of wild horses and burros on public lands*. Washington, DC: Bureau of Land Management.

Bureau of Land Management (BLM). 1993. Ninth report to Congress on the administration of the Wild Free-Roaming Horse and Burro Act. Washington, DC: BLM.

Boice, R., and Boice, C. 1970. Intraspecific competition in captive *Bufo marinus* and *Bufo americanus* toads. *Journal of Biological Psychology* 12:32–36.

Boydstun, Fuller, C. P., and Williams, J. D. 1995. Nonindigenous fish. In *Our living resources*. Washington, DC: U.S. Department of the Interior, National Biological Service.

Bradford, D. 1989. Allotopic distribution of native frogs and introduced fishes in high Sierra Nevada lakes of California: Implication of the negative effect of fish introduction. *Copeia* 989:775–78.

Brand, D. D. 1988. The honeybee in New Spain and Mexico. *Journal of Cultural Geography* 9:71–81.

Bridgman, H. A. 1994. *Global air pollution: Problems in the 1990s*. New York: John Wiley & Sons.

Broecker, W. 1987. Unpleasant surprises in the greenhouse? *Nature* 328:123–26.

Brown, P. 1997. *Global warming: Can civilization survive?* New York: Blandford Press.

Bruun, P. 1962. Sea level rise as a cause of shore erosion. *Proceedings of the American Society of Engineers and Journal Waterways Harbors Division* 88:117–30.

Bruun, P. 1986. *Worldwide impacts of sea level rise on shorelines, effects of changes in stratospheric ozone and global climate*. Vol. 4, 99–128. New York: UNEP/EPA.

Buchmann, S. L., O'Rourke, M. K., Shipman, C. W., Thoenes, S. C., and Schmidt, J. O. 1992. Pollen harvest by honeybees in Saguaro National Monument: potential effects on plant reproduction. In *Proceedings of the Symposium on Research in Saguaro National Monument, Tucson, AZ*, ed. C. P. Stone and E. S. Bellantoni, 149–56.

Bury, R. B. 1979. *Review of the ecology and conservation of the bog turtle, Clemmys muhlenbergii*. U.S. Fish and Wildlife Service Special Scientific Report Wildlife 219.

Bury, R. B., and Whelan, J. A. 1984. *Ecology and management of the bullfrog*. U.S. Fish and Wildlife Service Research Publication 155.

Capel, S., Crawford, J. A., Robel, R. J., Burger, Jr., L. W., and Southerton, N. W. 1993. Agricultural practices and pesticides. In *Quail symposium*, ed. K. E. Church and T. V. Donley, 172–73. Topeka: Kansas Department Wildlife and Parks.

Carpenter, S. R., Fisher, S. G., Grimm, N. B., and Kitchell, J. F. 1992. Global change and freshwater ecosystems. *Annual Review of Ecology and Systematics* 23:119–40.

Coddington, J., and Field, K. G. 1978. *Rate and endangered vascular plant species in Massachusetts*. Washington, DC: U.S. Fish and Wildlife Service.

Committee on Earth and Environmental Sciences (CEES). 1990. *Our changing planet: The FY1991 research plan of the U.S. Global Change Research Program*. Reston, VA: U.S. Geological Survey.

Courtenay, W. R., and Williams, J. D. 1992. Dispersal of exotic species from aquaculture sources, with emphasis on freshwater fish. In *Dispersal of living organisms into aquatic ecosystems*, ed. A. Rosenfield and R. Mann, 49–81. College Park: Maryland Sea Grant.

Crawford, M. 1976. *Air pollution control theory*. New York: McGraw-Hill.

Crowley, T. J., and North, G. R. 1988. Abrupt climate change and extinction events in Earth's history. *Science* 240:996.

Davis, M. L., and Cornwell, D. A. 1991. *Introduction to environmental engineering.* New York: McGraw-Hill.

Diamond, J. 1992. The arrow of disease. *Discover* 13, no. 910:64–73.

Dimitriades, B., and Whisman, M. 1971. Carbon monoxide in lower atmosphere reactions. *Environmental Science and Technology* 5:213.

Edgerton, L. 1991. *The rising tide: Global warming and world sea levels.* Washington, DC: Island Press.

Ellis, T. M. 1980. Caiman crocodiles: An established exotic in south Florida. *Copeia* 190:152–54.

Fellers, G. M., and Drost, C. A. 1993. Disappearance from the Cascades of *Rana cascadae* at the southern end of its range, California, USA. *Biological Conservation* 65:177–81.

Franck, I., and Brownstone, D. 1992. *The green encyclopedia.* New York: Prentice-Hall.

Friend, M. 1987. Avian cholera. In *Field guide to wildlife diseases.* Vol. 1. *General Field procedures and diseases of migratory birds,* ed. M. Friend and C. J. Laitman, 69–82. U.S. Fish and Wildlife Service Research Publication 167.

Friend, M. 1995. Increased avian diseases with habitat change. In *Our living resources.* Washington, DC: U.S. Department of the Interior, National Biological Service.

Friend, M., and G.L. Pearson. 1973. Duck plague: The present situation. *Proceedings of the Western Association of State Game and Fish commissioners* 53:315–325.

Fritts, T. H., and McCoid, J. J. 1991. Predation by the brown tree snake *Boiga irregularis* on poultry and other domesticated animals on Guam. *The Snake* 23:75–80.

Fritts, T. H., and McCoid, M. J., et al. 1994. Symptoms and circumstances associated with bites by the brown tree snake on Guam. *Journal of Herpetology* 28(1):27–33.

Fritts, T. H., McCoid, M. J., and Haddock, R. L. 1990. Risks to infants on Guam from bites of the brown tree snake. *American Journal of Tropical Medicine and Hygiene* 42:607–11.

Fritts, T. H., Scott, N. J., and Savidge, J. A. 1987. Activity of the arboreal brown tree snake on Guam as determined by electrical power outages. *The Snake* 19:51–58.

Gates, D. M. 1993. *Climate change and its biological consequences.* New York: Sinauer.

Gatter, W. 1992. Zuzellen und Sugmuster im Herbst: Einfluss des Treibhauseeffekts auf den Vogelzug. *Journal fur Ornithologue* 133:427–36.

Graedel, T. E., and Crutzen, P. J. 1989. The changing atmosphere. *Scientific American* (September): 58–68.

Graham, R. W., and Grimm, E. C. 1990. Effects of global climate change on the patterns of terrestrial biological communities. *Trends in Ecology and Evolution* 5, no. 9:289–92.

Guzman-Novoa, E., and Page, R. 1994. The impact of Africanized bees on Mexican beekeeping. *American Bee Journal* 134:101–106.

Hansen, J. E., et al. 1986. Climate sensitivity to increasing greenhouses gases. In *Greenhouse effect and sea level rise: A challenge for this generation,* ed., M. C. Barth and J. G. Titus. New York: Van Nostrand Reinhold.

Hansen, J. E., et al. 1989. Greenhouse effect of chlorofluorocarbons and other trace gases. *Journal of Geophysical Research* 94:16,417–16,421.

Hayes, M. P., and Jennings, M. R. 1986. Decline of ranid frog species in western North America: Are bullfrogs responsible? *Journal of Herpetology* 20:490–509.

Hayes, M. P., and Jennings, M. R. 1988. Habitat correlates of distribution of the California red-legged frog (*Rana aurora draytonii*) and the foothill yellow-legged frog (*Rana boylei*): Implications for management. In *Management of amphibians, reptiles, and small mammals in North America,* ed. R. C. Szaro, K. E. Severson, and D. R. Parton, 144–58. Fort Collins, CO: U.S. Forest Service General Technical Report RN-166.

Haynes, R. P. 1991. Science, technology, and the farm crisis. In *Ethics and agriculture: An anthology on current issues in world context*, ed. C. V. Blatz, 121–29. Moscow: University of Idaho Press.

Hight, S. D. 1990. Available feeding niches in populations of *Lythrum salicaria* L. (purple loosestrife) in the northeastern United States. *Proceedings of the International Symposium of Biological Control of Weeds* 7:269–78.

Holland, R. E. 1993. Changes in the planktonic diatoms and water transparency in Hatchery Bay, Bass Island area, western Lake Erie since the establishment of the zebra mussel. *Journal of Great Lakes Res.* 19:617–24.

Houghton, J. T., Jenkins, G. J., and Ephramus, J. J., eds. 1990. *Climate change: The IPCC scientific assessment.* Report prepared for the Intergovernmental Panel on Climate Change by Working Group 1. Cambridge: Cambridge University Press.

Interdepartmental Committee for Atmospheric Sciences. 1975. *The possible impact of fluoro-carbons and hydrocarbons on ozone.* Washington, DC: U.S. Government Printing Office.

Intergovernmental Panel on Climate Change (IPCC). 1990. *Climate change: The Intergovernmental Panel on Climate Change scientific assessment.* Geneva: World Meteorological Organization and United Nations Environment Programme.

Intergovernmental Panel on Climate Change (IPCC). 1992. *1992 Intergovernmental Panel on Climate Change supplement.* Geneva: World Meteorological Organization and United Nations Environment Programme.

Jennings, M. D. 1995. Habitat assessment. In *Our living resources.* Washington, DC: U.S. Department of the Interior, National Biological Service.

Jennings, M. D., and Reganold, J. P. 1991. A theoretical basis for managing environmentally sensitive areas. *Environmental Conservation* 18, no. 3:211–18.

Jennings, M. R., and Hayes, M. P. 1994. Decline of native ranids in the desert southwest. In *Proceedings of the Conference on the Herpetology of the North American Deserts*, ed. P. R. Brown and J. W. Wright. Los Angeles: Southwestern Association of Herpetologists Special Publication 5.

Kalmbach, E. R., and Gunderson, M. F. 1934. Western duck sickness: A form of botulism. *U.S. Department of Agriculture Technical Bulletin* 411.

Keniry, T., and Marsden, J. E. 1995. Zebra mussels in southwestern Lake Michigan. In *Our living resources.* Washington, DC: U.S. Department of the Interior, National Biological Service.

Kerr, W. E. 1967. The history of the introduction of African bees to Brazil. *South African Bee Journal* 39:3–5.

Knutson, R. D., Penn, J. B., and Boehm, W. T. 1990. *Agricultural and food policy.* Englewood Cliffs, NJ: Prentice Hall.

Kunzmann, M. R., Buchmann, S. L., Edwards, J. F., Thoenes, S. C., and Erickson, E. H. 1995. Africanized bees in North America. In *Our living resources.* Washington, DC: U.S. Department of the Interior, National Biological Service..

LaRoe, E. T. 1991. The effects of global climate change on fish and wildlife resources. *Transactions of the North American Wildlife and Natural Resources Conference* 56:171–76.

LaRoe, E. T. 1995. Global climate change. In *Our living resources.* Washington, DC: U.S. Department of the Interior, National Biological Service..

Lassuy, D. R. 1994. Aquatic nuisance organisms: Setting national policy. *Fisheries* 19, no. 4:14–17.

Lawson, W. J. 1987. *The cane toad Buffo martinis: A bibliography.* Queensland: Australian Environmental Studies Working Paper 1/87. Griffith University.

Leibovitz, L., and Hwang, J. 1968. Duck plague on the American continent. *Avian Diseases* 12:361–78.

Leonard, W. P., Brown, H. A., Jones, L. L. C., McAllister, K. R., and Storm, R. M. 1993. *Amphibians of Washington and Oregon*. Seattle, WA: Seattle Audubon Society.

Li, H. W. 1995. Non-native species. In *Our living resources*. Washington, DC: U.S. Department of the Interior, National Biological Service.

MacArthur, R. H., and Wilson, E. O. 1967. *The theory of island biogeography*. Princeton, NJ: Princeton University Press.

Mack, R. N., and Thompson, J. N. 1982. Evolution in step with few large, hooved mammals. *American Naturalist* 119:757–73.

Malecki, R. 1995. Purple loosestrife. In *Our living resources*. Washington, DC: U.S. Department of the Interior, National Biological Service.

Malecki, R., and Rawinski, T. J. 1985. New methods for controlling purple loosestrife. *New York Fish and Game Journal* 32:9–19.

Masters, G. M. 1991. *Introduction to environmental engineering and science*. Englewood Cliffs, NJ: Prentice-Hall.

McCoid, M. J. 1991. Brown tree snake on Guam: A worst case scenario of an introduced predator. *Micronesica Supplement* 3:63–69.

McCoid, M. J. 1993. The new herpetofauna of Guam, Mariana Islands. *Herpetological Review* 24:16–17.

McCoid, M. J., and Fritts, T. H. 1980a. Notes on the diet of a feral population of *Xenopus laevis* (Pipidae) in California. *Southwestern Naturalist* 25:272–75.

McCoid, M. J., and Fritts, T. H. 1980b. Observations of feral populations of *Xenopus laevis* (Pipidae) in southern California. *Bulletin of the Southern California Academy of Sciences* 79:82–86.

McCoid, M. J., and Fritts, T. H. 1993. Speculations of colonizing success of the African clawed frog, *Xenopus laevis* (Pipidae), in California. *South African Journal of Zoology* 28:59–61.

McCoid, M. J., Hensley, R. A., and Witteman, G. J. 1994. Factors in the decline of *Varanus indicus* on Guam, Marian Islands. *Herpetological Review* 25:60–61.

McCoid, M. J., and Kleberg, C. 1995. Non-native reptiles and amphibians. In *Our living resources*. Washington, DC: U.S. Department of the Interior, National Biological Service.

McCoid, M. J., Pregill, K., and Sullivan, R. M. 1993. Possible decline of *Xeopus* populations in southern California. *Herpetological Review* 24:29–30.

McKenna, W. R. 1992. Killer bees: What the allergist should know. *Pediatric Asthma, Allergy and Immunology* 6, no. 4:19–26.

Minckley, W. L., and Deacon, J. E., eds. 1991. *Battle against extinction: Native fish management in the American West*. Tucson: University of Arizona Press.

Molina, M. J., and Rowland, F. S. 1974. Stratospheric sink for chlorofluoromethanes: Chlorine atom catalyzed destruction of ozone. *Nature* 248:810–12.

Morse, L. E., Kutner, L. S., and Kartesz, J. T. 1995. Potential impacts of climate change on North American flora. In *Our living resources*. Washington, DC: U.S. Department of the Interior, National Biological Service.

Moyle, P. B. 1973. Effects of introduced bullfrogs, *Rana catesbeiana*, on the native frogs of San Joaquin Valley, California. *Copeia* 1973:18–22.

National Academy of Sciences. 1982. *Wild and free-roaming horses and burros: Final report of the Committee on Wild Free-Roaming Horses and Burros Board on Agriculture and Renewable Resources National Research Council*. Washington, DC: National Academy Press.

National Research Council. 1989. *Alternative agriculture*. Washington, DC: National Academy Press.

Oglesby, R.T., and Smith, C. R. 1995. Climate change in the Northeast. In *Our living resources*. Washington, DC: U.S. Department of the Interior, National Biological Service.

O'Neill, R.V., DeAngelis, D. L., Walde, J. B., and Allen, T. F. H. 1986. *A hierarchical concept of ecosystems.* Princeton, NJ: Princeton University Press.

Pellet, M. 1977. Purple loosestrife spreads down river. *American Bee Journal* 117:214–15.

Peters, R. L., and Lovejoy, T. L., eds. 1992. *Global warming and biological diversity.* New Haven, CT: Yale University Press.

Phillips, J. G., and Lincoln, F. C. 1930. *American waterfowl.* Cambridge, MA: Houghton Mifflin.

Pielou, E. C. 1991. *After the ice ages: The return of life to glaciated North America.* Chicago: University of Chicago Press.

Pogacnik, T. 1995. Wild horses and burros on public lands. In *Our living resources.* Washington, DC: U.S. Department of the Interior, National Biological Service.

Quortrup, E. R., Queen, F. B., and Merovka, L. T. 1946. An outbreak of *pasteurellosis* in wild ducks. *Journal of the American Veterinary Medical Association* 108:94–100.

Rawinski, T. J. 1982. *The ecology and management of purple loosestrife (Lythrum salicaria) in central New York.* M.S. thesis, Cornell University, Ithaca, NY.

Rawinski, T. J., and Malecki, R. A. 1984. Ecological relationships among purple loosestrife, cattail, and wildlife at the Montezuma National Wildlife Refuge. *New York Fish and Game Journal* 31:81–87.

Reeders, H. H., bij de Vaate, A., and Noordhuis, R. 1992. Potential of the zebra mussel for water quality management. In *Zebra mussels: Biology, impacts, and control,* ed. T. F. Nalepa and D. W. Schloesser, 439–51. Boca Raton, FL: Lewis.

Reichelderfer, K. 1990. Environmental Protection and agricultural support: Are trade-offs necessary? In *Agricultural policies in a new decade: Resources for the Future and National Planning Association,* ed. K. Allen, 201–30. Washington, DC.

Rhoades, K. R., and Rimler, R. B. 1991. Pasteurellosis. In *Disease of poultry,* 9th ed., ed. B. W. Calnck, H. J. Barnes, C. W. Beard, W. H. Reid and H. W. Yoder, Jr., 145–62. Ames: Iowa State University Press.

Ribaudo, M. O. 1989. *Water quality benefits from the Conservation Reserve Program.* U.S. Department of Agriculture, Resources and Technology Division, Economic Research Service. Agricultural Economic Report 606.

Robbins, C. S. 1995. Non-native birds. In *Our living resources.* Washington, DC: U.S. Department of the Interior, National Biological Service.

Rodda, G. H., and Fritts, T. H. 1992. The impact of the introduction of the colubrid snake *Boiga irregularis* on Guam's lizards. *Journal of Herpetology* 26:166–74.

Rodda, G. H., Fritts, T. H., and Conry, P. J. 1992. Origin and population growth of the brown tree snake, *Boiga irregularis,* on Guam. *Pacific Science* 46:195–210.

Root, T. L., and Weckstein, J. D. 1995. Changes in winter ranges of selected birds, 1901–1989. In *Our living resources.* Washington, DC: U.S. Department of the Interior, National Biological Service.

Rosen, P. C., and Schwalbe, C. R. 1995. Bullfrogs: Introduced predators in southwestern wetlands. In *Our living resources.* Washington, DC: U.S. Department of the Interior, National Biological Service.

Roubik, D. W. 1989. *Ecology and natural history of tropical bees.* Cambridge: Cambridge University Press.

Ruddiman, W. F., and Wright, Jr., H. E., eds. 1987. *North America and adjacent oceans during the last deglaciation.* Boulder, CO: Geological Society of America.

Savidge, J. A. 1987. Extinction of an island avifauna by an introduced snake. *Ecology* 68:660–68.

Schaffer, W. M., Zeh, D. W., Buchmann, S. L., Kleinhaus, S., Schaffer, M. V., and Antrim, J. 1983. Competition for nectar between introduced honeybees and native North American bees and ants. *Ecology* 64:564–77.

Schmalzel, R. J. 1980. *The diet breadth of Apis.* M.S. thesis, University of Arizona, Tucson.

Schneider, S. H., Mearns, L., and Gleick, P. H. 1992. Climate-change scenarios for impact assessment. In *Global warming and biological diversity,* ed. R. L. Peters and T. L. Lovejoy, 38–55. New Haven, CT: Yale University Press.

Schumacher, J. J., Schmidt, J. O., Egen, N. B., and Dillion, K. A. 1992. Biochemical variability of venoms from individual European and Africanized honeybees (*Apis mellifera*). *The Journal of Allergy and Clinical Immunology* 90:59–65.

Schwalbe, C. R., and Rosen, P. C. 1988. Preliminary report on effects of bullfrogs on wetland herpetofauna in southeastern Arizona. In *Management of amphibians, reptiles, and small mammals in North America,* ed. R. C. Szaro, K. E. Severson, and D. R. Patton, 166–73. Fort Collins, CO: U.S. Forest Service General Technical Report RM-166.

Schwartz, M. D. 1990. Detecting the onset of spring: A possible application of phenological models. *Climate Research* 1:23–29.

Spellman, F. R., and Whiting, N. E. 2006. *Environmental science and technology: Concepts and applications* (2nd ed.). Rockford, MD: Government Institutes Press.

Soil and Water Conservation Society. 1994. *When conservation and reserve program contracts expire: The policy options—Conference proceedings, February 10–11, 1994.* Arlington, VA: Soil and Water Conservation Society, Ankeny, IA.

Southwick, E. E., and Southwick, Jr., L. 1992. Economic value of honeybees (*Hymenoptera: Apidae*) in the United States. *Journal of Economic Entomology* 85, no. 3: 621.

St. Amant, J. A. 1975. Exotic visitor becomes permanent resident. *Terra* 13:22–23.

Stein, B. A., Breen, T., and Warner, R. 1995. Significance of federal lands for endangered species. In *Our living resources.* Washington, DC: U.S. Department of the Interior, National Biological Service.

Steirer, F. S., Jr. 1992. Historical perspective on exotic species. In *Introductions and transfers of marine species,* ed. M. R. DeVoe, 1–4. South Carolina: Carolina Sea Grant Consortium.

Stout, I. J., and Cornwell, G. W. 1976. Nonhunting mortality of fledged North American waterfowl. *Journal of Wildlife Management* 40:681–93.

Stuart, J. N., and Painter, C. W. 1993. *Rana catesbeiana* (bullfrog) cannibalism. *Herpetological Review* 24:103.

Stuckey, R. L. 1980. Distributional history of *Lythrum salicaria* (purple loosestrife) in North America. *Bartonia* 47:3–20.

Thompson, D. J. 1995. The seasons, global temperature, and precession. *Science* 268:59.

Thompson, D. Q., Stuckey, R. I., and Thompson, E. B. 1987. *Spread, impact, and control of purple loosestrife (Lythrum salicaria) in North American wetlands.* U.S Fish and Wildlife Service Research Report 2.

Toronto Conference Proceedings. 1988. *Climate change.* Toronto, Canada.

Tyler, M. 1989. *Australian frogs.* Ringwood, Victoria: Viking O Neill.

U.S. Congress, Office of Technology Assessment. 1993. *Harmful non-indigenous species in the United States.* Washington, DC: U.S. Government Printing Office OTA-F-565.

U.S. Department of Agriculture (USDA). 1994. *African honeybee fact sheet.* Washington, DC: U.S. Department of Agriculture.

U.S. Department of the Interior, National Biological Service. 1995. *Our living resources.* Washington, DC: U.S. Department of the Interior, National Biological Service.

U.S. Environmental Protection Agency (USEPA). 1995. *The probability of sea level rise.* Washington, DC: Environmental Protection Agency.

Vial, J. L., and Saylor, L. 1993. *The status of amphibian populations: A compilation and analysis.* International Union for Conservation of Nature and Natural Resources, Species Survival Commission, Declining Amphibian Population Task Force Working Document 1.

Villanueva, R. 1984. Plantas de importancia apicola en el ejido de Plan del Rio: Veracruz, Mexico. *Biotica* 9:279–340.

Welling, C. H., and Becker, R. I. 1990. Seed bank dynamics of *Lythrum salicaria L.*: Implications for control of this species in North America. *Aquatic Botany* 8:303–309.

Wiegand, K., and A. J. Eames. 1925. *The flora of the Cayuga Lake basin, New York*. Ithaca, NY: Cornell University Agricultural Experiment Station Memoir 92.

Wigley, T. M., Jones, P. D., and Kelly P. M. 1986. *Empirical climate studies: Warm world scenarios and the detection of climatic change induced by radioactively active gases, the greenhouse effect, climatic change, and ecosystems*, ed. B. Bolin et al. New York: Wiley.

Willford, W. 1995. Human influences. In *Our living resources*. Washington, DC: U.S. Department of the Interior, National Biological Service.

Williams, T. T. 2002. *Red: Passion and patience in the desert*. New York: Vintage Books.

Wilson, L. D., and Porras, L. 1983. *The ecological impact of man on the south Florida herpetofauna*. University of Kansas Museum of Natural History Special Publication 9.

Winston, M. L. 1987. *The biology or the honeybee*. Cambridge, MA: Harvard University Press.

Zug, G. R., and Zug, P. B. 1979. The marine toad, *Bufo marinus:* A natural history resume of native populations. *Smithsonian Contributions to Zoology* 284.

Zurer, P. S. 1988. Studies on ozone destruction expand beyond Antarctic, *C & E News* (May): 18–25.

Appendix: Answers to Chapter Review Questions

Chapter 1 Review Questions

1.1 Ecology
1.2 A major subdivision of ecology that studies the individual organism or a species
1.3 A major subdivision of ecology that studies groups of organisms associated together as a unit
1.4 A niche is the role that an organism plays in its natural ecosystem, including its activities, resource use, and interaction with other organisms.
1.5 An adverse alteration to the environment by a pollutant
1.6 Temperature; rainfall; light; mineral; wind; humidity; elevation; predominant land forms; tide; medium upon which the organisms exist (water, sand, mud, rock)

Matching:

(1)	c	(8)	x	(15)	o	(22)	s
(2)	f	(9)	k	(16)	u	(23)	l
(3)	n	(10)	r	(17)	y	(24)	b
(4)	q	(11)	v	(18)	w	(25)	g
(5)	i	(12)	a	(19)	j	(26)	e
(6)	t	(13)	d	(20)	m		
(7)	z	(14)	h	(21)	p		

Label the pond:

A. Primary consumers
B. Producers
C. Tertiary Consumer
D. Decomposers
E. Secondary Consumer

Chapter 2 Review Questions

2.1   Describes the tendency of chemical elements, including all the essential elements of protoplase, to circulate in the biosphere in characteristic paths from environment to organism and back to the environment
2.2   Reservoirs
2.3   Residence time
2.4   Mean residence time
2.5   Meteorological; geological; biological
2.6   Local; global
2.7   Gaseous; sedimentary
2.8   Hydrosphere; lithosphere; atmosphere
2.9   Carbon dioxide
2.10   Photosynthesis
2.11   Carbon dioxide; beneficial
2.12   Nitrogen
2.13   Nitrate
2.14   Ammonia
2.15   Nitrification
2.16   Denitrification
2.17   Rock
2.18   Fertilizer; algal bloom
2.19   One-problem/one solution syndrome
2.20   Sulfur

Chapter 3 Review Questions

3.1   Energy may not be created or destroyed.
3.2   Energy; materials
3.3   Flow of energy is from a greater to lesser amount only
3.4   Food chain
3.5   Food web
3.6   Decomposers
3.7   Used to estimate the amount of energy transferred through a food chain
3.8   Graphical representation of the number of organisms at various trophic levels in a food chain, represented by several levels placed one above the other with the base formed by producers and the apex formed by the final consumer
3.9   Energy, biomass, numbers
3.10   The energy accumulated by plants
3.11   Net community productivity
3.12   Net primary production

Chapter 4 Review Questions

4.1   The branch of ecology that studies that structure and dynamics of population.

4.2   Organisms in a population are ecologically equivalent.

4.3   Species; density

4.4   Immigration

4.5   Community ecology

4.6   Clumped distribution

4.7   Carrying capacity

4.8   Environmental carrying capacity

4.9   Population controlling

4.10 Ecological succession

4.11 Bare rocks exposed to the elements; rocks become colonized by lichens; masses replace the lichens; grasses and flowering plants replace the mosses; woody shrubs begin replacing the grasses and flowering plants; a forest eventually grows where bare rocks once existed.

4.12 Pioneer community

4.13 Biotic; abiotic

4.14 We can never do merely one thing. Any intrusion into nature has numerous effects, many of which are unpredictable.

Chapter 5 Review Questions

5.1   Variety of different species, genetics variability among individuals within each species, variety of ecosystems, and functions such as energy flow and matter cycling needed for survival of species and biological communities

5.2   Ruin

5.3   Fungi, bacteria and protozoans

5.4   Birth rate has fallen below replacement levels

5.5   Temperate zone

5.6   Species-level

5.7   Persistence, constancy, resilience

Chapter 6 Review Questions

6.1   Wapiti

6.2   Decline

6.3   Losses of breeding

6.4   Depending on where they feed

6.5   Wading birds

6.6   Natural resources
6.7   Mourning dove
6.8   Common raven
6.9   Sirenians
6.10   Gray wolves
6.11   Grizzly bear
6.12   Black-footed ferret
6.13   Amphibians
6.14   Loggerhead, green, Kemp's Ridley, leatherback, hawksbill
6.15   Natures-study, fishing, philosophy
6.16   White sturgeon
6.17   Invertebrates
6.18   Grasshopper
6.19   Moths
6.20   Oxygen sag curve

Chapter 7 Review Questions

7.1   Biomes
7.2   North Dakota
7.3   Shenandoah
7.4   Natural
7.5   Pocosin
7.6   Fire
7.7   Fire regime
7.8   Fuel consumption, intensity, severity, frequency, seasonality
7.9   Dwarf mistletoe
7.10   Whitebark pine nuts

Chapter 8 Review Questions

8.1   Nature continuously strives to maintain a stream in a clean, healthy, normal state. This is accomplished by maintaining the stream's flora and fauna in a balanced state. Nature balances stream life by maintaining the number and types of species present in any one part of the stream. Nature structures the steam environment so that interdependency is maintained by a balance between plants and animals.
8.2   Medium
8.3   20
8.4   Lentic; lotic
8.5   Seasonal
8.6   Littoral

8.7   Plankton
8.8   Profundal
8.9   Benthic
8.10   Riffle, run, pool
8.11   Neuston
8.12   A condition or a substance that limits the presence or success of an organism or a group of organism in an area
8.13   Water quality; temperature; turbidity; DO; acidity; acidity; alkalinity; organic or inorganic chemicals; heavy metals; toxic substances; habitat structure; substrate types; water depth and velocity; spatial and temporal complexity of physical habitats; flow regime; water volume; temporal distribution of flows; energy sources; type, amount, and particle size of organic material entering stream; seasonal pattern of energy availability; biotic interactions; competition; predation, parasitism; mutualism
8.14   Lowered light penetration
8.15   Carbon dioxide
8.16   Carbon dioxide
8.17   A lake's natural aging process that results in organic material being produced in abundance due to a ready supply of nutrients. It causes a lake to turn into a bog and, eventually, into a terrestrial ecosystem.
8.18   Mesotrophic lake
8.19   Marl
8.20   Open; closed

Chapter 9 Review Questions

9.1   Answers will vary.
9.2   Answers will vary.
9.3   Answers will vary.
9.4   Answers will vary.
9.5   Answers will vary.

Index

About the Author

Frank R. Spellman is assistant professor of Environmental Health at Old Dominion University, Norfolk, Virginia.

Spellman's 54 published book titles range from *Concentrated Animal Feeding Operations* (CAFOs) to several topics in all areas of environmental science and occupational health. Many of Spellman's texts are listed on Amazon.com and Barnes and Noble. Several of his texts have been adopted for classroom use at major universities throughout the United States of America, Canada, Europe, and Russia; two are currently being translated into Spanish for South American markets.

Spellman has been cited in more than 400 publications; serves as a professional expert witness for three law groups; accident investigator for a northern Virginia law firm; and consults on homeland security vulnerability assessments (VAs) for critical infrastructure, including water/wastewater facilities nationwide.

Spellman receives numerous requests to co-author with well-recognized experts in several scientific fields. For example, he is a contributing author for the prestigious text *The Engineering Handbook, 2nd Ed.* (CRC Press).

Spellman lectures on homeland security and health and safety topics throughout the country and teaches water/wastewater operator short courses at Virginia Tech.

Spellman holds a BA in Public Administration, BS in Business Management, MBA, Master of Science in Environmental Engineering, and PhD in Environmental Engineering.